ASTRONOMY AND ASTROPHYSICS LIBRARY

Series Editors: G. Börner, Garching, Germany
A. Burkert, München, Germany
W. B. Burton, Charlottesville, VA, USA and
   Leiden, The Netherlands
M. A. Dopita, Canberra, Australia
A. Eckart, Köln, Germany
T. Encrenaz, Meudon, France
B. Leibundgut, Garching, Germany
J. Lequeux, Paris, France
A. Maeder, Sauverny, Switzerland
V. Trimble, College Park, MD, and Irvine, CA, USA

Peter Hoyng

Relativistic Astrophysics and Cosmology

A Primer

With 114 Figures, 16 in color and 12 Tables

Peter Hoyng
SRON Netherlands Institute for Space Research
Sorbonnelaan 2
3584 CA Utrecht, The Netherlands
p.hoyng@sron.nl

Cover picture: 'Zwarte Kom op Geel Vlak' 2002 by Olav Cleofas van Overbeek, picture Galerie Lieve Hemel, Amsterdam.

Library of Congress Control Number: 2006920111

ISSN 0941-7834
ISBN-10 1-4020-4521-2 Springer Berlin Heidelberg New York
ISBN-13 978-1-4020-4521-9 Springer Berlin Heidelberg New York

This work is subject to copyright. All rights are reserved, whether the whole or part of the material is concerned, specifically the rights of translation, reprinting, reuse of illustrations, recitation, broadcasting, reproduction on microfilm or in any other way, and storage in data banks. Duplication of this publication or parts thereof is permitted only under the provisions of the German Copyright Law of September 9, 1965, in its current version, and permission for use must always be obtained from Springer. Violations are liable to prosecution under the German Copyright Law.

Springer is a part of Springer Science+Business Media

springer.com

© Springer-Verlag Berlin Heidelberg 2006
Printed in The Netherlands

The use of general descriptive names, registered names, trademarks, etc. in this publication does not imply, even in the absence of a specific statement, that such names are exempt from the relevant protective laws and regulations and therefore free for general use.

Typesetting by author and SPI Publisher Services using a Springer LATEX macro package

Cover design: *design & production* GmbH, Heidelberg

Printed on acid-free paper   SPIN: 11301257   3144/SPI - 5 4 3 2 1 0

Preface

This textbook offers a succinct and self-contained introduction into general relativity and its main areas of application: compact objects, gravitational waves and cosmology. It has evolved from lecture courses I have taught at the University of Utrecht since 1990. The book is intended for advanced undergraduate and beginning graduate students in physics and astrophysics.

The past decades have seen spectacular new developments in our knowledge of cosmology, the physics of compact objects and in high precision gravity experiments. As a result, relativistic astrophysics and cosmology have become a very attractive element in the (astro)physics curriculum, and there is a variety of excellent textbooks. But most of these are either too advanced, too elementary, or too voluminous for my purpose, or they do not cover all topics. My object in writing this book has been to provide a concise text that addresses general relativity and its applications homogeneously, at an intermediate level, conveying a maximal physical insight with a minimal amount of formalism. It is often a revelation for students to see that it is possible, at least for the range of subjects addressed here, to cut down the usual tangle of math to manageable proportions without watering down the discussion. My guiding principle has been to keep only what is really useful, but that does not mean that it is always the more difficult topics that have been eliminated. For example, I kept very little formal tensor calculus as it is not really needed – only the basics are indispensable. But variational calculus is used extensively because it is by far the simplest way to compute Christoffel symbols and therefore very useful.

The approach is theoretical, but the text is interlaced with discussions of observational, instrumental and historical aspects where appropriate. The book is divided into (1) *preparatory material:* special relativity, geometry of Riemann spaces, and general relativity, (2) *Schwarzschild metric and applications:* classical tests, binary pulsars, gravitational lenses, neutron stars and black holes, (3) *experimental gravity:* gravitational waves and their detectors, Gravity Probe B, and finally (4) *cosmology:* Robertson-Walker metric, evolution of the universe, observational cosmology, and inflation. Due to the self-imposed restrictions several topics had to be skipped. But in view of their current interest, extra attention has been given to the operation of interferometer detectors for gravitational waves, to the Gravity Probe B mission, and to structure formation in relation to the results of the Wilkinson Microwave Anisotropy Probe (WMAP).

The reader is supposed to be familiar with linear algebra and calculus, ordinary differential equations, and with elementary thermal physics, electrodynamics, special relativity and quantum mechanics - in other words, the basic education of advanced physics undergraduates. Prior knowledge of differential geometry, general relativity and astrophysics is helpful but not required. The necessary mathematical techniques are introduced informally, following geometrical intuition as much as possible. The admirable texts of Dirac (1975), Price (1982) and Schutz (1985) have been a source of inspiration for me in this regard. And the astrophysical concepts are likewise briefly introduced to a level where they should be intelligible for physics students. There are about 145 exercises with hints for their solution. These exercises are an indispensable element in helping students to come to grasp with the subject matter, and to train them to solve elementary problems independently. In my experience 40 – 45 lectures (45 min.) of oral instruction would suffice to expound all material, excluding tutorials for exercises.

References to the literature are eclectic rather that complete, and appear as footnotes in the text. General references (mostly textbooks) are given in Appendix A. The finiteness of the alphabet did cause some problems of notation. The reader is alerted to my propensity for the symbol a. There are many different constants a in the text, but confusion is unlikely as they have only a local meaning. Likewise h has three different meanings, ($H_0/100$, the constant of the motion $r^2\dot{\varphi}$ in the Schwarzschild metric, and Planck's constant).

The cover picture is a still life by the Dutch artist Olav Cleofas van Overbeek entitled *Black bowl on yellow plane* (2002). Its simplicity and well-balanced design epitomize the rotational and translational symmetries that are so ubiquitous in physics, and in this book embodied in the Schwarzschild metric and the Robertson-Walker metric, respectively. The cartoons opposite to the chapter headings have been drawn by Roeland van Oss, and I am grateful for his permission to reproduce them here. The drafting of the figures reflects the technical developments of the period, and began on rice paper, to proceed entirely by electronic means in the end. I wish to thank Hans Braun, Arjan Bik, and in particular Artur Pfeifer for their assistance in this area. There are instances where we have been unable to trace or contact the copyright holder of some of the reproduced figures. If notified the publisher will be pleased to rectify any errors or omissions at the earliest opportunity.

I want to express my gratitude to Jan van der Kuur for his help in solving my Latex problems, and to Constance Jansen who generously provided library assistance. Lucas van der Wiel has helped me with the first English translation. In the course of the years that led up to this book I have benefitted from discussions and correspondence with many colleagues. I cannot name them all, but I do wish to thank Bram Achterberg and Ed van den Heuvel and several unknown referees who read sections of the manuscript. I am in particular indebted to my friend and colleague John Heise who since many years is my discussion partner on matters relativistic and other. His influence is pervasive throughout the book. And last but not least, I should thank all the students who continually forced me to improve the presentation of the material, from my first notes in 1990 (to which I think in slight embarrassment), to the present text which is, I hope, of some use to the reader.

Utrecht, *Peter Hoyng*
February 2005

Contents

1 Introduction 1
   1.1 Special relativity (SR) 3
   1.2 General relativity (GR) 10
   1.3 The need for GR in astrophysics.......................... 14

2 Geometry of Riemann Spaces 19
   2.1 Definition 19
   2.2 The tangent space 21
   2.3 Tensors 23
   2.4 Parallel transport and Christoffel symbols 26
   2.5 Geodesics 30
   2.6 The covariant derivative.................................. 33
   2.7 Riemann tensor and curvature 35

3 General Relativity 43
   3.1 Co-ordinates, metric and motion 43
   3.2 Weak fields (1) 47
   3.3 Conservation of mass 50
   3.4 The field equations 52
   3.5 Weak fields (2) 56
   3.6 Discussion 59

4 The Schwarzschild Metric 65
   4.1 Preliminary calculations 65
   4.2 The Schwarzschild metric................................. 69
   4.3 Geodesics of the Schwarzschild metric 72
   4.4 The classical tests of GR 77
   4.5 Gravitational lenses...................................... 82

5 Compact Stars 89
5.1 End products of stellar evolution 89
5.2 The maximum mass M_c 95
5.3 The Tolman-Oppenheimer-Volkoff equation 97
5.4 A simple neutron star model 101
5.5 Realistic neutron star models 103

6 Black Holes 109
6.1 Introduction 109
6.2 Observations 110
6.3 Elementary properties 113
6.4 Kruskal-Szekeres co-ordinates 121
6.5 Rotating black holes: the Kerr metric 125
6.6 Hawking radiation 128

7 Gravitational waves 133
7.1 Small amplitude waves 133
7.2 The effect of a gravitational wave on test masses 136
7.3 Generation of gravitational radiation 138
7.4 Bar detectors 143
7.5 Interferometer detectors 145

8 Fermi-Walker Transport 155
8.1 Transport of accelerated vectors 155
8.2 Thomas precession 159
8.3 Geodesic precession 161
8.4 Gravity Probe B 164

9 The Robertson-Walker Metric 169
9.1 Observations 169
9.2 Definition of co-ordinates 174
9.3 Metric and spatial structure 177
9.4 Equations of motion 181
9.5 The cosmological constant 183
9.6 Geodesics 185

10 The Evolution of the Universe 189
10.1 Equation of state 189
10.2 The matter era 191
10.3 The radiation era 197
10.4 The formation of structure 203

11 Observational Cosmology 213
  11.1 Redshift and distance 213
  11.2 The visible universe and the horizon 218
  11.3 Luminosity distance and Hubble relation 224
  11.4 The microwave background 228
  11.5 Light-cone integrals 231

12 The Big Bang 237
  12.1 Nuclear reactions 237
  12.2 The first 100 seconds 240
  12.3 The synthesis of light elements 246

13 Inflation 253
  13.1 The horizon problem 254
  13.2 Evolution of a universe with a scalar field 258
  13.3 Chaotic inflation 261
  13.4 Discussion 266

1

Introduction

From the earliest days of history mankind has shown an avid interest in the heavenly phenomena, and astronomers have good reasons to claim that theirs is the oldest profession of the world but one. This interest arose largely from practical needs. In a differentiated society where rituals play an important role it is useful to know the direction of the North and to be able to predict the turn of the seasons, days of festivities, and so on. Astronomy was still tightly interwoven with religion and astrology. The Babylonians had an extensive knowledge of practical mathematics and astronomy. The two important issues were the *calendar* (i.e. the question of the relative length of the year, months, days and the time of important feast-days), and the *ephemeris* (the positions of the Sun, Moon and the planets, lunar and solar eclipses, etc., as a function of time). In parallel to this practical knowledge, a whole variety of mythological ideas developed about the origin of the world around us. It is a peculiar coincidence that the Hindus arrived at time scales close to what we now think to be the age of the universe. The Hindus believed in a cyclic universe. It was created by Brahma, and exists in an orderly state for a period of one Brahma day (4.32×10^9 year).[1] At the end of the day Brahma will go to rest, and the universe will turn into chaos. Light, orderly motion and life only return when Brahma wakes up again. Ultimately Brahma himself will die, and the universe and the Hindu pantheon will perish with him. A new Brahma will then be born, and the endless cycle of creation and destruction will repeat itself.

The Greek were the first to develop rational concepts about the world. According to Pythagoras and his followers (ca. 500 B.C.) the Earth is spherical. The Sun, Moon and planets reside on concentric spheres revolving around the central fire Hestia. The stars are located on the outermost sphere. The idea that the Earth is not at the centre of the universe is therefore very old. Eudoxus (about 408-355) and Aristotle (384 - 322) developed a spherical world model consisting of a great number of concentric spheres with the

---

[1] Thomas, P.: 1975, *Hindu religion, customs and manners*, Taraporevala Sons & Co, Bombay.

Earth located at the centre. Each celestial body (Sun, Moon, and the five known planets) has a set of spheres associated with it, and is located on the innermost sphere of its own set. Each sphere of a set revolves around an axis attached to the sphere directly within. Because the axes of the spheres are not aligned, the apparent motions of the planets could be reproduced approximately. To the Greek, esthetic considerations played an important role, and this trend has persisted in physics to this day because it is often productive ('a theory is plausible *because* it is elegant'). Religious aspects played a role as well, and this has also lingered on for a very long time (cf. for example Newton). And haven't we all at times been overwhelmed by the beauty of the night sky – a strong emotional experience bordering to a religious experience? In a letter to his brother Theo, Vincent van Gogh wrote '.. It does not prevent me from having a terrible need of – shall I say the word – of religion, then I go outside in the night to paint the stars ..' [2]

Based on Babylonian observations Hipparchus (ca. 190 - 125) catalogued some 850 stars and their positions. He also invented the concept of epicycles to explain the brightness variations associated with the apparent motion of the planets. It should be kept in mind that in those days stars and planets were regarded as independent light sources of a divine nature, and that only the Earth and the Moon were thought to be lit by the Sun. The insight that the Earth and the planets are actually comparable objects came much later. Geocentric world models with epicycles were gradually refined. Ptolemy (87 – 150) recorded his version in the Almagest[3], a summary of ancient astronomy and one of the most influential texts in the development of Western thinking. Much earlier, Aristarchus (ca. 310 – 230) had proposed a simpler, truly heliocentric model with the Earth rotating around its axis and around the Sun. He was therefore 1800 years ahead of his time, but his ideas did not prevail. The history of astronomy would arguably have been quite different if they had, and this example may serve as a consolation for those who feel that the world does not hear their voice. The heliocentric theory became gradually accepted only after the publication of the work of Copernicus in 1543. For more information on these matters see Koestler (1959), Dijsterhuis (1969), Pannekoek (1989), Evans (1998), and Bless (1995).

The transition from a geocentric to a heliocentric world model meant that mankind had to give up its privileged position at the centre of the universe. This development continued well into the last century, one might say, until Hubble proved in 1924 that the spiral nebulae are actually galaxies located far outside our own galaxy, as Kant had already postulated in 1755. As a result, our galaxy became one among many. This led to the formulation of the *cosmological principle*, which says that our position in the universe is in no way special – the complete antithese of the geocentric view.

---

[2] J. van Gogh-Bonger (ed.), *Verzamelde brieven van Vincent van Gogh*, Wereldbibliotheek, Amsterdam (1973), Vol III, letter 543, p. 321.
[3] From the Arabic-Greek word Kitab al-megiste, the Great Book.

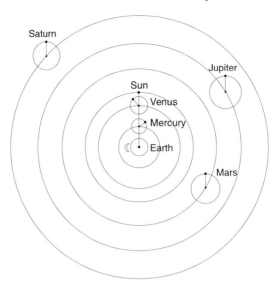

Fig. 1.1. Ptolemy's world model, very much simplified and not to scale. The centres of the epicycles of the inner planets are on the Sun-Earth line, while the radii of the epicycles of the outer planets run parallel to this line. The innermost sphere around the Earth (the 'sublunary') belongs to the Moon. The stars are located on an outermost sphere (not shown). The whole system operates like a clockwork as the Sun moves around the Earth. To the modern eye, a strange aspect of the model is that the motion of the other planets is connected with the motion of the Sun around the Earth. This coincidence is removed in Copernicus's heliocentric model. After Dijksterhuis (1969).

1.1 Special relativity (SR)

Modern cosmology is based on the theory of general relativity (GR), which is a natural generalisation of the theory of special relativity (SR). This section recapitulates the main ideas of special relativity, that is, physics in the absence of gravity. For a more thorough discussion we refer to Schutz (1985). We consider space and time to be a 4-dimensional continuum, called *Minkowski spacetime*. A (global) co-ordinate system in Minkowski spacetime is usually called a *reference frame* or just a *frame*. A point P with co-ordinates $\{x^\alpha\}$ is called an *event*. The motion of a particle can be represented by its *worldline*, Fig. 1.2. SR is based on two postulates:

- The *principle of relativity*, which states that the laws of physics must have the same form in every inertial frame.

- The *speed of light* has a constant value c in all inertial frames.

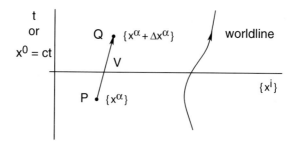

Fig. 1.2. The Minkowski spacetime, with events P and Q, a vector $\mathbf{V}$ connecting these events, and the worldline of a particle.

An inertial frame is a rigid system of spatial co-ordinates with synchronised clocks to measure t, in which test particles on which no forces are exerted move uniformly with respect to each other. An example of an inertial frame is a frame that does not move (no rotation, no translation) with respect to the distant galaxies. Inertial frames in SR are global, and they all move uniformly with respect to each other. In this section we admit only inertial frames. The principle of relativity is very old and goes back to Galilei. The second postulate is Einstein's innovative step, which he based, among other things, on Michelson and Morley's experiment which demonstrated the impossibility of measuring the velocity of the Earth with respect to the ether. The consequence is that invariance for Galilean transformations, as e.g. Newton's laws possess, no longer applies.

Simultaneity exit

SR often evokes major conceptual problems due to the fact that some very deeply rooted (Newtonian) ideas about space and time are not consistent with observations. Paramount among these is the fact that simultaneity has no longer an invariant meaning. Consider an inertial observer W, who tries to locate the events in his co-ordinate system (x,t) that are simultaneous with the origin $x = t = 0$, see Fig. 1.3, left. W argues: all events P that reflect light such that the moments of emission and detection are symmetrical with respect to $t = 0$ (emission at $t = -t_0$, detection at $t = t_0$ for all t_0). W's conclusion is: all events on the x-axis. Now consider observer $\overline{W}$ who moves uniformly to the right in W's frame, Fig. 1.3, right. At $t = 0$, W and $\overline{W}$ are both at the origin. $\overline{W}$'s worldline serves as the $\bar{t}$-axis of his frame, and $\bar{t} = 0$ is chosen at the common origin. $\overline{W}$ repeats W's experiment, but since the value of c is frame-independent, $\overline{W}$ identifies a different set of events, effectively his $\bar{x}$-axis, as being simultaneous with the origin. The $\bar{x}$-axis lies tilted in W's frame, and the tilt angle depends on $\overline{W}$'s velocity. Different observers $\overline{W}$ will

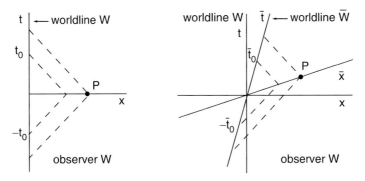

Fig. 1.3. As explained in the text, an invariant definition of simultaneity is impossible in SR.

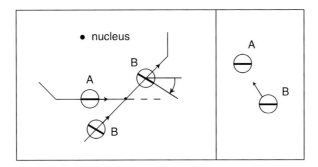

Fig. 1.4. Thomas precession of an electron orbiting a nucleus explained in the spirit of Fig. 1.3. After Taylor and Wheeler (1966).

therefore disagree as to which events are simultaneous with the origin.

Inaccurate reasoning in SR has led to many paradoxes (clock paradox, car-in-garage paradox). A vivid illustration of how drastically SR turns our perception of space and time upside down is the Thomas precession of the spin of an electron in an atom, a purely special-relativistic effect. Fig. 1.4 shows the classical orbit, approximated by a polygon. The heavy line is the projection of the spin axis on the plane of the orbit. After the electron has rounded a corner, its spin axis has turned. An analysis of what happens during the acceleration at the corner can be avoided by replacing electron A there by electron B, demanding that the spin vectors are aligned in a frame moving with A (A's rest-frame; right figure). But in the laboratory frame these orientations are different – this is a consequence of the relative meaning of simultaneity as explained in Fig. 1.3. Note that the electron is subject to

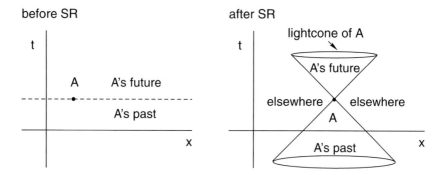

Fig. 1.5. The causal structure of Minkowski space. In SR every event A has its own invariant light-cone that divides Minkowski space into a past, a future and an elsewhere.

an additional precession due to electromagnetic interaction with the nucleus. The question arises whether a gyroscope in orbit around the Earth will also exhibit a precession. At the time of writing, the Gravity Probe B mission is performing the experiment, see further Ch. 8.

Lorentz metric

An important concept in SR is the *interval* Δs^2 between two events P and Q with co-ordinates x^α and $x^\alpha + \Delta x^\alpha$:

$$\Delta s^2 = c^2 \Delta t^2 - \Delta x^i \Delta x^i = \eta_{\alpha\beta} \Delta x^\alpha \Delta x^\beta ; \tag{1.1}$$

$$\eta_{\alpha\beta} = \begin{pmatrix} 1 & & & \emptyset \\ & -1 & & \\ & & -1 & \\ \emptyset & & & -1 \end{pmatrix}. \tag{1.2}$$

Notation:
$$x^0 = ct, \qquad \Delta t^2 \equiv (\Delta t)^2. \tag{1.3}$$

Relation (1.1) defines the *metric*, i.e. the distance between two events in Minkowski space, and is called the *Lorentz metric*. Here and everywhere else: summation convention ; Roman indices run from 1 to 3 and Greek indices from 0 to 3. Note that we adopt the signature : $1, -1, -1, -1$.[4]

---

[4] The sign convention is important as it leads to sign differences everywhere, but it has of course no influence on the physics. The advantage of the present choice is that for timelike geodesics the curve parameter p, the interval length s and proper time τ are proportional, see § 2.5.

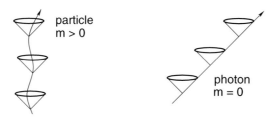

Fig. 1.6. The worldline of a particle with nonzero mass is located inside the light-cone, that of a photon is tangent to it.

Two events connected by a light ray have $\Delta s^2 = 0$, irrespective of their spatial distance. No matter how enormous the distance of some objects in the universe may be, the interval Δs^2 between them and the telescope is zero. The value of Δs^2 is also invariant: if some other observer $\overline{W}$ computes $\Delta \overline{s}^2 \equiv \eta_{\alpha\beta} \Delta \overline{x}^\alpha \Delta \overline{x}^\beta$ in his rest-frame (i.e. in a comoving inertial frame), the value he finds is equal to Δs^2 (proof: e.g. Schutz (1985), p. 11). This leads to an important relation between events, see Fig. 1.5. Prior to the advent of SR, all events were located either in the future, in the past, or they were simultaneous with a given event A. In SR there is A's *light-cone* $\Delta s^2 = 0$ that divides Minskowski space into a past and a future (with which A can have causal relations), and an 'elsewhere' (with which A cannot have any interaction). This division is independent of the reference frame because Δs^2 is invariant. Hence we can speak of *the* light-cone. The worldline of a particle with non-zero mass is always located inside the light-cone, see Fig. 1.6. Depending on the value of Δs^2, the vector connecting events P and Q in Fig. 1.2 is called a

$$\left. \begin{array}{ll} \text{timelike vector}: & \text{when } \Delta s^2 > 0\,; \\ \text{null vector}: & \text{when } \Delta s^2 = 0\,; \\ \text{spacelike vector}: & \text{when } \Delta s^2 < 0\,. \end{array} \right\} \quad (1.4)$$

The *proper time* interval $\Delta \tau$ between two (timelike connected) events on the worldline of a particle is defined as:

$$c^2 \Delta \tau^2 \equiv \Delta s^2 = c^2 \Delta t^2 - \Delta x^i \Delta x^i\,. \quad (1.5)$$

For positive Δs^2 we may define $\Delta s \equiv (\Delta s^2)^{1/2}$ and proper time intervals as $\Delta \tau = \Delta s / c$. Proper time intervals are invariant because Δs^2 is. By transforming to the rest-frame of the observer $\overline{W}$, so that $\Delta \overline{x}^i = 0$, we find that $\Delta \tau^2 = \Delta \overline{t}^2$, which shows that the proper time is just the time of a clock moving with the observer (his own wristwatch). Now substitute $\Delta x^i = (\Delta x^i / \Delta t) \Delta t = v^i \Delta t$ in (1.5) and compute the limit:

$$d\tau = \sqrt{1 - (v/c)^2}\, dt\,, \quad (1.6)$$

8   1 Introduction

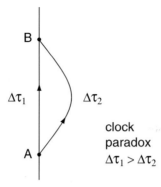

Fig. 1.7. The clock paradox. Two clocks moving from event A to B along different worldlines indicate different readings $\Delta\tau$ for the duration of the trip.

where v is the speed of the particle (the ordinary 3-velocity). The proper time $\Delta\tau$ elapsed between two events can be found by integrating (1.6) along the worldline connecting the events. The answer will depend on the shape of the worldline, which leads to the famous clock paradox, Fig. 1.7, explained in detail in Schutz (1985), § 1.13.

Lorentz transformations

The co-ordinates x^α and $\bar{x}^\alpha$ of an event with respect to two different inertial frames can be expressed into each other by means of a *Lorentz transformation*:

$$\bar{x}^\alpha = L^\alpha{}_\nu x^\nu . \tag{1.7}$$

The $L^\alpha{}_\nu$ are constants that depend only on the relative velocity $\boldsymbol{v}$ of the two frames. Relation (1.7) is a linear transformation that leaves Δs^2 invariant. If the co-ordinate axes (x, t) and $(\bar{x}, \bar{t})$ are defined as in Fig. 1.3 the transformation is

$$L^\alpha{}_\nu = \begin{pmatrix} \gamma & -\beta\gamma & 0 & 0 \\ -\beta\gamma & \gamma & 0 & 0 \\ 0 & 0 & 1 & 0 \\ 0 & 0 & 0 & 1 \end{pmatrix} \tag{1.8}$$

with $\beta = v/c$ and $\gamma = (1 - \beta^2)^{-1/2}$. The mathematical formulation of SR proceeds in terms of 4-vectors and tensors, that transform according to a Lorentz transformation. The trick is to try and write the laws of physics as relations between scalars, vectors and tensors *only*, because in that case they are automatically invariant for Lorentz transformations.

Lorentz transformations are global. In GR we allow arbitrary curvilinear

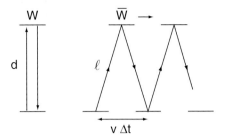

Fig. 1.8. An Einstein clock consists of photons traveling between two parallel mirrors at a distance d; the time for a round trip $\Delta t = 2d/c$ serves as the time unit. This clock will run slower if it moves with respect to the observer because the photons traverse a distance $\ell > d$ while c is constant. The merit of this example is that the time dilation is immediately obvious, but it is not so evident that it is impossible to eliminate the effect by using another clockwork. However, it can be shown that the effect is quite general and independent of the way the clock is constructed.

reference frames. As we shall see in § 2.3, the effect is that the global Lorentz transformation is replaced by a mesh of local Lorentz transformations that are different at each position in spacetime.

Exercise 1.1: Explain the time dilation with the help of Einstein's clock, Fig. 1.8:

$$(\Delta t)_{\text{measured by W}} = \frac{(\Delta \bar{t})_{\text{measured by }\overline{W}}}{\sqrt{1 - v^2/c^2}}. \tag{1.9}$$

Hint: W observes $\overline{W}$'s clock as it travels to the right at velocity v. W measures $\Delta t = 2\ell/c$, and $\ell^2 = d^2 + (v\Delta t/2)^2 = d^2 + (v\ell/c)^2$, from which $\ell = d/\sqrt{1 - (v/c)^2}$, and $(\Delta \bar{t})_{\text{measured by }\overline{W}} = 2d/c$.

Exercise 1.2: Below relation (1.6) it was said that the proper time elapsed between events depends on the worldline connecting the two events. Doesn't that contradict the fact that $d\tau$ is invariant?

Hint: In a given set of $d\tau$, each $d\tau$ is invariant under co-ordinate transformations, but another integration path simply implies a different set of $d\tau$.

1.2 General relativity (GR)

If we extend SR to arbitrarily moving reference frames, we would be able to do physics from the point of view of an accelerated observer. There is, however, another important motivation. Since in doing so apparent forces appear that are closely related to gravity, we may perhaps also be able to address gravity. And this turns out to be true. But if we are only after gravity, it would seem more straightforward to try and incorporate gravity in the framework of SR. Unfortunately, that doesn't work. Newtonian gravity may be summarised by $\nabla^2 \Phi = 4\pi G\rho$ and $K = -m\nabla\Phi$. It follows that gravity operates instantaneously – a change in ρ alters Φ everywhere at the same moment. This is inconsistent with SR because what is instantaneous in one frame is no longer so in another. This theory holds therefore only in one preferred frame. The problem might be overcome by replacing the equation for the potential by $\Box\Phi = (c^{-2}\partial^2/\partial t^2 - \nabla^2)\Phi = -4\pi G\rho$, for example, but then other difficulties appear, see e.g. Robertson and Noonan (1969) and Price (1982). Special relativistic theories of gravity using a flat spacetime and a single global reference frame don't work because they cannot accommodate the gravitational redshift and the weak equivalence principle. A different approach is needed.

Weak equivalence

At this point we need to be more precise about the concept of mass. A force K acting on a particle with *inertial mass* m_i causes an acceleration a given by Newton's law $K = m_i a$. The inertial mass expresses the fact that objects resist being accelerated. To compute the force K we need the field(s) in which the particle moves, and the charge(s) that couple to those field(s). For example, $K = q(E + v \times B/c)$ for a particle with electric charge q moving with speed v in an electric field E and magnetic field B. For a particle with a gravitational charge m_g, usually called the *gravitational mass*, we have $K = -m_g \nabla\Phi$. It follows that $a = -(m_g/m_i)\nabla\Phi$.

It is an experimental fact that materials of different composition and mass experience exactly the same acceleration in a gravitational field. Eötvös verified that with an accuracy of 10^{-8} in 1896, and Dicke attained 2×10^{-11} in 1962. Both experiments used a torsion balance. Presently, torsion balance and free-fall experiments achieve an accuracy of $\sim 10^{-12}$.[5][6] Hence m_g/m_i is

---

[5] Chen and Cook (1993) § 4.8; Will (1993) Ch. 14. With the help of lunar laser-ranging an accuracy of 7×10^{-13} has been achieved (Dickey, J.O. et al., *Science* 265 (1994) 482). The idea is that the lunar orbit as a whole must be displaced along the Earth-Sun line in case the Moon and the Earth experience a slightly different acceleration with respect to the Sun.

[6] The gravitational constant G, however, is only known with a precision of a few times 10^{-4}.

a universal constant, taken to be unity in classical mechanics. This is called the *weak principle of equivalence*. It follows that the concept of gravity loses its meaning, as the field can be made to vanish by transforming to a freely falling reference frame. For an electromagnetic field this is impossible as q/m_i is most certainly not a universal constant. From this Einstein (and others before him) concluded that light must be deflected by a gravitational field, because it moves along a straight line in a freely falling frame where there is no gravity. This trick of transforming gravity away works only locally. In a frame that moves with a freely falling particle, neighbouring particles will initially move uniformly with respect to each other, but not after some time. In the famous elevator thought experiment it is impossible to distinguish *locally* gravity from an externally imposed acceleration. But a distinction is possible by observing two test particles at some distance from each other, because the latter is homogenous while the former is not. The so-called *tidal forces* cannot be transformed away, because a 'real' gravitational field is inhomogeneous.

The fact that inertial and gravitational mass are identical is an unexplained coincidence, in some sense comparable to the unexplained coincidence in Ptolemy's world model, Fig. 1.1. Einstein took that as a basis for a new theory. The fact that motion in a gravitational field depends neither on the composition nor on the mass of the particles suggests that the particle orbits might perhaps be determined by the structure of spacetime. In SR the worldlines of free particles are straight, independent of the nature of the particles. If we now switch on gravity, maybe a more general formulation is possible, in which the worldlines remain 'straight' (i.e. geodesics) in a *curved* spacetime.[7] In that case gravity would no longer be a force, but rather a consequence of the curvature of spacetime. The elaboration of this idea is what we now know as the theory of General Relativity (GR). Global inertial frames no longer exist, only local inertial frames do. For according to GR there are no forces working on freely falling particles, while it is at the same time not possible to define a reference frame in which two freely falling particles move uniformly with respect to each other.

Curvature

That curvature is the way to go ahead may be gleaned, for instance, from the experiments of Pound, Rebka and Snider.[8] Photons moving vertically in the Earth's gravity field turn out to be slightly redshifted, see Fig. 1.9. The

---

[7] A space is said to be *flat* when Euclides's 5th postulate on the existence of a single parallel holds (in metric terms: the Riemann tensor is zero). A space is said to have Euclidean geometry if the metric can be cast in the form $ds^2 = dx^\alpha dx^\alpha$. The Minkowski spacetime of SR is flat but not Euclidean.

[8] Pound, R.V. and Rebka, G.A. *Phys. Rev. Lett. 4* (1960) 337; Pound, R.V. and Snider, J.L. *Phys. Rev. B 140* (1965) 788.

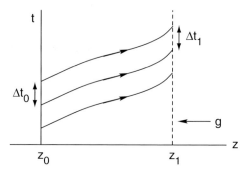

Fig. 1.9. The Pound-Rebka-Snider experiment. Photons move vertically upwards over a distance of $z_1 - z_0 = 22.5$ meters and get redshifted.

required precision could be attained with the help of the Mössbauer effect. The worldlines of subsequent wave crests in the Minkowski diagram must be congruent because the gravity field does not depend on time. Therefore Δt_0 should be equal to Δt_1, regardless of the shape of the worldlines, but the experiment shows that $\Delta t_1 > \Delta t_0$ (a redshift). This suggests (but does not prove) that one can no longer assume that the Minkowski spacetime is globally flat in the presence of gravity. A curved spacetime is descibed by a *local* metric:

$$c^2 \mathrm{d}\tau^2 = \mathrm{d}s^2 = g_{\alpha\beta}\,\mathrm{d}x^\alpha \mathrm{d}x^\beta \;, \tag{1.10}$$

and $\mathrm{d}s^2$ is the interval ('distance') between two events at x^α and $x^\alpha + \mathrm{d}x^\alpha$; $g_{\alpha\beta}$ is the *metric tensor*. The relation $\mathrm{d}s^2 = c^2 \mathrm{d}\tau^2$ between interval and proper time is taken to remain valid (for particles with mass), but the relation between $\mathrm{d}t$ and $\mathrm{d}\tau$ is no longer as simple as in (1.6) because $g_{\alpha\beta} \neq \eta_{\alpha\beta}$. The possibility of transforming gravity away locally amounts to the following requirement: at any point $\{x^\mu\}$ of spacetime there should exist a transformation that casts (1.10) into the SR form $\mathrm{d}s^2 = \eta_{\alpha\beta}\,\mathrm{d}x^\alpha \mathrm{d}x^\beta$. In doing so we have constructed a local inertial (i.e. freely falling) frame in $\{x^\mu\}$ where gravity does not exist[9] – provided the frame is not too big, otherwise we will notice the effect of curvature in the form of tidal forces. Sufficiently small sections of spacetime are flat, 'small' meaning small compared to the typical dimension of the system (the Schwarzschild radius, the scale factor S of the universe, etc.). Spacetime curvature and tidal forces will be the hallmark of a real gravitational field. Weight is merely a pseudo-force caused by being in the wrong (not freely falling) frame, just as centrifugal and Coriolis forces are pseudo-forces caused by being in a wrong (rotating) frame.

---

[9] The terms 'local inertial frame' and 'local freely falling frame' will be used interchangeably. A local rest-frame is a local inertial frame in which a particle or an observer is instantaneously at rest.

Strong equivalence and general covariance

In order to generalise existing physical laws to GR we broaden the scope of the weak equivalence principle, and assume that it is impossible to detect locally any effect of gravity in a freely falling frame, whatever other forces may be acting. In other words, in a freely falling frame *all* laws of physics have the form they have in SR in the absence of gravity. This is called the *strong principle of equivalence*. These laws / equations are then generalised by replacing the tensors that appear in them by tensors that are invariant for *arbitrary* co-ordinate transformations instead of only for Lorentz transformations. This is called the *principle of general covariance*. The application of this principle is somewhat arbitrary, as we shall see, but the obvious way out of adopting the simplest possible generalisation has sofar proven to be effective. The term 'principle of general covariance', incidentally, is misleading in that it has nothing to do with the covariant form of tensors. Principle of general *in*variance (for arbitrary co-ordinate transformations) would have been a much better name. Note also that general covariance has no deeper significance of its own (Friedman, 1983). It is a self-imposed regime of great heuristic value in finding physically correct equations, in some way comparable to checking the correct dimension of an expression.

Mach's principle

A number of ideas, collectively known today under the name *Mach's principle*, have strongly influenced Einstein in his formulation of GR. Mach rejected the Newtonian concept of absolute space, as Leibniz had done earlier. Mach was struck by the fact that the frame defined by the distant matter in the universe happens to be an inertial frame, and that inertia manifests itself only if masses are accelerated with respect to this frame. He argued that this cannot be just a coincidence, and that the inertial mass may somehow be 'induced' by the gravitational mass of all matter in the universe. This led Einstein to seek a theory in which the geometry of spacetime, i.e. $g_{\alpha\beta}$, is determined by the mass distribution. The frame-dragging effect near rotating massive objects, for example (Ch. 6), may be seen as a manifestation of Mach's principle. However, Gödel's solution[10] of the field equations indicates that Mach's principle is only partially contained in GR as it is presently formulated, see Friedman (1983) for more information.

Exercise 1.3: GR and cosmology are fields of many principles. Formulate in your own words the meaning of these principles: relativity, strong and weak equivalence principle, Mach, general covariance, cosmological and the anthropic principle (§ 13.4).

---

[10] Gödel, K., *Rev. Mod. Phys. 21* (1949) 447.

1.3 The need for GR in astrophysics

While SR was born out of the need to resolve a major conflict, namely the failure to measure the velocity (of the Earth) with respect to the ether, GR was created rather for esthetic reasons: the wish to have a relativistic theory of gravity. But there was no compelling conflict with observations that called for a solution. The problem of the perihelium precession of Mercury was known at the time, but was considered to be a nut for the astronomers to crack – not as a stumble block to progress in physics. Consequently, after its conception, GR remained for a long time what is was: an elegant but rather inconsequential theory that was accepted by the physics community precisely because of its elegance. After the correct prediction of the perihelium shift and the spectacular confirmation of the deflection of starlight in 1919, there weren't many other things that could be measured. The technology of the day, for example, was inadequate to detect the gravitational redshift in the solar spectrum. SR on the other hand, led to many observable consequences and was soon completely integrated in the framework of physics as an indispensable basic element. It was recognised that GR was relevant for cosmology,[11] but in the first half of 20th century cosmology was very much a slightly esoteric field that a decent physicist did not touch, because there were very few observations that could show the way. Notions such as a hot big bang, light element synthesis and structure formation were as yet unheard of. And so GR remained outside the mainstream of physics. That state of affairs began to change only in the second half of the 20th century. In particular the 60ies saw a rapid succession of novel developments and discoveries. Technological advances led to a demonstration of the gravitational redshift in the laboratory (1960), soon followed by a measurement in the solar spectrum (1962). Radar reflections from Venus (1964) showed that the travel time of light increases when it moves closely past the Sun. This effect had been predicted by GR as a consequence of the warping of spacetime near a massive object, causing distances to be generally longer.

Astrophysics, too, began to profit from several new developments. Most important were the emergence of radio astronomy, and the possibility to deploy instruments in space which opened up the field of X-ray astronomy. Non-solar X-rays were first detected in 1962 and led to the discovery of X-ray binaries. The X-ray emission is believed to be due to accretion of matter onto a neutron star or black hole, two objects whose existence is predicted by GR. The energy released per unit mass by accretion on such a compact object depends on various parameters, and is of the order of 10% of the infalling rest mass energy – a factor 10-20 more than hydrogen fusion. As the matter falls into the deep potential well, it is heated to X-ray temperatures and serves as

---

[11] In particular the work of Lemaître was influential in this regard (Lemaître, G., *Ann. Soc. Sci. Bruxelles 47A* (1927) 49 and *M.N.R.A.S. 91* (1931) 483).

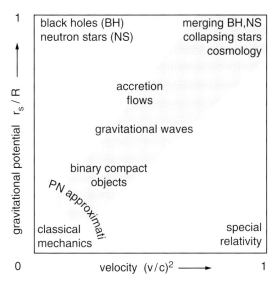

Fig. 1.10. A classification of some applications of General Relativity. For weak fields there is only the horizontal axis. The world of GR unfolds as we move upward to stronger gravitational potential, measured by $r_s/R = 2GM/Rc^2 \sim |\Phi|/c^2$, where $\Phi \sim -GM/R$ and R = typical size of the object (r_s = its Schwarzschild radius). The post-Newtonian approximation gives first order corrections to classical mechanics. Neutron stars and black holes are in the strong field corner. Binary objects have $\Phi \sim -v^2$ and are approximately on the grey diagonal, slowly moving up to their eventual merger and generating gravitational waves as they do so. The latter may also be generated to the left of the diagonal (oscillating / rotating neutron stars) or to the right (close encounters). To position cosmology the universe is considered to be a compact object with expansion velocities approaching c near the horizon (though fields and velocities are locally small).

a bright probe of conditions very close to the compact object. In some cases the mass of the object could be shown to be larger than $3M_\odot$. Since this is larger than the theoretical maximum mass of a neutron star, the object is, in all likelihood, a black hole. Accretion flows thus provide an important diagnostic tool of these compact systems, but it is not the only one. Direct proof of the existence of neutron stars came in 1967 with the discovery of pulsars. It was soon realised (1968) that pulsars are spinning neutron stars equipped with a radio beacon, a feat no one had ever dreamt of. Neutron stars had been hypothesized by Baade and Zwicky (1934) following the discovery of the neutron (1932). They suggested that neutron stars are formed in a supernova explosion, a gravitational collapse of a heavy, evolved star that has run out of nuclear fuel. In 1939 Oppenheimer and Volkoff calculated the structure of a neutron star and showed that it is completely determined by

GR. Now, after 33 years, it turned out that these objects actually did exist. And if stellar evolution, that great creator, can make neutron stars, it may very well produce black holes too. These and other developments led to a revival of theoretical studies in GR which had been stagnant for years. The properties of these mysterious black holes and the generation of gravitational waves, for example, drew much attention. Experimental gravity received a boost as well, leading to the development of detectors for gravitational waves and Gravity Probe B, a space mission for detecting relativistic precession effects – to name only two.

The first binary pulsar was discovered by Hulse and Taylor in 1975. This system turned out to be a perfect cosmic experiment featuring two neutron stars in a tight orbit, one of which is a precision clock. Since the system is clean, application of GR permitted determination of all system parameters. In 1979 it was shown that the system loses energy at a rate that is consistent with energy loss by gravitational waves. This is a strong if indirect argument for the existence of gravitational waves. Several of these binaries have now been found, and there should be many more out there that we cannot see because they contain no pulsar. However, the gravitational waves they emit should be detectable. As the binary loses energy it shrinks and moves slowly along the diagonal in Fig. 1.10 until the components merge in a gigantic explosion, unleashing a final burst of gravitational radiation and γ-rays into space which should be visible throughout the universe. Perhaps this is the explanation of the so-called short-duration γ-ray bursts, whose nature is still not understood. And the hunt for gravitational waves is on: detectors for gravitational waves are in an advanced state of development and several are operating in science mode.

The discovery of quasi-stellar objects or quasars (1963) showed that there are distant objects that are typically 100 times brighter than ordinary galaxies in our neighbourhood. It was gradually understood that these and other objects (Seyferts, BL Lac objects,..) are different visual manifestations of active galactic nuclei (AGNs) with a huge power release, up to 10^{48} erg s^{-1}. Rapid variability pointed to a small gravitational powerhouse casting as the main actors a black hole of $10^6 - 10^9$ $M_\odot$, a surrounding disc swallowing matter (in some cases as much as $10 - 100$ $M_\odot$ per year), and collimated bipolar outflows. Another line of evidence for the existence of massive black holes comes from galactic rotation curves which demonstrate that many galaxies contain heavy objects ($10^6 - 10^9$ $M_\odot$) within a small radius at the centre, very likely a black hole. And there is very strong evidence that a $\sim 3.6 \times 10^6$ $M_\odot$ black hole is lurking at the centre of our own galaxy, which is currently not accreting any appreciable amount of mass.

The gravitational deflection of light by the Sun discovered in 1919 received a spectacular follow-up in 1979 when the quasars Q0957+561 A and

B were identified as two images of the same object whose light is deflected by an intervening galaxy. Many gravitational lenses have been found since then. In principle this opens the possibility to weigh the lens including the dark matter it contains, and to study magnified images of very distant objects. There have been many other advances in cosmology, but there are two that outshone all others. The first is the cosmic microwave background (CMB), discovered in 1965, with suggestions as to its existence dating back to 1946. The CMB was a monumental discovery that marked the beginning of cosmology as a quantitative science. It put an end to the so-called steady state model and permitted for example a quantitative prediction of the synthesis of the light elements in the universe (1967), which has been confirmed by observations. The latest highlight is the WMAP mission which has measured the tiny fluctuations in the temperature of the CMB across the sky. This has resulted in a determination of the basic parameters that fix the structure and evolution of our universe. The second very important development was of a theoretical nature and took place in 1981: the discovery of the possibility of an inflation phase right after the Big Bang. The inflation concept repairs some basic defects of the classical Friedmann-Robertson-Walker cosmology that had to do with causality. The inflation paradigm is very powerful but speculative. Pending some unsettled 'fine-tuning' it seems to explain why the universe expands, why it is homogeneous and flat, as well as the origin of the density fluctuations out of which galaxies evolve later.

This overview illustrates that GR is nowadays being studied in all corners of the diagram of Fig. 1.10. The field has really opened up and there is a great sense of anticipation and promise of new results every day. Particle physicists turn to cosmology in the hope to find answers to questions that particle accelerators seem unable to address. This symbiosis of cosmology and particle physics has sparked off the new field of astroparticle physics. And although it may take years before the detectors for gravitational waves currently in operation actually observe a wave, it may also be tomorrow! This element of suspense and impending surprise renders GR and its application to astrophysics and cosmology a highly attractive field, and some of the thrill, it is hoped, will transpire in the following chapters.

2

Geometry of Riemann Spaces

The fact that the geometry of the space in which we live is Euclidean is a very basic daily experience. This may explain why it took so long before it was realised that this may actually not be correct, and that the question of the geometry of the space around us is a matter of empirical assessment. Early in the 19th century Gauss studied the geometry of curved surfaces, and showed that all references to a flat embedding space could be eliminated. In the same way Riemann formulated in 1854 the geometry of 3D spaces. He found that Euclidean geometry is merely one possibility out of many. Riemann's method could be generalized to spaces of arbitrary dimension. The geometry of these curved Riemann spaces is wholly described within the space itself, by the use of co-ordinates and the metric tensor. No embedding is required. These geometrical concepts gradually spread beyond the mathematical incrowd, and in the last quarter of the 19th century the idea that a fourth (spatial) dimension might exist had mesmerized the public's imagination, perhaps even more so than black holes did a century later. One of the products of that period was Abbott's famous Flatland.[1] The flatland analogy is nowadays a standard technique of teachers to explain some of the intricacies of curved spaces.

The theoretical framework of Riemann spaces is also the starting point for the mathematical formulation of GR. In this chapter we discuss the tools that any student should master in order to be able to deal with GR beyond the level of handwaving. In doing so we have deliberately chosen to stay close to intuition as that outweighs the merits of rigour, certainly on first acquaintance.

2.1 Definition

A Riemann space has the following properties:

---

[1] Abbott, E.A.: 1884, *Flatland: A Romance of many Dimensions, by a Square*, Seeley & Co. (London).

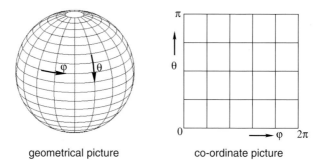

Fig. 2.1. A geometrical picture and the corresponding co-ordinate picture of the space defined by (2.2). Co-ordinate pictures will be frequently used.

1. Any point can be identified by a set of co-ordinates $\{x^\mu\}$; the number of independent x^μ is called the dimension.

2. It is possible to define continuously differentiable functions of $\{x^\mu\}$, in particular one-to-one co-ordinate transformations $\{x^\mu\} \leftrightarrow \{\bar{x}^\nu\}$.

3. There is a metric that specifies the distance ds^2 between two nearby points x^μ and $x^\mu + dx^\mu$:

$$ds^2 = g_{\alpha\beta}\, dx^\alpha dx^\beta \; ; \qquad g_{\alpha\beta} = g_{\beta\alpha} \; . \tag{2.1}$$

An antisymmetric part of $g_{\alpha\beta}$ does not contribute to ds^2. Example: a spherical surface with radius 1 and co-ordinates θ, φ:

$$ds^2 = d\theta^2 + \sin^2\theta\, d\varphi^2 \; . \tag{2.2}$$

Notation: $d\theta^2 \equiv (d\theta)^2$, $d\varphi^2 \equiv (d\varphi)^2$, but $ds^2 = (ds)^2$ only if $ds^2 > 0$ as in (2.2). But the metric is in general not positive definite! In this simple case the geometrical structure may be visualised through embedding in an Euclidean space of one higher dimension, but for Riemann spaces of higher dimension this is no longer possible. Moreover, a Riemann space of dimension D cannot always be embedded in a flat space of dimension $D + 1$. It is often useful to draw a *co-ordinate picture* of a suitably chosen subspace, even though it contains no information on the geometry, see Fig. 2.1.

An important point is that the metric determines the local structure of the space, but reveals nothing about its global (topological) structure. A plane, a cone and a cylinder all have the same metric $ds^2 = dx^2 + dy^2$, but entirely different global structures.

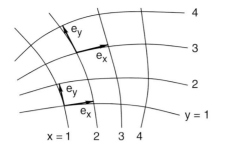
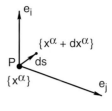

Fig. 2.2. Co-ordinate lines and base vectors spanning the tangent space. The choice of the co-ordinates is entirely free, and in practice dictated by the question which co-ordinates are the most expedient to use.

2.2 The tangent space

In each point we construct a set of base vectors tangent to the co-ordinate lines, as in Fig. 2.2. The arrow points towards increasing x^i. The base vectors span the flat tangent space, which has the same dimension as Riemann space. This construction evidently requires the existence of a flat embedding space, but that can be avoided as follows. Consider the curves $\{x^\alpha(p)\}$ through a point P in Riemann space (p = curve parameter), and construct $A^\sigma = [\mathrm{d}x^\sigma/\mathrm{d}p]_P$. These vectors A^σ span the abstract tangent space of P, which exists independent of any embedding. Usually, however, the abstract tangent space may be identified with the tangent space constructed in Fig. 2.2. For our discussion there is no real advantage in making the distinction and we shall work with the intuitive picture of Fig. 2.2.

We may use any metric we like in the tangent space, but there exists a preferred metric. Consider an infinitesimal section of Riemann space. This section is flat and virtually coincides with the tangent space. To an infinitesimal vector $\mathrm{d}\boldsymbol{s} = \mathrm{d}x^\alpha \boldsymbol{e}_\alpha$ *in the tangent space* we may therefore assign the length of the line element ds *in Riemann space*, i.e. we require $\mathrm{d}\boldsymbol{s} \cdot \mathrm{d}\boldsymbol{s} = \mathrm{d}s^2$:

$$\begin{aligned}\mathrm{d}\boldsymbol{s} \cdot \mathrm{d}\boldsymbol{s} &= (\mathrm{d}x^\alpha \boldsymbol{e}_\alpha) \cdot (\mathrm{d}x^\beta \boldsymbol{e}_\beta) = \boldsymbol{e}_\alpha \cdot \boldsymbol{e}_\beta \, \mathrm{d}x^\alpha \mathrm{d}x^\beta \\ &= g_{\alpha\beta} \, \mathrm{d}x^\alpha \mathrm{d}x^\beta \;, \end{aligned} \qquad (2.3)$$

and it follows that

$$g_{\alpha\beta} \equiv \boldsymbol{e}_\alpha \cdot \boldsymbol{e}_\beta \;. \qquad (2.4)$$

Here · represents the vector inner product. This may be the usual inner product, for example when we deal with 2D surfaces embedded in a flat R_3. But in case of the Minkowski spacetime of SR, and in GR, the inner product is not positive definite, and we may have that $\boldsymbol{A} \cdot \boldsymbol{A} < 0$ (for spacelike vectors). By taking $\mathrm{d}x^\alpha = 1$ in (2.3) and all other $\mathrm{d}x^\beta = 0$ we see that $\boldsymbol{e}_\alpha \cdot \boldsymbol{e}_\alpha = \mathrm{d}\boldsymbol{s} \cdot \mathrm{d}\boldsymbol{s}$ (no summation). It follows that the 'length' of $\boldsymbol{e}_\alpha$ corresponds to a jump $\Delta x^\alpha = 1$, at constant value of the other co-ordinates. Due to the curvature this is of course only approximately correct. These base vectors are called a co-ordinate basis because they are defined entirely by the co-ordinates and the metric. The length of the base vectors depends on the choice of the co-ordinates, and is in general a function of position. Consider for example polar co-ordinates in a plane, Fig. 2.3. The length of $\boldsymbol{e}_r$ is constant, while $|\boldsymbol{e}_\varphi| \propto r$:

$$\mathrm{d}s^2 = 1 \cdot \mathrm{d}r^2 + r^2 \mathrm{d}\varphi^2 \ . \qquad (2.5)$$
$$\uparrow \qquad \uparrow$$
$$\boldsymbol{e}_r \cdot \boldsymbol{e}_r \quad \boldsymbol{e}_\varphi \cdot \boldsymbol{e}_\varphi$$

Now that we have defined the basis we may construct finite vectors $\boldsymbol{A} = A^\alpha \boldsymbol{e}_\alpha$ in the tangent space through the usual parallelogram construction. These so called contravariant components A^α are the components of $\boldsymbol{A}$ along the basis.

The next step is to define another (covariant) representation A_α of $\boldsymbol{A}$ by demanding that $\boldsymbol{A} \cdot \boldsymbol{A} = A_\alpha A^\alpha$, for every $\boldsymbol{A}$:

$$\boldsymbol{A} \cdot \boldsymbol{A} = (A^\alpha \boldsymbol{e}_\alpha) \cdot (A^\beta \boldsymbol{e}_\beta) = g_{\alpha\beta} A^\beta A^\alpha \equiv A_\alpha A^\alpha \ , \qquad (2.6)$$

which leads to:

$$A_\alpha = g_{\alpha\beta} A^\beta \ . \qquad (2.7)$$

In a more advanced treatment a distinction is made between tensors as geometrical objects, their contravariant representation located in an abstract tangent space, and the dual tangent space, in which the covariant representations reside. In the current, more primitive context the following interpretation suggests itself. Since $A_\gamma = g_{\gamma\beta} A^\beta = \boldsymbol{e}_\gamma \cdot \boldsymbol{e}_\beta A^\beta = (A^\beta \boldsymbol{e}_\beta) \cdot \boldsymbol{e}_\gamma = \boldsymbol{A} \cdot \boldsymbol{e}_\gamma$, it follows that A_γ is the projection of $\boldsymbol{A}$ on $\boldsymbol{e}_\gamma$. Hence, the contravariant components A^β are the components of $\boldsymbol{A}$ along the base vectors $\boldsymbol{e}_\beta$ (parallelogram construction), while the covariant component A_α is the projection of $\boldsymbol{A}$ on the base vector $\boldsymbol{e}_\alpha$, Fig. 2.3, right:

$$\text{contravariant } (A^\beta): \ \boldsymbol{A} = A^\beta \boldsymbol{e}_\beta \ , \qquad (2.8)$$
$$\text{covariant } (A_\alpha): \ A_\alpha = \boldsymbol{A} \cdot \boldsymbol{e}_\alpha \ . \qquad (2.9)$$

Finally, the concept of *index raising and lowering*. We can lower an index with the help of (2.7). The inverse operation of raising is defined as:

$$A^\gamma = g^{\gamma\alpha} A_\alpha \ . \qquad (2.10)$$

The meaning of $g^{\gamma\alpha}$ can be gleaned from:

$$A^\gamma = g^{\gamma\alpha} A_\alpha = g^{\gamma\alpha} g_{\alpha\nu} A^\nu \ , \qquad (2.11)$$

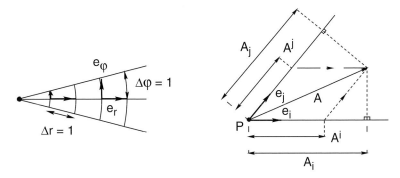

Fig. 2.3. Left: polar co-ordinates and the base vectors e_r and e_φ. Right: interpretation of the contravariant and covariant representation of a vector $\boldsymbol{A}$.

so that $g^{\gamma\alpha}g_{\alpha\nu} = \delta^\gamma_\nu$, i.e. $\{g^{\gamma\alpha}\}$ is the inverse of $\{g_{\alpha\nu}\}$. In summary:

$$\left.\begin{aligned} \text{index lowering}: \quad & A_\alpha = g_{\alpha\beta}A^\beta\,,\\ \text{index raising}: \quad & A^\gamma = g^{\gamma\nu}A_\nu\,,\\ & \{g^{\gamma\nu}\} = \{g_{\alpha\beta}\}^{-1}\,. \end{aligned}\right\} \qquad (2.12)$$

We have silently adopted the *summation convention*: if an index occurs twice, once as a lower and once as an upper index, summation over that index is implied. Note that the rules for index raising and lowering are always valid, and have nothing to do with the question whether one is dealing with a tensor or not. The tensor concept is related to behaviour under co-ordinate transformations, which was not an issue above, and to which we turn our attention now.

2.3 Tensors

We are now in a position to do linear algebra in the tangent space, but we leave that aside and study the effect of co-ordinate transformations. Consider two overlapping sets of co-ordinates $\{x^\mu\}$ and $\{x^{\mu'}\}$. The notation is sloppy – it would be more appropriate to write $\{\bar{x}^\mu\}$ instead of $\{x^{\mu'}\}$, but $\{x^{\mu'}\}$ is much more expedient if used with care. A displacement $\delta x^{\mu'}$ is related to a displacement δx^ν through:

$$\delta x^{\mu'} = \frac{\partial x^{\mu'}}{\partial x^\nu}\delta x^\nu \equiv x^{\mu'}_{,\nu}\delta x^\nu\,. \qquad (2.13)$$

Notation:
$$X_{,\nu} \equiv \frac{\partial X}{\partial x^\nu} \; ; \qquad X_{,\nu\rho} \equiv \frac{\partial^2 X}{\partial x^\nu \partial x^\rho} \; , \qquad \text{etc.} \qquad (2.14)$$

where X can be anything (A_α, $g^{\alpha\beta}$, ...). We may freely interchange indices behind the comma: $X_{,\alpha\beta\gamma} = X_{,\alpha\gamma\beta} = X_{,\gamma\alpha\beta}$ etc.

Any set A^μ transforming according to (2.13) is called a contravariant tensor of rank 1:
$$A^{\mu'} = x^{\mu'}_{,\nu} A^\nu \quad \leftrightarrow \quad A^\nu \quad \text{contravariant.} \qquad (2.15)$$

Hence δx^ν is a contravariant tensor. Tensors of rank 1 are often referred to as vectors, and henceforth we shall use the word vector in this sense only. A function such as the temperature distribution $T(x)$ is called a *scalar*, a tensor of rank zero. Its value in a point is independent of the co-ordinate system, i.e. invariant for co-ordinate transformations: $T'(x') = T(x)$, where T' is the new function prescription. The derivative of a scalar Q,
$$B_\mu = \frac{\partial Q}{\partial x^\mu} \equiv Q_{,\mu} \qquad (2.16)$$

transforms like $B_{\mu'} = Q'_{,\mu'} = Q_{,\nu} x^\nu_{,\mu'} = x^\nu_{,\mu'} B_\nu$. Every B_ν that transforms in this way is called a covariant vector or tensor of rank 1:
$$B_{\mu'} = x^\nu_{,\mu'} B_\nu \quad \leftrightarrow \quad B_\nu \quad \text{covariant.} \qquad (2.17)$$

From two covariant vectors we can form $T_{\mu\nu} = A_\mu B_\nu$, a covariant tensor of rank 2. More general tensors can be constructed through summation, $T_{\mu\nu} = A_\mu B_\nu + C_\mu D_\nu + ...$ This process may be continued: $T_{\alpha\beta} C^\gamma$ and $A_\mu C^\nu B_\rho$ are mixed tensors of rank 3 (provided T, A, B and C are tensors themselves). The indices of tensors of higher rank transform according to (2.15) resp. (2.17), for example:
$$T^{\alpha'}{}_{\beta'\gamma'}{}^{\delta'} = x^{\alpha'}_{,\mu} \, x^\nu_{,\beta'} \, x^\sigma_{,\gamma'} \, x^{\delta'}_{,\tau} \, T^\mu{}_{\nu\sigma}{}^\tau \; . \qquad (2.18)$$

There is no other choice because (2.18) must hold for the special tensor $T^\alpha{}_{\beta\gamma}{}^\delta = P^\alpha Q_\beta R_\gamma S^\delta$, and the transformation rules for vectors have already been fixed! Note that we get a glimpse here of how the Lorentz transformations of SR will be generalised in GR: relation (1.7) of SR will be replaced by (2.15). This transformation is still locally linear, but different in each point of Riemann space as the $\{x^{\mu'}_{,\nu}\}$ are functions of position. The single global Lorentz transformation will be replaced by a mesh of local Lorentz transformations.

The horizontal position of the indices is important: $T^\mu{}_\nu$ is different from $T_\nu{}^\mu$! The summation over double indices is called *contraction*. It lowers the rank by two. For example $T^\mu{}_\mu$, $T^\alpha{}_{\beta\alpha}{}^\gamma$, $P_{\alpha\beta} Q^{\beta\gamma}$, $T^\alpha{}_{\beta\alpha}{}^\beta$ (double contraction). Double indices are *dummies*: $T^\alpha{}_\alpha = T^\mu{}_\mu$, dummies may occur only twice, once as an upper and once as a lower index. If you encounter expressions like $C^{\mu\mu}$, $P^{\alpha\beta}Q^\alpha_\gamma$ or $P_{\alpha\beta}Q^{\alpha\gamma}R_{\delta\alpha}$ then you have made a mistake somewhere!

Index raising and lowering, finally, is done by factors $g_{\alpha\beta}$ or $g^{\mu\nu}$ for each upper/lower index, e.g.:

$$\left.\begin{aligned} T^{\mu\nu} &= g^{\mu\alpha}T_\alpha{}^\nu\;, \\ T^\alpha{}_\beta{}^{\gamma\delta} &= g^{\alpha\mu}g_{\beta\nu}g^{\delta\sigma}T_\mu{}^{\nu\gamma}{}_\sigma\;,\quad \text{etc.} \end{aligned}\right\} \tag{2.19}$$

Again, like in (2.18), we have hardly any other choice here, because (2.19) must hold for the special tensors $T^{\mu\nu} = P^\mu Q^\nu$ and $T^\alpha{}_\beta{}^{\gamma\delta} = P^\alpha Q_\beta R^\gamma S^\delta$, and the rules for index raising and lowering for vectors have already been fixed. We are now in a position that we can raise and lower indices at liberty. We emphasise once more that the rules (2.12) and (2.19) for index gymnastics are generally valid, also for non-tensors. For example, $Q_{\mu\nu} = A_{\mu,\nu}$ is not a tensor (exercise 2.4), and yet $Q^\mu{}_\nu = g^{\mu\alpha}Q_{\alpha\nu}$.

Exercise 2.1: The unit tensor is defined as $\delta^\alpha{}_\beta = 1$ for $\alpha = \beta$, otherwise 0. Prove that $\delta^\alpha{}_\beta$ is a tensor, and that $\delta^\alpha{}_\beta = \delta_\beta{}^\alpha$, so that we may write δ^α_β without risk of confusion. Show that $\delta_{\alpha\beta} = g_{\alpha\beta}$. Is $\eta_{\alpha\beta}$ a tensor? And $g_{\alpha\beta}$? One could define $\delta_{\alpha\beta} = 1$ for $\alpha = \beta$, and 0 otherwise, but then $\delta_{\alpha\beta}$ is not a tensor.

Hint: $\delta^{\alpha'}{}_{\beta'}$ must be equal to $x^{\alpha'}{}_{,\nu}\,x^\mu{}_{,\beta'}\,\delta^\nu{}_\mu$, or $\delta^{\alpha'}{}_{\beta'} = x^{\alpha'}{}_{,\nu}\,x^\nu{}_{,\beta'} = x^{\alpha'}{}_{,\beta'}$ (chain rule) $= 1$ for $\alpha' = \beta'$ otherwise 0. Hence $\delta^\alpha{}_\beta$ is tensor. And $\delta_\beta{}^\alpha = g_{\beta\mu}g^{\alpha\nu}\delta^\mu{}_\nu = g_{\beta\mu}g^{\alpha\mu} = g^{\alpha\mu}g_{\mu\beta} = 1$ for $\alpha = \beta$, otherwise 0, i.e. identical to $\delta^\alpha{}_\beta$; $\delta_{\alpha\beta} = g_{\alpha\nu}\delta^\nu{}_\beta = g_{\alpha\beta}$; $\eta_{\alpha\beta}$ is a tensor in SR only, i.e. under Lorentz transformations; $g_{\alpha\beta}$ tensor: use (2.1), require that ds^2 is also tensor in GR (invariant scalar), and dx^α is tensor, then exercise 2.3. Other definition $\delta_{\alpha\beta}$: $\delta_{\alpha'\beta'} = x^\nu{}_{,\alpha'}\,x^\mu{}_{,\beta'}\,\delta_{\nu\mu}$? No, because the chain rule can no longer be used.

Exercise 2.2: If $T^{\alpha\beta}$ and $P^\mu{}_\nu$ are tensors then $P^\mu{}_\mu$ is a scalar, but $T^{\alpha\alpha}$ is not. The inner product $A_\nu B^\nu$ of two vectors is a scalar.

Hint: $P^{\mu'}{}_{\mu'} = x^{\mu'}{}_{,\alpha}\,x^\beta{}_{,\mu'}\,P^\alpha{}_\beta$, then the chain rule.

Exercise 2.3: Quotient theorem: If $A^\lambda P_{\lambda\mu\nu}$ is a tensor for arbitrary vector A^λ, then $P_{\lambda\mu\nu}$ is a tensor; $\mu\nu$ may be replaced with an arbitrary sequence of upper / lower indices.

Hint: $A^\lambda P_{\lambda\mu\nu}$ is a tensor, i.e. $A^{\lambda'} P_{\lambda'\mu'\nu'} = x^\alpha_{,\mu'} x^\beta_{,\nu'} A^\sigma P_{\sigma\alpha\beta}$ (λ' and σ are dummies!), then substitute $A^\sigma = x^\sigma_{,\lambda'} A^{\lambda'}$, etc.

Exercise 2.4: The derivative $A_{\mu,\nu}$ of a covariant vector A_μ is not a tensor, as it transforms according to:

$$A_{\mu',\nu'} = A_{\alpha,\beta}\, x^\alpha_{,\mu'} x^\beta_{,\nu'} + A_\alpha\, x^\alpha_{,\mu'\nu'}\ . \tag{2.20}$$

The problem is in the second term of (2.20). In SR only linear (Lorentz) transformations are allowed. In that case the second term is zero and $A_{\mu,\nu}$ is a tensor.

Hint: Start from $A_{\mu',\nu'} = (x^\alpha_{,\mu'} A_\alpha)_{,\nu'}$, then use the product rule.

Exercise 2.5: Prove $T_\alpha{}^\nu A_\nu = T_{\alpha\nu} A^\nu$; $T_\alpha{}^\alpha = T^\alpha{}_\alpha$; $g^\nu{}_\nu = 4$; $\eta^\nu{}_\nu = g^{00} - g^{11} - g^{22} - g^{33}$.

Hint: We know that $g^\nu{}_\nu = g^{\nu\alpha} g_{\alpha\nu} = \delta^\nu_\nu = 4$. The following may be illuminating: the scalar $g^\nu{}_\nu$ is invariant, compute in a freely falling frame: $g^\nu{}_\nu = \eta^\nu{}_\nu$, SR holds in that frame: $\eta^\nu{}_\nu = \eta^{\nu\alpha} \eta_{\alpha\nu} = 4$. But in GR: $\eta^\nu{}_\nu = g^{\nu\alpha} \eta_{\alpha\nu} =$ etc.

2.4 Parallel transport and Christoffel symbols

Consider a particle at position P in Riemann space, Fig. 2.4. The vectors associated with it (velocity, spin, ..) reside in the tangent space of P. At some later time the particle has moved to position Q, but the tangent space of Q does not coincide with that of P. To be able to do dynamics, we must develop a way to compare vectors in the different tangent spaces along the worldline of the particle. In other words, we need something against which to gauge the concept of 'change'. This is what parallel transport in GR is about.

Fig. 2.4 shows the curve $x^\sigma(p)$ in Riemann space. The vector $\boldsymbol{A}$ is always in the tangent space, but the tangent spaces of P, Q, R, .. are disjunct, and comparison of $\boldsymbol{A}(P)$ with $\boldsymbol{A}(Q)$ or $\boldsymbol{A}(R)$ is not possible. To this end we define a *connection* between tangent spaces, that is, a mathematical prescription telling us how a vector $\boldsymbol{A}(P)$ lies in the tangent space of Q if we 'transport' it along a given path from P to Q. This can be done in a variety of ways, but much of the mathematical freedom that we have is eliminated by the physical

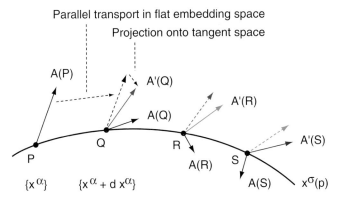

Fig. 2.4. Conceptual definition of parallel displacement of a vector along a curve $x^\sigma(p)$ in Riemann space: first an ordinary parallel displacement in the flat embedding space (resulting in the dashed arrows) followed by projection on the local tangent space. The process is repeated in infinitesimal steps.

requirement that we recover what we ordinarily do when we transport a vector parallel to itself in a flat space. Imagine the Riemann space embedded in a flat space of higher dimension. We know how to move $\boldsymbol{A}(P)$ around parallel to itself in this embedding space, because it is flat. Having arrived in Q, the result is projected onto the local tangent space. To order $O(\mathrm{d}x^\alpha)$ projection does not change the length of the vector: the projection angle γ is $O(\mathrm{d}x^\alpha)$, but $\cos\gamma = 1$ up to $O(\mathrm{d}x^\alpha)$. This process is now repeated with infinitesimal steps, and generates the coloured vector field $\boldsymbol{A}'$ in Fig. 2.4, starting from $\boldsymbol{A}(P)$. In this way we have generalized the concept of parallel transport to curved spaces, in such a way that it reduces to normal parallel transport for flat spaces. Not surprisingly, it is also the definition that turns out to work in GR. The result of the transport operation depends on the path, see Fig. 2.5. However, when $\boldsymbol{e}$ in Fig. 2.5 is parallel-transported along a small curve on the sphere there is virtually no change, because there is hardly any curvature felt (exercise 2.17).

We now formalise our intuitive approach. The difference $\mathrm{d}\boldsymbol{A} = \boldsymbol{A}(Q) - \boldsymbol{A}(P)$ is not defined, but up to order $O(\mathrm{d}x^\alpha)$ we have that $\mathrm{d}\boldsymbol{A} \simeq \boldsymbol{A}(Q) - \boldsymbol{A}'(Q)$, and this is useful as both vectors lie in the same tangent space. The vector $\mathrm{d}\boldsymbol{A}$ may be interpreted as the intrinsic change of $\boldsymbol{A}$, after correction for the 'irrelevant' change in the orientation of the tangent space:

$$\mathrm{d}\boldsymbol{A} \simeq \boldsymbol{A}(Q) - \boldsymbol{A}'(Q) \tag{2.21}$$

$$= \mathrm{d}(A^\mu \boldsymbol{e}_\mu) = (\mathrm{d}A^\mu)\boldsymbol{e}_\mu + A^\mu(\mathrm{d}\boldsymbol{e}_\mu). \tag{2.22}$$

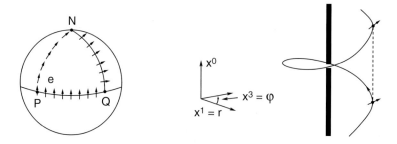

Fig. 2.5. Left: Parallel displacement of the vector e along PNQ and along PQ produces entirely different results. To the right, the geodesic precession of a top in orbit around a central mass, see text.

Here, $\mathrm{d}\boldsymbol{A}$ has been split into two contributions: the change $\mathrm{d}A^\mu \equiv A^\mu(Q) - A^\mu(P)$ of the contravariant components of $\boldsymbol{A}$, and a contribution from the change of the base vectors. On general grounds we anticipate $\mathrm{d}\boldsymbol{e}_\mu$ to be proportional to both $\{\mathrm{d}x^\beta\}$ and $\{\boldsymbol{e}_\alpha\}$:

$$\mathrm{d}\boldsymbol{e}_\mu = \Gamma^\alpha_{\mu\beta}\, \mathrm{d}x^\beta\, \boldsymbol{e}_\alpha\ . \tag{2.23}$$

$\Gamma^\alpha_{\mu\beta}$ is called the *Christoffel symbol of the second kind*, and as may be expected it is intimately related to the metric tensor:

$$\Gamma^\mu_{\nu\sigma} = \tfrac{1}{2} g^{\mu\lambda}\left(g_{\lambda\nu,\sigma} + g_{\lambda\sigma,\nu} - g_{\nu\sigma,\lambda}\right) \equiv g^{\mu\lambda}\Gamma_{\lambda\nu\sigma}\ . \tag{2.24}$$

The $=$ sign is proved in § 2.5. The $\equiv$ sign defines the *Christoffel symbol of the first kind*, simply by raising one index with $g^{\mu\lambda}$. According to (2.23) the Christoffel symbols define the connection between the base vectors of the tangent spaces at different positions. As pointed out above, there exist more general connection coefficients than (2.24), but these play no role in GR.

Insert (2.23) in (2.22) and rename the dummy-indices:

$$\mathrm{d}\boldsymbol{A} = (\mathrm{d}A^\mu + \Gamma^\mu_{\nu\sigma} A^\nu\, \mathrm{d}x^\sigma)\,\boldsymbol{e}_\mu \equiv (\mathrm{D}A^\mu)\,\boldsymbol{e}_\mu\ . \tag{2.25}$$

The right hand side defines the intrisic change $\mathrm{D}A^\mu$, which apparently obeys the following equation:

$$\frac{\mathrm{D}A^\mu}{\mathrm{D}p} = \frac{\mathrm{d}A^\mu}{\mathrm{d}p} + \Gamma^\mu_{\nu\sigma} A^\nu \frac{\mathrm{d}x^\sigma}{\mathrm{d}p} \qquad \text{(contravariant)}; \tag{2.26}$$

$$\frac{\mathrm{D}A_\mu}{\mathrm{D}p} = \frac{\mathrm{d}A_\mu}{\mathrm{d}p} - \Gamma^\nu_{\mu\sigma} A_\nu \frac{\mathrm{d}x^\sigma}{\mathrm{d}p} \qquad \text{(covariant)}. \tag{2.27}$$

For the second relation (2.27) see exercise 2.8. We may apply these equations in two ways. For a *given* vector field we may compute DA^μ or DA_μ for a displacement dp along $x^\sigma(p)$. On the other hand, one may solve $DA^\mu/Dp = 0$ or $DA_\mu/Dp = 0$ starting from an initial value $A^\mu(P)$ or $A_\mu(P)$, and construct a vector field along $x^\sigma(p)$ for which $d\boldsymbol{A} = \boldsymbol{A} - \boldsymbol{A}' = 0$. Parallel transport of a vector along $x^\sigma(p)$ is therefore described by the differential equation

$$\frac{DA_\mu}{Dp} = 0 \quad \text{or} \quad \frac{DA^\mu}{Dp} = 0 \ . \tag{2.28}$$

We mention a few properties of the Christoffel symbols. They are symmetrical in the last two indices:

$$\Gamma^\mu_{\nu\sigma} = \Gamma^\mu_{\sigma\nu} \ ; \quad \Gamma_{\lambda\nu\sigma} = \Gamma_{\lambda\sigma\nu} \ . \tag{2.29}$$

By interchanging the indices in (2.24) we may infer $\Gamma_{\nu\lambda\sigma}$, and on adding that to $\Gamma_{\lambda\nu\sigma}$ one obtains

$$\Gamma_{\lambda\nu\sigma} + \Gamma_{\nu\lambda\sigma} = g_{\lambda\nu,\sigma} \ . \tag{2.30}$$

The Christoffel symbol transforms according to

$$\Gamma^{\mu'}_{\nu'\sigma'} = \Gamma^\rho_{\alpha\beta} \, x^{\mu'}_{,\rho} \, x^\alpha_{,\nu'} \, x^\beta_{,\sigma'} + x^{\mu'}_{,\rho} \, x^\rho_{,\nu'\sigma'} \ . \tag{2.31}$$

The proof is for diehards (see literature). The first term is what we would expect if the Christoffel symbol were a tensor, but the second term makes that it is actually not a tensor. The concept of parallel transport will be used in § 2.5 to define geodesics.

In SR the velocity and spin vector of a particle on which no forces are exerted are constant. They are transported parallel along the 'straight' orbit of the particle. The idea of GR is that a particle under the influence of gravity moves freely in a *curved* spacetime. A natural generalisation is that velocity and spin vector of the particle can be found by parallel transport along the orbit in spacetime. In this way we are able to understand the geodesic precession of a top. Fig. 2.5 shows a co-ordinate picture, with $x^0 = ct$ on the vertical axis and polar co-ordinates $x^1 = r$ and $x^3 = \varphi$ in the horizontal plane. The worldline of the top orbiting the central object (vertical bar) is a spiral. The spin 4-vector (whose spatial part is directed along the spin axis) is parallel-transported along the worldline. After one revolution the top has returned to same spatial position, but because spacetime is not flat – not visible in a co-ordinate picture – the spin vector has changed its direction. At this point one may wonder how the effect is related to the Thomas precession. We refer to Ch. 8 for a more general treatment, from which both Thomas precession and geodesic precession emerge in the appropriate limit.

Exercise 2.6: The length of a vector remains constant under parallel transport:
$$\mathrm{d}\, A^\nu A_\nu = \mathrm{d}\,(g_{\mu\nu} A^\mu A^\nu) = 0\,.$$
Hint: First attempt: $\mathrm{d} = \mathrm{D} =$ intrinsic change: $\mathrm{D} A^\nu A_\nu = (\mathrm{D} A^\nu) A_\nu + A^\nu (\mathrm{D} A_\nu) = 0$, because $\mathrm{D} A^\nu = (\mathrm{D} A^\nu/\mathrm{D} p)\,\mathrm{d} p = 0$, etc. But (2.27) must still be proven, and for that we need $\mathrm{d}\, A^\nu A_\nu = 0$. Second attempt: $\mathrm{d} =$ total change: $\mathrm{d}\, g_{\mu\nu} A^\mu A^\nu = 2 A_\nu \mathrm{d} A^\nu + A^\mu A^\nu g_{\mu\nu,\sigma}\,\mathrm{d} x^\sigma$; (2.26): $\mathrm{d} A^\nu = -\Gamma^\nu_{\mu\sigma} A^\mu \mathrm{d} x^\sigma$; exercise 2.5 and (2.30): $2 A_\nu \mathrm{d} A^\nu = -2\Gamma_{\nu\mu\sigma} A^\nu A^\mu \mathrm{d} x^\sigma = -g_{\nu\mu,\sigma} A^\nu A^\mu \mathrm{d} x^\sigma$.

Exercise 2.7: Prove that $\mathrm{d}\, A^\nu B_\nu = 0$ under parallel transport.

Hint: The length of $A^\nu + B^\nu$ is constant.

Exercise 2.8: For parallel transport of a covariant vector:
$$\mathrm{d} B_\mu = \Gamma^\nu_{\mu\sigma} B_\nu\, \mathrm{d} x^\sigma\,. \qquad (2.32)$$

Hint: $0 = \mathrm{d}\, A^\mu B_\mu = A^\mu\, \mathrm{d} B_\mu + B_\mu\, \mathrm{d} A^\mu$, and $\mathrm{d} A^\mu$ is known.

Exercise 2.9: Prove that
$$\Gamma^\mu_{\nu\mu} = g_{,\nu}/2g = \tfrac{1}{2}\bigl(\log|g|\bigr)_{,\nu}\,;\qquad g = \det\{g_{\alpha\beta}\}\,. \qquad (2.33)$$

Hint: (2.24): $\Gamma^\mu_{\nu\mu} = \tfrac{1}{2} g^{\lambda\mu} g_{\lambda\mu,\nu}$. For a matrix M we have that $\mathrm{Tr}\,(M^{-1} M_{,\nu}) = (\mathrm{Tr}\,\log M)_{,\nu} = (\log \det M)_{,\nu}$. Take $M = \{g_{\alpha\beta}\}$.

2.5 Geodesics

Intuitively, a geodesic is a line that is 'as straight as possible' on a curved surface. We say that a curve $x^\mu(p)$ is a geodesic when the tangent vector $\mathrm{d} x^\mu/\mathrm{d} p$ remains a tangent vector under parallel transport along $x^\mu(p)$. Therefore $\dot{x}^\mu \equiv \mathrm{d} x^\mu/\mathrm{d} p$ must satisfy (2.28), and we arrive at the geodesic equation:
$$\frac{\mathrm{D}}{\mathrm{D} p}\left(\frac{\mathrm{d} x^\mu}{\mathrm{d} p}\right) = 0 \quad\to\quad \ddot{x}^\mu + \Gamma^\mu_{\nu\sigma}\dot{x}^\nu \dot{x}^\sigma = 0\,, \qquad (2.34)$$

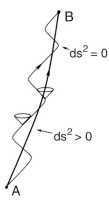

Fig. 2.6. A timelike geodesic connecting events A and B is the curve with the *maximum* possible interval length between A and B, see text.

with $\dot{} = d/dp$. For timelike geodesics[2] the parameter p in (2.34) is proportional to the interval length s. Proof: according to exercise 2.6 the length of $\dot{x}^\alpha = dx^\alpha/dp$ is constant along $x^\mu(p)$, i.e. $\dot{x}^\alpha \dot{x}_\alpha = g_{\alpha\beta} \dot{x}^\alpha \dot{x}^\beta \equiv (ds/dp)^2$ is constant. For timelike geodesics $ds^2 > 0$, and we may take the square root to conclude that $ds = \text{const} \cdot dp$. Later, when GR is cast into the geometrical framework developed here, this result will be connected to proper time (a physical concept that does not yet exist here): $ds = c d\tau$, so that

$$dp \propto ds \propto d\tau \quad \text{for timelike geodesics.} \quad (2.35)$$

This is important as it implies that we may, for timelike geodesics, replace the curve parameter p in (2.34) by the interval length s or the proper time τ.

Intuitively, a geodesic is also the shortest possible route between two points. For a positive definite metric this is indeed the case, but ds^2 can be positive as well as negative in GR. Assuming that the interval $\int ds = \int \dot{s} \, dp$ of a timelike geodesic is an extremum (see below), it is easy to see that it should be a maximum: there always exists an arbitrarily nearby worldline that has a smaller $\int ds$, by letting it jump more or less from light-cone to light-cone, as in Fig. 2.6 (see e.g. Wald (1984) § 9.3). The construction of Fig. 2.6 fails for spacelike geodesics.

---

[2] In an analogy with (1.4) we speak of a timelike (spacelike) worldline or geodesic when $ds^2 > 0$ ($ds^2 < 0$). A null worldline or null geodesic has $ds^2 = 0$. For spacelike and null geodesics p can no longer be interpreted as an interval length.

Eq. (2.34) may also be derived from a variational principle.[3] The simplest is $\delta \int \dot{s}\,dp = 0$, and this is equivalent to $\delta \int F(\dot{s})\,dp = 0$ provided F is monotonous, $F' \neq 0$. We choose $\delta \int \dot{s}^2\,dp = 0$, or

$$\delta \int L\,dp = 0 ; \quad L(x^\alpha, \dot{x}^\beta) = (ds/dp)^2 = g_{\alpha\beta}\,\dot{x}^\alpha\,\dot{x}^\beta . \tag{2.36}$$

The solution is determined by the Euler-Lagrange equations (Appendix C)

$$\frac{\partial L}{\partial x^\lambda} = \frac{d}{dp}\left(\frac{\partial L}{\partial \dot{x}^\lambda}\right) . \tag{2.37}$$

Now, $\partial L/\partial x^\lambda = g_{\alpha\beta,\lambda}\,\dot{x}^\alpha\,\dot{x}^\beta$ because only $g_{\alpha\beta}$ depends on $\{x^\mu\}$. By using $\partial \dot{x}^\alpha/\partial \dot{x}^\lambda = \delta^\alpha_\lambda$ one gets $\partial L/\partial \dot{x}^\lambda = 2 g_{\alpha\lambda}\,\dot{x}^\alpha$. Substitute this in (2.37):

$$g_{\alpha\beta,\lambda}\,\dot{x}^\alpha\,\dot{x}^\beta = 2(g_{\alpha\lambda}\,\dot{x}^\alpha)^{\cdot}$$

$$= 2(g_{\alpha\lambda,\beta}\,\dot{x}^\beta\,\dot{x}^\alpha + g_{\alpha\lambda}\,\ddot{x}^\alpha) ,$$

or

$$g_{\alpha\lambda}\,\ddot{x}^\alpha + \tfrac{1}{2}(2 g_{\lambda\alpha,\beta} - g_{\alpha\beta,\lambda})\,\dot{x}^\alpha\,\dot{x}^\beta = 0 . \tag{2.38}$$

Now comes a frequently used trick: renaming of dummy indices: $2 g_{\lambda\alpha,\beta} \cdot \dot{x}^\alpha\,\dot{x}^\beta = g_{\lambda\alpha,\beta}\,\dot{x}^\alpha\,\dot{x}^\beta + g_{\lambda\beta,\alpha}\,\dot{x}^\beta\,\dot{x}^\alpha = (g_{\lambda\alpha,\beta} + g_{\lambda\beta,\alpha})\,\dot{x}^\alpha\,\dot{x}^\beta$. Substitution in (2.38) and multiplication with $g^{\mu\lambda}$ gives:

$$\ddot{x}^\mu + \tfrac{1}{2} g^{\mu\lambda}\left(g_{\lambda\alpha,\beta} + g_{\lambda\beta,\alpha} - g_{\alpha\beta,\lambda}\right)\dot{x}^\alpha\,\dot{x}^\beta = 0 . \tag{2.39}$$

This is of the form of (2.34) and the factor multiplying $\dot{x}^\alpha\,\dot{x}^\beta$ must be equal to $\Gamma^\mu_{\alpha\beta}$, which proves (2.24). Variational calculus is a very efficient tool for this type of problem. Without much difficulty, it permits us to find the geodesic equation directly from the metric, and from this equation one may just read the Christoffel symbols $\Gamma^\mu_{\nu\sigma}$. This is usually a lot faster than calculating them from (2.24), and this method is therefore highly recommended.

The following result is very helpful when analysing the dynamics of a test particle in GR (assuming that its orbit is a geodesic), because it allows us to find *constants of the motion*. From the text below (2.37) we see that $\partial L/\partial x^\lambda$ vanishes if $g_{\alpha\beta,\lambda} = 0$. And then eq. (2.37) says that $\partial L/\partial \dot{x}^\lambda = 2 g_{\alpha\lambda}\dot{x}^\alpha$ is constant. In terms of the *4-velocity* $u^\mu = dx^\mu/dp$ we have found that the covariant 4-velocity $u_\lambda = g_{\lambda\alpha} u^\alpha$ is constant:

$$g_{\alpha\beta,\lambda} = 0 \quad \rightarrow \quad u_\lambda = g_{\lambda\nu}\,\dot{x}^\nu = \text{constant} \tag{2.40}$$

with $\cdot = d/dp$. The fact that u_λ is a constant along a geodesic if the metric is independent of x^λ – doesn't that ring a bell?

---

[3] Here we switch to another definition of geodesics without proving its equivalence with (2.34).

Exercise 2.10: Show that the geodesics of the Lorentz metric ($g_{\alpha\beta} = \eta_{\alpha\beta}$) are straight lines.

Exercise 2.11: Show that the variational problem (2.36) is equivalent to $\delta \int F(L) \, \mathrm{d}p = 0$ if F is monotonous, $F' \neq 0$.

Hint: Write down (2.37) with $L \to F(L)$; use $\partial F(L)/\partial x^\lambda = F' \partial L/\partial x^\lambda$, and $(F' \partial L/\partial \dot{x}^\lambda)\dot{} = (F')\dot{}\, \partial L/\partial \dot{x}^\lambda + F'(\partial L/\partial \dot{x}^\lambda)\dot{}$. But $(F')\dot{} = F'' \mathrm{d}L/\mathrm{d}p = 0$ (L is constant on $x^\mu(p)$ because $\dot{x}^\alpha \dot{x}_\alpha$ is).

2.6 The covariant derivative

For a given vector field A^μ that is not restricted to the curve $x^\sigma(p)$ we can elaborate $\mathrm{d}A^\mu/\mathrm{d}p$ in (2.26) as $\mathrm{d}A^\mu/\mathrm{d}p = A^\mu{}_{,\sigma}\, \dot{x}^\sigma$, because we are able to compute derivatives in other directions than along the curve. This leads to the introduction of the covariant derivative

$$\frac{\mathrm{D}A^\mu}{\mathrm{D}p} = \left(A^\mu{}_{,\sigma} + \Gamma^\mu{}_{\nu\sigma} A^\nu\right) \dot{x}^\sigma \equiv A^\mu{}_{:\sigma} u^\sigma , \qquad (2.41)$$

where $u^\sigma = \dot{x}^\sigma = \mathrm{d}x^\sigma/\mathrm{d}p$ and

$$A^\mu{}_{:\sigma} \equiv A^\mu{}_{,\sigma} + \Gamma^\mu{}_{\nu\sigma} A^\nu \qquad (2.42)$$

is the *covariant derivative* of A^μ. It may be regarded as the 'intrinsic derivative', the derivative after correction for the meaningless change in orientation of the base vectors. In a similar way we may obtain the covariant derivative of a covariant vector from (2.27):

$$A_{\mu:\sigma} = A_{\mu,\sigma} - \Gamma^\nu{}_{\mu\sigma} A_\nu . \qquad (2.43)$$

Important is that both $A^\mu{}_{:\sigma}$ and $A_{\mu:\sigma}$ are tensors if A^μ is a vector, even though neither of the two terms on the right hand sides of (2.42) and (2.43) are tensors themselves. The proof is a matter of combining relations (2.20) and (2.31), and is left to the reader.

Next follow a few definitions. The covariant derivative of a product XY of two tensors is:

$$(XY)_{:\sigma} = X_{:\sigma} Y + X Y_{:\sigma} . \qquad (2.44)$$

For example:

$$(A_\mu B_\nu)_{:\sigma} = (A_{\mu,\sigma} - \Gamma^\alpha_{\mu\sigma} A_\alpha) B_\nu + A_\mu (B_{\nu,\sigma} - \Gamma^\alpha_{\nu\sigma} B_\alpha)$$

$$= (A_\mu B_\nu)_{,\sigma} - \Gamma^\alpha_{\mu\sigma} A_\alpha B_\nu - \Gamma^\alpha_{\nu\sigma} A_\mu B_\alpha \,. \tag{2.45}$$

Accordingly, we define the covariant derivative of a covariant second rank tensor as:

$$T_{\mu\nu:\sigma} = T_{\mu\nu,\sigma} - \Gamma^\alpha_{\mu\sigma} T_{\alpha\nu} - \Gamma^\alpha_{\nu\sigma} T_{\mu\alpha} \,. \tag{2.46}$$

The recipe for tensors of higher rank should be clear by now. For example, if we need an expression for $T_\alpha{}^\beta{}_{\gamma:\sigma}$, then we merely have to work out $(P_\alpha Q^\beta R_\gamma)_{:\sigma}$ as in (2.44) and (2.45). The general pattern is $T^*_{*:\sigma} = T^*_{*,\sigma} \pm \Gamma$-term for every index. For a scalar:

$$Q_{:\sigma} = Q_{,\sigma} \,. \tag{2.47}$$

Covariant derivatives do not commute, unlike normal derivatives ($X_{,\alpha\beta} = X_{,\beta\alpha}$ for every X). We calculate $B_{\mu:\nu:\sigma}$ by substituting $T_{\mu\nu} = B_{\mu:\nu}$ in (2.46):

$$B_{\mu:\nu:\sigma} = B_{\mu:\nu,\sigma} - \Gamma^\alpha_{\mu\sigma} B_{\alpha:\nu} - \Gamma^\alpha_{\nu\sigma} B_{\mu:\alpha} \,, \tag{2.48}$$

which should be elaborated further with (2.43). After that, interchange ν and σ and subtract. The result of a somewhat lengthy calculation is:

$$B_{\mu:\nu:\sigma} - B_{\mu:\sigma:\nu} = B_\alpha R^\alpha{}_{\mu\nu\sigma} \tag{2.49}$$

with

$$R^\alpha{}_{\mu\nu\sigma} = \Gamma^\alpha_{\mu\sigma,\nu} - \Gamma^\alpha_{\mu\nu,\sigma} + \Gamma^\tau_{\mu\sigma} \Gamma^\alpha_{\tau\nu} - \Gamma^\tau_{\mu\nu} \Gamma^\alpha_{\tau\sigma} \,. \tag{2.50}$$

$R^\alpha{}_{\mu\nu\sigma}$ is called the RIEMANN tensor. It is a tensor because (2.49) is valid for every vector B_α and because the left hand side is a tensor. Then apply the quotient theorem. Apparently, covariant derivatives commute only if $R^\alpha{}_{\mu\nu\sigma} = 0$. The Riemann tensor plays a crucial role in GR because it contains all information about the curvature of space. Note the remarkable fact that according to (2.49) the difference of two consecutive covariant differentiations is proportional to the vector itself. The explanation is given in the next section.

Exercise 2.12: Show that

$$T^{\mu\nu}{}_{:\sigma} = T^{\mu\nu}{}_{,\sigma} + \Gamma^\mu_{\alpha\sigma} T^{\alpha\nu} + \Gamma^\nu_{\alpha\sigma} T^{\mu\alpha} \,. \tag{2.51}$$

Great care is needed in using these relations. For example, let $T^{\mu\nu}$ be diagonal. Then it seems evident that $T^{1\mu}{}_{:\mu} = T^{11}{}_{:1}$, but that is not the case. Why not?

Hint: Write out $(A^\mu B^\nu)_{:\sigma}$ as in (2.45). It is due to the action of the invisible dummy index α.

Exercise 2.13: An important property is that the metric tensor behaves as a constant under covariant differentiation:

$$g_{\mu\nu:\sigma} = 0 . \qquad (2.52)$$

Hint: Use (2.46) and (2.30).

Exercise 2.14: Prove the following compact form of the geodesic equation:

$$u^\sigma u_{\mu:\sigma} = 0 \quad \text{or} \quad u^\sigma u^\mu_{\;:\sigma} = 0 . \qquad (2.53)$$

Hint: The last relation is just $0 = Du^\mu/Dp$ = (2.41); the first relation with (2.52): $0 = g_{\lambda\mu} u^\mu_{\;:\sigma} u^\sigma = (g_{\lambda\mu} u^\mu)_{:\sigma} u^\sigma =$ etc.

Exercise 2.15: A reminder of the linear algebra aspects of tensor calculus. Given a 2D Riemann space with co-ordinates x, y, a metric and two vectors in the tangent space of the point (x, y):

$$ds^2 = dx^2 + 4dxdy + dy^2 \;;\quad A^\alpha = \begin{pmatrix} 1 \\ 4 \end{pmatrix} \;;\quad B_\alpha = \begin{pmatrix} y \\ x \end{pmatrix} .$$

Write down $g_{\mu\nu}$ and $g^{\mu\nu}$ and show that all Christoffel symbols are zero. Compute A_ν and $B^\nu_{\;:\nu}$.

Hint: $g_{11} = g_{22} = 1$; $g_{12} = g_{21} = 2$, use (2.24) for the Christoffel symbols;

$$g^{\mu\nu} = \frac{1}{3}\begin{pmatrix} -1 & 2 \\ 2 & -1 \end{pmatrix} \;;\quad A_\mu = \begin{pmatrix} 9 \\ 6 \end{pmatrix} \;;\quad B^\nu_{\;:\nu} = \frac{4}{3} .$$

The Γ's being zero we have $B^\nu_{\;:\nu} = B^\nu_{\;,\nu}$.

2.7 Riemann tensor and curvature

The metric tensor does not tell us whether a space is flat, because the use of 'strange' co-ordinates is not prohibited. For example $ds^2 = dr^2 + r^2 d\varphi^2$

(planar polar co-ordinates) defines a flat space, but (2.2) defines a curved space. The metric tensor contains apparently a mix of information on co-ordinates and curvature. The intrinsic curvature properties are determined by the Riemann tensor. We shall illustrate this by transporting a vector A^μ parallel to itself along two different paths to the same final position, see Fig. 2.7. According to (2.26), $\mathrm{d}A^\mu = -f_\sigma(x)\mathrm{d}x^\sigma$ with $f_\sigma(x) = \Gamma^\mu_{\nu\sigma}A^\nu$ (the upper index μ is omitted for brevity as it does not change). The difference of the two final vectors is:

$$\begin{aligned}
\mathrm{d}A^\mu &= A_1^\mu - A_2^\mu \\
&= -f_\sigma(x)\mathrm{d}\xi^\sigma - f_\sigma(x+\mathrm{d}\xi)\mathrm{d}\eta^\sigma + f_\sigma(x)\mathrm{d}\eta^\sigma + f_\sigma(x+\mathrm{d}\eta)\mathrm{d}\xi^\sigma \\
&\simeq -f_\sigma \mathrm{d}\xi^\sigma - f_\sigma \mathrm{d}\eta^\sigma - f_{\sigma,\lambda}\,\mathrm{d}\xi^\lambda \mathrm{d}\eta^\sigma + f_\sigma \mathrm{d}\eta^\sigma + f_\sigma \mathrm{d}\xi^\sigma + f_{\sigma,\lambda}\,\mathrm{d}\eta^\lambda \mathrm{d}\xi^\sigma \\
&= (f_{\sigma,\lambda} - f_{\lambda,\sigma})\,\mathrm{d}\xi^\sigma \mathrm{d}\eta^\lambda \;. \quad (2.54)
\end{aligned}$$

Now substitute $f_\sigma = \Gamma^\mu_{\nu\sigma}A^\nu = A^\mu_{;\sigma} - A^\mu_{,\sigma}$. The terms $A^\mu_{,\sigma}$ cancel, and after some index gymnastics we arrive at (exercise 2.16):

$$\begin{aligned}
\mathrm{d}A^\mu &= (A^\mu_{;\sigma,\lambda} - A^\mu_{;\lambda,\sigma})\,\mathrm{d}\xi^\sigma \mathrm{d}\eta^\lambda \\
&= g^{\mu\nu}(A_{\nu;\sigma,\lambda} - A_{\nu;\lambda,\upsilon})\,\mathrm{d}\xi^\sigma \mathrm{d}\eta^\lambda \\
&= g^{\mu\nu}(A_{\nu;\sigma:\lambda} - A_{\nu;\lambda:\sigma})\,\mathrm{d}\xi^\sigma \mathrm{d}\eta^\lambda \\
&= g^{\mu\nu} R_{\alpha\nu\sigma\lambda}\,A^\alpha\,\mathrm{d}\xi^\sigma \mathrm{d}\eta^\lambda \\
&= g^{\mu\nu} R_{\nu\alpha\lambda\sigma}\,A^\alpha\,\mathrm{d}\xi^\sigma \mathrm{d}\eta^\lambda \\
&= R^\mu{}_{\alpha\lambda\sigma}\,A^\alpha\,\mathrm{d}\xi^\sigma \mathrm{d}\eta^\lambda \;. \quad (2.55)
\end{aligned}$$

On account of (2.24) the Christoffel symbols vanish identically in a flat space with rectangular co-ordinates, since $g_{\mu\nu}$ has only constant elements. Therefore the Riemann tensor (2.50) is zero as well. The transformation properties of a tensor then ensure that $R^\alpha{}_{\mu\nu\sigma}$ is zero in a flat space for any choice of the co-ordinates.[4] In that case parallel transport along a closed path leaves a vector unchanged.[5] But in a curved space the orientation of the vector will have

---

[4] Contrary to the Christoffel symbols, which are not tensors. For example, the Christoffel symbols vanish in rectangular co-ordinates in a plane, but *not* in polar co-ordinates.

[5] Conversely, if the Riemann tensor is zero, it can be proven that there exist co-ordinates so that $g_{\mu\nu}$ is constant which implies that the space is flat, see e.g. Dirac (1975) § 12.

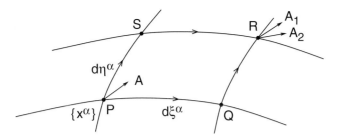

Fig. 2.7. Parallel transport of the vector $\boldsymbol{A}$ from P to R along path 1 (PQR) and path 2 (PSR) produces a different result.

changed. Once this is accepted intuitively, it is clear that the difference dA^μ must be proportional to the length of the vector, which explains the factor A^α in (2.55). The derivation in (2.55) shows that the difference dA^μ is also proportional to the difference of two consecutive covariant differentiations, and this explains why this difference is proportional to the vector itself, as in (2.49).

There are several other ways to illustrate the relation between the Riemann tensor and curvature. One is the equation for the *geodesic deviation*, see exercise 2.18. Another is the relation between *Gaussian curvature* and the Riemann tensor. Gaussian curvature refers to surfaces embedded in a flat 3D space. The curvature κ in a point P of a curve on the surface is defined as the inverse radius of the osculating circle at P. Each point has two principal curvatures κ_1 and κ_2, and the Gaussian curvature $K \equiv \kappa_1\kappa_2$ is an invariant determined by the geometry of the surface, which has several interesting properties.[6] Turning now to Riemann spaces, take two orthogonal unit vectors $\boldsymbol{e}_1$ and $\boldsymbol{e}_2$ in the tangent space of a point P which are not null. Now consider those geodesics in Riemann space that are tangent in P to the plane spanned by $\boldsymbol{e}_1$ and $\boldsymbol{e}_2$. These geodesics subtend, locally around P, a 2D curved subspace of Riemann space. The Gaussian curvature of this 2D space at P is $R_{\alpha\mu\nu\sigma}e_1^\alpha e_2^\mu e_1^\nu e_2^\sigma$, apart from the sign.[7]

The Riemann tensor obeys several symmetry relations that reduce the number of independent components from n^4 to $n^2(n^2-1)/12$ (see literature). In 4 dimensions $R^\alpha{}_{\nu\rho\sigma}$ has only 20 independent components, and all contractions of $R^\alpha{}_{\nu\rho\sigma}$ are either zero or equal, apart from the sign. We choose

---

[6] E.g. Gauss's theorem on integral curvature: the sum of the three interior angles of a geodesic triangle (bounded by 3 geodesics) equals π plus the surface integral of K.

[7] For a proof of these statements see e.g. Robertson and Noonan (1969) p. 216.

the *Ricci tensor*:[8]

$$R_{\mu\nu} \equiv R^{\alpha}{}_{\mu\nu\alpha} \qquad \text{(RICCI)}. \qquad (2.56)$$

The explicit expression follows from (2.50):

$$R_{\mu\nu} = \Gamma^{\alpha}{}_{\mu\alpha,\nu} - \Gamma^{\alpha}{}_{\mu\nu,\alpha} - \Gamma^{\alpha}{}_{\mu\nu}\Gamma^{\beta}{}_{\alpha\beta} + \Gamma^{\alpha}{}_{\mu\beta}\Gamma^{\beta}{}_{\nu\alpha}. \qquad (2.57)$$

We infer from (2.33) that $\Gamma^{\alpha}{}_{\mu\alpha,\nu} = \frac{1}{2}(\log|g|)_{,\mu\nu}$ so that all terms in (2.57) are symmetric in μ and in ν. Hence $R_{\mu\nu}$ is symmetric:

$$R_{\mu\nu} = R_{\nu\mu}. \qquad (2.58)$$

We may contract once more:

$$R \equiv R^{\nu}{}_{\nu} = g^{\nu\mu}R_{\mu\nu} = R^{\alpha\beta}{}_{\beta\alpha}. \qquad (2.59)$$

R is called the *total curvature*. Finally we introduce the *Einstein tensor* $G_{\mu\nu}$:

$$G_{\mu\nu} = R_{\mu\nu} - \frac{1}{2}g_{\mu\nu}R \qquad \text{(EINSTEIN)}. \qquad (2.60)$$

The Einstein tensor will be useful later because its divergence is zero:

$$G^{\mu\nu}{}_{:\nu} = (R^{\mu\nu} - \frac{1}{2}g^{\mu\nu}R)_{:\nu} = 0. \qquad (2.61)$$

Riemann, Ricci en Einstein tensor contain at most second derivatives of $g_{\alpha\beta}$. By substituting (2.24) in (2.50) we get:

$$R^{\alpha}{}_{\mu\nu\sigma} = \frac{1}{2}g^{\alpha\beta}\left(g_{\beta\sigma,\mu\nu} - g_{\mu\sigma,\beta\nu} - g_{\beta\nu,\mu\sigma} + g_{\mu\nu,\beta\sigma}\right)$$
$$+ g^{\alpha\beta}\left(\Gamma^{\tau}{}_{\beta\sigma}\Gamma^{\tau}{}_{\mu\nu} - \Gamma^{\tau}{}_{\beta\nu}\Gamma^{\tau}{}_{\mu\sigma}\right). \qquad (2.62)$$

The corresponding expressions for $R_{\mu\nu}$ and for $G_{\mu\nu}$ can be found from this by contraction. The first term contains all second-order derivatives. The first-order derivatives are in the second term. The proofs of (2.61) and (2.62) can be found in the literature, but are not important here.

Exercise 2.16: Provide the missing details of the derivation of (2.55).

Hint: Second = sign: $A^{\mu}{}_{:\sigma,\lambda} = (g^{\mu\nu}A_{\nu})_{:\sigma,\lambda} = (g^{\mu\nu}A_{\nu:\sigma})_{,\lambda} = g^{\mu\nu}{}_{,\lambda}A_{\nu:\sigma} + g^{\mu\nu}A_{\nu:\sigma,\lambda}$, but A^{μ} is parallel transported, hence $A_{\nu:\sigma} = 0$, etc. Third = sign: $A_{\nu:\sigma:\lambda} = A_{\nu:\sigma,\lambda}$ from (2.48). Fifth = sign: $R_{\alpha\nu\sigma\lambda} = R_{\nu\alpha\lambda\sigma}$ is a symmetry relation of the Riemann tensor.

---

[8] Other authors define $R_{\mu\nu} = R^{\alpha}{}_{\mu\alpha\nu}$, another source of sign differences. For a complete classification of all sign conventions see the red pages in Misner et al. (1971). In terms of this classification we follow the $- + -$ convention.

2.7 Riemann tensor and curvature   39

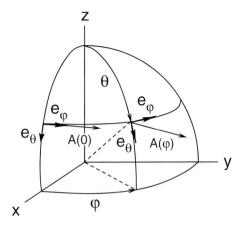

Fig. 2.8. Parallel transport of a vector A over the surface of a sphere with radius $r = 1$, see exercise 2.17.

Exercise 2.17: Consider a 2D spherical surface with radius $r = 1$, see Fig. 2.8. Calculate the Christoffel symbols and the total curvature R. Convince yourself that $R \propto r^{-2}$. Show that a vector A will rotate in the tangent space as it is parallel-transported along a circle $\theta = \theta_0$. Try to understand this with the intuitive definition of parallel transport in § 2.4. Start in $(\theta, \varphi) = (\theta_0, 0)$ with $(A^\theta, A^\varphi) = (0, 1/\sin\theta_0)$. Show that $A^i A_i$ is always 1, i.e. $|A| \equiv 1$, and that after one full revolution A has rotated over an angle $2\pi \cos\theta_0$. Discuss the limiting cases $\theta_0 = \pi/2$ (geodesic!) and $\theta_0 \ll 1$.

Hint: (2.2): $g_{11} = 1$, $g_{22} = \sin^2\theta$ ($\theta = 1$, $\varphi = 2$). Do *not* use (2.24), but rather (2.37) with $L(\theta, \dot\theta, \dot\varphi) = \dot\theta^2 + \sin^2\theta\, \dot\varphi^2$:

$$\frac{\partial L}{\partial \theta} = \left(\frac{\partial L}{\partial \dot\theta}\right)^\cdot \rightarrow \ddot\theta - \sin\theta\cos\theta\, \dot\varphi^2 = 0\,;$$

$$\frac{\partial L}{\partial \varphi} = \left(\frac{\partial L}{\partial \dot\varphi}\right)^\cdot \rightarrow \ddot\varphi + 2\cot\theta\, \dot\theta\dot\varphi = 0\,.$$

By comparing with (2.34) we may just read the Γ's: $\Gamma^1_{22} = -\sin\theta\cos\theta$; $\Gamma^2_{12} = \cot\theta$ (double product!). All other Γ's are zero. (2.33) $\rightarrow$ $\Gamma^\alpha_{\mu\alpha,\nu} = (\log\sin\theta)_{,\mu\nu}$ $\rightarrow$ $\Gamma^\alpha_{1\alpha,1} = -1/\sin^2\theta$. And $\Gamma^\alpha_{11,\alpha} = 0$; $\Gamma^\alpha_{22,\alpha} = -(\sin\theta\cdot\cos\theta)_{,\theta} = \sin^2\theta - \cos^2\theta$. Algebra: $R_{11} = -1$ and $R_{22} = -\sin^2\theta$. Finally $R = g^{\mu\nu}R_{\mu\nu} = g^{11}R_{11} + g^{22}R_{22} = R_{11} + (1/\sin^2\theta)R_{22} = -2$. For a sphere with radius r: $R = -2/r^2$ (minus sign due to sign convention).
Parallel transport: p is proportional to the arc length (why?), so choose $p = \varphi$; (2.28)+(2.26): $A^\mu_{,\varphi} + \Gamma^\mu_{\nu\sigma} A^\nu x^\sigma_{,\varphi} = 0$ with $x^1_{,\varphi} = d\theta/d\varphi = 0$ and $x^2_{,\varphi} = d\varphi/d\varphi = 1$:

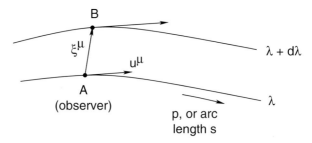

Fig. 2.9. The geodesic deviation.

$$A^\theta_{,\varphi} = \sin\theta_0 \cos\theta_0 \, A^\varphi \; ; \qquad A^\varphi_{,\varphi} = -\cot\theta_0 \, A^\theta \; .$$

Eliminate A^φ: $A^\theta_{,\varphi\varphi} + \cos^2\theta_0 \, A^\theta = 0$, same equation holds for A^φ. Harmonic oscillator with frequency $\cos\theta_0$. Solution for given initial value:

$$A^\theta = \sin(\varphi\cos\theta_0) \; ; \qquad A^\varphi = \cos(\varphi\cos\theta_0)/\sin\theta_0 \; .$$

$\boldsymbol{A}$ rotates clockwise when looking down on the tangent space from outside; $\theta_0 = \pi/2$: $A^\theta \equiv 0$ and $A^\varphi \equiv 1/\sin\theta_0 = 1$, therefore $\boldsymbol{A}$ remains a tangent vector; $\theta_0 \ll 1$ (small circle around the north pole): in that case the tangent space is always almost parallel to the equatorial plane, with base vectors $\boldsymbol{x}$ en $\boldsymbol{y}$, and $\boldsymbol{e}_\theta \simeq \boldsymbol{x}\cos\varphi + \boldsymbol{y}\sin\varphi$ and $\boldsymbol{e}_\varphi \simeq (\boldsymbol{y}\cos\varphi - \boldsymbol{x}\sin\varphi)\sin\theta_0$. For $\theta_0 \ll 1$ it follows that $\boldsymbol{A} = A^\theta \boldsymbol{e}_\theta + A^\varphi \boldsymbol{e}_\varphi \simeq \boldsymbol{y}$, so that $\boldsymbol{A}$ remains virtually unchanged with respect to a fixed frame.

Exercise 2.18: Given a set of geodesics $x^\mu(p,\lambda)$ where p is the curve parameter and λ labels different geodesics (λ is constant along one geodesic). Consider two neighbouring geodesics λ and $\lambda + \delta\lambda$. The points A and B are connected by the vector $\xi^\mu = x^\mu(p,\lambda+\delta\lambda) - x^\mu(p,\lambda) \simeq (\partial x^\mu/\partial\lambda)\delta\lambda \equiv e^\mu \delta\lambda$. Prove that:

$$\frac{\mathrm{D}^2 \xi^\mu}{\mathrm{D}p^2} = R^\mu{}_{\alpha\beta\nu} u^\alpha u^\beta \xi^\nu \; ; \qquad u^\alpha = \dot{x}^\alpha = \frac{\partial x^\alpha}{\partial p} \; . \tag{2.63}$$

This is the equation for the geodesic deviation, that will play an important role later. In a flat space the Riemann tensor is zero, and then ξ^μ is a linear function of p, and for timelike geodesics also a linear function of the arc length s, as expected. In a curved space however this is no longer the case. For example, on a sphere $\xi^\mu(s)$ will be something like a sine-function.

Hint: The proof comes in three steps:

(a) $\dfrac{\partial e^\mu}{\partial p} = \dfrac{\partial^2 x^\mu}{\partial p\,\partial\lambda} = \dfrac{\partial u^\mu}{\partial \lambda} = u^\mu{}_{,\alpha}\dfrac{\partial x^\alpha}{\partial \lambda} = u^\mu{}_{,\alpha}\,e^\alpha\,;$

(b) $e^\mu{}_{:\alpha}\,u^\alpha \equiv \dfrac{De^\mu}{Dp} = \dfrac{\partial e^\mu}{\partial p} + \Gamma^\mu{}_{\alpha\beta}\,e^\alpha u^\beta$

$\qquad = u^\mu{}_{,\alpha}\,e^\alpha + \Gamma^\mu{}_{\alpha\beta}\,e^\alpha u^\beta = u^\mu{}_{:\alpha}\,e^\alpha\,;$

(c) $\dfrac{D^2 e^\mu}{Dp^2} \equiv (e^\mu{}_{:\alpha}\,u^\alpha)_{:\beta}\,u^\beta = (u^\mu{}_{:\alpha}\,e^\alpha)_{:\beta}\,u^\beta$

$\qquad = u^\mu{}_{:\alpha}\,e^\alpha{}_{:\beta}\,u^\beta + u^\mu{}_{:\alpha:\beta}\,e^\alpha u^\beta$

$\qquad = u^\mu{}_{:\alpha}\,u^\alpha{}_{:\beta}\,e^\beta + u^\mu{}_{:\alpha:\beta}\,u^\alpha e^\beta + \left(u^\mu{}_{:\alpha:\beta} - u^\mu{}_{:\beta:\alpha}\right) u^\beta e^\alpha$

$\qquad = \left(u^\mu{}_{:\alpha}\,u^\alpha\right)_{:\beta}\,e^\beta + g^{\mu\nu}\left(u_{\nu:\alpha:\beta} - u_{\nu:\beta:\alpha}\right) u^\beta e^\alpha$

$\qquad = g^{\mu\nu}\,u_\sigma\,R^\sigma{}_{\nu\alpha\beta}\,u^\beta e^\alpha$

$\qquad = R_\sigma{}^\mu{}_{\alpha\beta}\,u^\sigma u^\beta e^\alpha$

$\qquad = R^\mu{}_{\sigma\beta\alpha}\,u^\sigma u^\beta e^\alpha\,.$

In (c) we have twice used (b), next $u^\mu{}_{:\beta:\alpha}\,e^\alpha u^\beta = u^\mu{}_{:\alpha:\beta}\,e^\beta u^\alpha$ is added and substracted again, and then (2.53) and (2.49). The last $=$ sign is a symmetry relation of the Riemann tensor. Because $\delta\lambda$ is constant, the equation also holds for $\xi^\mu = e^\mu \delta\lambda$.

Exercise 2.19: Be aware of some inconsistencies in the notation. We encountered one in exercise 2.12. Meet two more here. In § 2.2 and § 2.3 it was stressed that the rules for index raising and lowering are always valid. Does that mean that

$$g^{\mu\alpha} g_{\alpha\lambda,\nu} \stackrel{?}{=} g^\mu{}_{\lambda,\nu}\,; \qquad (2.64)$$

$$g_{\mu\alpha}\dot{u}^\alpha \stackrel{?}{=} \dot{u}_\mu\,. \qquad (2.65)$$

Hint: In exercise 2.12 the trouble was caused by a hidden index; here we discover that the symbols without derivative had already been defined; one way to see that (2.64) cannot be correct is to note that $g^\mu{}_{\lambda,\nu} \equiv \delta^\mu{}_{\lambda,\nu} = 0$, and since $\det\{g^{\mu\alpha}\} \neq 0 \to g_{\alpha\lambda,\nu} = 0 \to g_{\alpha\lambda} = \mathrm{const}$. Instead, $0 = (g^{\mu\alpha} g_{\alpha\lambda})_{,\nu} = g^{\mu\alpha}{}_{,\nu}\,g_{\alpha\lambda} + g^{\mu\alpha} g_{\alpha\lambda,\nu}$, etc. Likewise, u_μ is *defined* as $g_{\mu\alpha} u^\alpha$ so that $\dot{u}_\mu = (g_{\mu\alpha} u^\alpha)^\cdot = g_{\mu\alpha,\sigma}\,u^\sigma u^\alpha + g_{\mu\alpha}\dot{u}^\alpha$. Also correct is $\dot{u}_\mu = u_{\mu,\alpha}\dot{x}^\alpha = u_{\mu,\alpha}u^\alpha$.

3

General Relativity

We shall now put the ideas of GR on solid footing by casting them into the framework of Riemann spaces. From now on we deal again with a 4-dimensional *spacetime* in which every *event* is determined by the co-ordinates $x^0, .. , x^3$ ($x^0 = ct$). First we say a few words about the meaning of these co-ordinates and their relation to the metric. Then we discuss the field equations for the metric tensor, and the classical limit for weak fields.

3.1 Co-ordinates, metric and motion

It is important to understand that the co-ordinates serve merely as *labels* that identify events in spacetime. They can be chosen arbitrarily, as long as they are well-behaved (continuous, one-to-one,..), but they have usually no physical meaning. In particular, differences in time or spatial co-ordinates are meaningless because they are not invariant. In GR, measurable quantities such as lengths and times are always expressed in terms of the co-ordinates and the metric tensor, so that the result is invariant for a co-ordinate transformation. Consider for example radial distances in the Schwarzschild metric. The difference $r_2 - r_1$ of two radial positions r_1 and r_2 is not invariant and not equal to the measured distance. If we travel radially from r_1 to r_2 and measure the distance with a measuring rod, the result is equal to $\int_{r_1}^{r_2} \sqrt{-g_{\rm rr}(r)}\, dr$. This strange expression will become clear in a moment. The point is that the outcome of a measurement is always given by an invariant expression (invariant for co-ordinate transformations) involving the metric tensor. These two functions of labelling and measuring are frequently confused in daily life, for example, in the case of cartesian co-ordinates (think of millimetre paper), but in GR they are strictly separated.

Even though the choice of the co-ordinates is free, some co-ordinates are much easier to use than others. It is not very wise to use rectangular co-ordinates for a spherically symmetric system, and this is also very much true

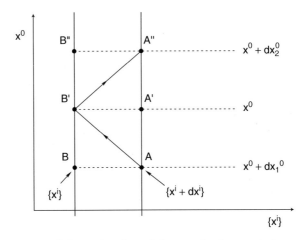

Fig. 3.1. Experimental determination of the metric of space in terms of the metric of spacetime. After Landau and Lifshitz (1971), § 84.

in GR. By 'natural selection' a few standard co-ordinate systems have emerged for frequently occurring physical situations that everybody uses because it saves a lot of work.

Time and distance measurements

To begin with the co-ordinate time t, one way to define t is to count light flashes of a beacon, for example a pulsar. The co-ordinate time interval Δt between n flashes has the same value everywhere in space (namely n), but the proper time interval does not. Their relation is determined by the metric:

$$c^2 \mathrm{d}\tau^2 = \mathrm{d}s^2 = g_{\alpha\beta} \, \mathrm{d}x^\alpha \mathrm{d}x^\beta \; . \qquad (3.1)$$

For timelike worldlines ($\mathrm{d}s^2 > 0$) we interpret $\mathrm{d}s \equiv (\mathrm{d}s^2)^{1/2}$ as $c \times$ the *proper time* interval $\mathrm{d}\tau$, like in SR. An observer *at rest* ($\mathrm{d}x^i = 0$, $i = 1, 2, 3$) has a timelike worldline[1] and hence $(c\mathrm{d}\tau)^2 = \mathrm{d}s^2 = g_{00}(c\mathrm{d}t)^2$:

$$\mathrm{d}\tau = \sqrt{g_{00}} \, \mathrm{d}t \; . \qquad (3.2)$$

$\mathrm{d}\tau$ is the interval read from the clock of the observer at rest, while $\mathrm{d}t$ is the co-ordinate time interval. It follows that g_{00} must be positive.

---

[1] This need not be a geodesic. In the Schwarzschild metric an observer needs a rocket to remain at rest. But 'rest' ($\mathrm{d}x^i = 0$) is not an invariant concept. If one drops a stone into a black hole once every second, the radial position of any point can be expressed in terms of the fractional stone number. These co-ordinates are not stationary, but perfectly legitimate. In these co-ordinates 'an observer at rest' is a freely falling observer. Co-ordinates in which a freely falling observer is at rest are used in cosmology.

3.1 Co-ordinates, metric and motion

For the spatial co-ordinates, too, there should exist properly defined *measuring procedures* to determine their value. As an example we illustrate in the next chapter how the values of the Schwarzschild co-ordinates may be determined. Another issue is this: an observer A (co-ordinates $x^\mu + dx^\mu$) may determine the metric of the *space* in his neighbourhood empirically. How is this metric related to the metric of *spacetime*? A places a mirror at B at a distance dl, Fig. 3.1, measures on his clock the time $d\tau$ it takes a light signal to travel from A to B and back again, and argues: $dl = cd\tau/2$. Light signals must travel along null worldlines.[2] The interval ds^2 between the *events* A, B' and B', A'' is zero, or, from (3.1):

$$g_{ij}\, dx^i dx^j + 2g_{0i}\, dx^i dx^0 + g_{00}\, (dx^0)^2 = 0 \,. \tag{3.3}$$

Roman indices run from 1 to 3. Solve for dx^0:

$$dx^0_{1,2} = \frac{1}{g_{00}} \left\{ -g_{0i}\, dx^i \pm \sqrt{(g_{0i}\, dx^i)(g_{0j}\, dx^j) - g_{00}\, g_{ij}\, dx^i dx^j} \right\} \tag{3.4}$$

($1 = -$; $2 = +$). The co-ordinate time interval dt between the events A and A'' is

$$cdt = dx^0_2 - dx^0_1 = \frac{2}{g_{00}} \sqrt{(g_{0i}\, g_{0j} - g_{00}\, g_{ij})\, dx^i dx^j} \,. \tag{3.5}$$

With the help of (3.2) we infer $dl^2 = (cd\tau/2)^2 = g_{00}(cdt)^2/4$. The spatial metric now follows from (3.5):

$$dl^2 = \left(\frac{g_{0i}\, g_{0j}}{g_{00}} - g_{ij}\right) dx^i dx^j \,. \tag{3.6}$$

Frequently $g_{0i} = 0$, in which case the metric of space simplifies to[3]

$$dl^2 = -g_{ij}\, dx^i dx^j \,. \tag{3.7}$$

As an application consider a curve along the x_1-axis, so that $dx_2 = dx_3 = 0$, and $dl = \sqrt{-g_{11}}\, dx_1$. The distance between $x_1 = a$ and $x_1 = b$ is $l = \int_a^b \sqrt{-g_{11}}\, dx_1$, and this explains the formula used earlier.

Strong equivalence

The metric $g_{\alpha\beta}$ is determined by the mass distribution and the choice of co-ordinates, through the field equations that are yet to come. When other co-ordinates are used, the metric becomes different as well, in such a way that ds^2 and all other physical quantities remain invariant. The metric tensor $g_{\alpha\beta}$

---

[2] We know that $ds^2 = 0$ in a local freely falling frame since SR holds there. But the way ds^2 is written in (3.1) makes it a scalar, hence it is zero in any frame.
[3] But not always. In the Kerr metric (rotating black holes) $g_{t\varphi} \neq 0$, and then weird things may happen.

is symmetric and may therefore be diagonalized locally by a transformation. Subsequently, the (real) eigenvalues may all be rescaled to ± 1 by redefining the units. Further analysis (omitted here) shows that the metric can always be brought into the following form, in the neighbourhood of any event $x_0 = \{x_0^\mu\}$, by a transformation of co-ordinates:

$$g_{\alpha\beta}(x_0 + \mathrm{d}x) = \eta_{\alpha\beta} + O(\mathrm{d}x^\mu \mathrm{d}x^\nu) \,, \tag{3.8}$$

see e.g. Schutz (1985, p. 154) or Kenyon (1990, p. 24). In x_0 the metric has approximately the SR-form. These co-ordinates define a local freely falling frame in x_0. It follows that the possibility to apply the strong equivalence principle is properly built into the theory.

Geodesic motion

In § 1.2 we anticipated that test masses on which only gravity acts move along geodesics. An elegant argument due to Weinberg (1972, p. 72) shows that geodesic motion is an almost inevitable consequence of the principles of weak equivalence and general covariance. In a freely falling frame a test mass moves as a free particle in SR. If its co-ordinates are $\{x^{\mu'}\}$ we have

$$\frac{\mathrm{d}^2 x^{\mu'}}{\mathrm{d}\tau^2} = 0 \,, \quad \text{and} \quad c^2 \mathrm{d}\tau^2 = \eta_{\mu'\nu'} \, \mathrm{d}x^{\mu'} \mathrm{d}x^{\nu'} \,, \tag{3.9}$$

where τ is the proper time of the mass. We now transform to co-ordinates x^λ. Denoting $\dot{} = \mathrm{d}/\mathrm{d}s$ with $\mathrm{d}s = c\mathrm{d}\tau$ we have $\mathrm{d}x^{\mu'}/\mathrm{d}s = x^{\mu'}_{,\lambda} \dot{x}^\lambda$, or

$$0 = (x^{\mu'}_{,\lambda} \dot{x}^\lambda)^{\cdot} = x^{\mu'}_{,\lambda} \ddot{x}^\lambda + x^{\mu'}_{,\lambda\sigma} \dot{x}^\lambda \dot{x}^\sigma \,. \tag{3.10}$$

On multiplying with $x^\alpha_{,\mu'}$ and summing over μ':

$$\ddot{x}^\alpha + \Gamma^\alpha_{\lambda\sigma} \dot{x}^\lambda \dot{x}^\sigma = 0 \quad \text{with} \quad \Gamma^\alpha_{\lambda\sigma} = x^\alpha_{,\mu'} \, x^{\mu'}_{,\lambda\sigma} \,. \tag{3.11}$$

This looks like the geodesic equation. Likewise,

$$\mathrm{d}s^2 = g_{\lambda\sigma} \, \mathrm{d}x^\lambda \mathrm{d}x^\sigma \quad \text{with} \quad g_{\lambda\sigma} = \eta_{\mu'\nu'} \, x^{\mu'}_{,\lambda} \, x^{\nu'}_{,\sigma} \,. \tag{3.12}$$

These $\Gamma^\alpha_{\lambda\sigma}$ and $g_{\lambda\sigma}$ have as yet nothing to do with the Christoffel symbols and the metric tensor, but Weinberg goes on to prove that the quantities thus defined obey relation (2.24)! This illustrates how neatly GR fits into the framework of Riemann geometry, and that the metric tensor is the basic quantity that determines everything else. And it is of course no coincidence that (3.11) for $\Gamma^\alpha_{\lambda\sigma}$ is equal to the second term in (2.31)!

We know that the geodesic is timelike because $\mathrm{d}s^2$ is invariant and positive in SR. According to (2.35) we may choose s for the curve parameter

p according and then (2.34) tells us that the 4-velocity $u^\mu = \mathrm{d}x^\mu/\mathrm{d}s$ obeys $\mathrm{D}u^\mu/\mathrm{D}s = 0$, or, in terms of the *4-momentum* $p^\mu = m_0 c u^\mu$:

$$\frac{\mathrm{D}p^\mu}{\mathrm{D}\tau} = 0, \qquad (3.13)$$

and this is a suggestive generalisation of the equations of motion $\mathrm{d}p^\mu/\mathrm{d}\tau = 0$ of a free particle in SR.

What about null geodesics? We concluded earlier that particles with zero rest mass must move along null worldlines, because of the invariance of $\mathrm{d}s^2$. We now assume that those lines are null geodesics. Here, too, we have hardly any choice because the worldlines are already null geodesics in SR.

Exercise 3.1: Show that the $-$ sign in (3.7) is a consequence of the adopted signature.

Hint: If we take $- + + +$ in (1.1), then $-c^2 \mathrm{d}\tau^2 = \mathrm{d}s^2 = g_{\alpha\beta}\,\mathrm{d}x^\alpha \mathrm{d}x^\beta$ in (3.1). How does (3.2) look in that case, and how does (3.5) follow from (3.4)?

Exercise 3.2: In a local frame in free fall all Christoffel symbols are zero, but the Riemann tensor is not.

Hint: (3.8) implies that $g_{\alpha\beta} = \eta_{\alpha\beta} + A_{\alpha\beta\mu\nu}(x^\mu - x_0^\mu)(x^\nu - x_0^\nu)$ in the neighbourhood of x_0 with constant $A_{\alpha\beta\mu\nu}$. First-order derivatives of $g_{\alpha\beta}$ are zero in $x = x_0$, second-order derivatives not. Diehards should try to express $A_{\alpha\beta\mu\nu}$ in terms of the Riemann tensor $R_{\beta\mu\nu\sigma}$ with (2.62).

3.2 Weak fields (1)

Assume now that we are dealing with weak, time-independent gravity fields and that the relevant velocities are non-relativistic, $\beta = v/c \ll 1$. Spacetime is then nearly flat, and it makes sense to do the substitution $g_{\mu\nu} = \eta_{\mu\nu} + \gamma_{\mu\nu}$ with $\gamma_{\mu\nu}$ small, and $\gamma_{\mu\nu,0} = 0$. We take once more $p = s$ and settle first the relation between $\mathrm{d}s$ and $\mathrm{d}t$ by using the metric (3.1), which we may write as $\mathrm{d}s^2 = (\mathrm{d}x^0)^2 - \mathrm{d}x^i \mathrm{d}x^i + \gamma_{\mu\nu}\mathrm{d}x^\mu \mathrm{d}x^\nu$. After 'division' by $\mathrm{d}t^2$:

$$\left(\frac{\mathrm{d}s}{\mathrm{d}t}\right)^2 \simeq c^2 - v^2 + \gamma_{00}c^2 + \text{terms } O(\gamma v c) \text{ or } O(\gamma v^2)$$

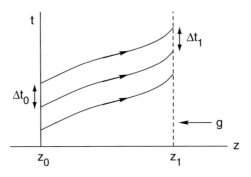

Fig. 3.2. The Pound-Rebka-Snider experiment revisited.

$$= c^2\{1 - \beta^2 + \gamma_{00} + O(\gamma\beta)\} \simeq c^2 , \qquad (3.14)$$

plus terms of order β^2. To order β we may put $d/ds = c^{-1}d/dt$ in the geodesic equation (2.34):

$$0 \simeq \frac{d^2 x^\mu}{dt^2} + \Gamma^\mu_{\nu\sigma} \frac{dx^\nu}{dt} \frac{dx^\sigma}{dt} \simeq \frac{d^2 x^\mu}{dt^2} + c^2 \Gamma^\mu_{00} , \qquad (3.15)$$

since due to $dx^\mu/dt \simeq (c, v^i)$ the summand $\nu = \sigma = 0$ is at least a factor c/v larger than all others. Now drop terms of order γ^2 and use $\gamma_{\mu\nu,0} = 0$ in (2.24):

$$\Gamma^\mu_{00} \simeq \tfrac{1}{2}\eta^{\mu\lambda}(2\gamma_{\lambda 0,0} - \gamma_{00,\lambda}) = -\tfrac{1}{2}\eta^{\mu\lambda}\gamma_{00,\lambda} . \qquad (3.16)$$

Hence, $\Gamma^0_{00} = 0$ and $\Gamma^i_{00} = \tfrac{1}{2}\gamma_{00,i}$. Relation (3.15) produces an identity for $\mu = 0$, and for $\mu = i$:

$$\frac{d^2 x^i}{dt^2} \simeq -\tfrac{1}{2}c^2\gamma_{00,i} = -\nabla_i \Phi . \qquad (3.17)$$

The second $=$ sign follows from the classic equation of motion of a test particle in a gravitational potential. The obvious choice is: $\Phi = \tfrac{1}{2}c^2\gamma_{00}$. We may now draw two conclusions for weak time-independent gravity fields:

1. Since the classical field equation is $\nabla^2 \Phi = \Phi_{,ii} = 0$, the field equation of GR must imply that

$$\nabla^2 \gamma_{00} = \gamma_{00,ii} = 0 . \qquad (3.18)$$

2. We know one component of the metric tensor:

$$g_{00} = 1 + \gamma_{00} = 1 + \frac{2\Phi(\mathbf{r})}{c^2} . \qquad (3.19)$$

Table 3.1. Parameters of characteristic objects.

Object	M	R (km)	r_s	$-\Phi/c^2$
Earth	5.98×10^{27} g	6370	0.89 cm	7×10^{-10}
Sun	1.99×10^{33} g	6.96×10^5	2.95 km	2×10^{-6}
Procyon B	$0.43\, M_\odot$	8800	1.3 km	7×10^{-5}
neutron star	$1.5\, M_\odot$	~ 10	4.4 km	~ 0.2

Gravitational redshift

At this point we sidestep to a related issue: the gravitational redshift. It follows from (3.19) that light passing through a gravitational field is subject to a frequency shift, see Fig. 3.2. In fact we compare proper time intervals $d\tau$ at two locations z_0 and z_1 in the gravitational field. Physically, the proper time is defined by identical oscillators at z_0 and z_1 (atoms of one species). With the help of (3.2) and (3.19) we obtain

$$\frac{d\tau(z_0)}{d\tau(z_1)} = \left\{\frac{g_{00}(z_0)}{g_{00}(z_1)}\right\}^{1/2} \simeq 1 + \frac{\Phi(z_0) - \Phi(z_1)}{c^2}. \qquad (3.20)$$

Use was made of $dt_0 = dt_1$ because the geodesics must be congruent as the gravity field is time-independent. Let n waves be emitted at z_0 and be detected at z_1. The measured frequencies ν_0 at z_0 and ν_1 at z_1 follow from $n = \nu_0 d\tau(z_0) = \nu_1 d\tau(z_1)$. Consequently

$$\frac{\Delta\nu}{\nu} = \frac{\nu_1 - \nu_0}{\nu_1} = 1 - \frac{d\tau(z_1)}{d\tau(z_0)} \simeq \frac{\Phi(z_0) - \Phi(z_1)}{c^2}. \qquad (3.21)$$

Note the difference between the SR and the GR point of view. In § 1.2 we tried to fit the Pound-Rebka-Snider experiment into the framework of SR, that is, in one global frame in which the source and the detector are both at rest. Under these circumstances we expect no differences in proper time, i.e. $dt(z_0) = dt(z_1)$, but this is refuted by the experiment. In GR the reasoning is different. We now interpret Fig. 3.2 as a co-ordinate picture, a picture that displays the co-ordinates but conveys no information about the geometry, see Fig. 2.1. The null geodesics must of course be congruent because the field is time-independent, i.e. $dt(z_0) = dt(z_1)$. However, a measurement refers to proper time, which in GR follows from (3.2).

Exercise 3.3: Show that the relative redshift in the Pound-Rebka-Snider experiment is about 10^{-15} ($h = 22.5$ m). For details on the experimental confirmation see also Adler et al. (1965, p. 129) and Misner et al. (1971,

p. 1056).

Hint: (3.21): $\Delta\nu/\nu = \Delta\Phi/c^2 = -gh/c^2$.

Exercise 3.4: Henceforth, 4-velocity and 4-momentum of a particle with rest mass m_0 are defined by $u^\mu = dx^\mu/ds$ and $p^\mu = m_0 c u^\mu$. Prove that

$$u^\mu \text{ is a vector} ; \quad u^\mu u_\mu = 1 ; \quad p^\mu p_\mu = (m_0 c)^2 . \quad (3.22)$$

Show that in the SR limit and for small velocity $v^i/c \ll 1$ ($E = \gamma m_0 c^2$; $\gamma = 1/\sqrt{1-\beta^2}$):

$$u^\mu = (\gamma, \gamma v^i/c) \simeq (1, v^i/c) ; \quad p^\mu = (E/c, p^i) \quad (3.23)$$

Hint: vector: $u^{\mu'} = dx^{\mu'}/ds = x^{\mu'}{}_{,\nu} dx^\nu/ds$ etc.; $u^\mu u_\mu = 1$: 'divide' (3.1) by ds^2; SR: $u^0 = dx^0/ds = dt/d\tau = \gamma$ according to (1.6); $u^i = dx^i/ds = c^{-1}(dx^i/dt)(dt/d\tau)$, etc.

Exercise 3.5: Estimate the values of Φ/c^2 in Table 3.1.

Hint. $\Phi/c^2 = -r_s/2R$ with $r_s = 2GM/c^2 =$ Schwarzschild radius.

3.3 Conservation of mass

In preparation for § 3.4 we analyse how conservation of mass is formulated in GR. The volume element $d^4x = dx^0 dx^1 dx^2 dx^3$ transforms according to

$$d^4x' = |J| d^4x , \quad (3.24)$$

where $J = \det\{x^{\mu'}{}_{,\alpha}\}$. From $g_{\alpha\beta} = x^{\mu'}{}_{,\alpha} x^{\nu'}{}_{,\beta} g_{\mu'\nu'}$ and $g \equiv \det\{g_{\alpha\beta}\}$ it follows that $g = J^2 g'$, or $\sqrt{-g} = |J|\sqrt{-g'}$,[4] because g and $g' < 0$. As a result,

$$\sqrt{-g}\, d^4x = \sqrt{-g'}\, d^4x' \equiv \text{proper volume element.} \quad (3.25)$$

This is important for integrations. It is physically not very meaningful to integrate a scalar S over a section of spacetime, because $\int S d^4x$ is not invariant, even though $S(x) = S'(x')$. But $\int S\sqrt{-g}\, d^4x = \int S'\sqrt{-g'}\, d^4x'$ is

---
[4] It follows, incidentally, that g is not a scalar.

3.3 Conservation of mass

invariant. The proper volume element is the physical volume element corresponding to the meaningless (i.e. not invariant) co-ordinate volume element $\mathrm{d}^4 x$. As an example consider spherical co-ordinates in a flat R_3: $\mathrm{d}s^2 = \mathrm{d}r^2 + r^2 \mathrm{d}\theta^2 + r^2 \sin^2\theta\, \mathrm{d}\varphi^2$ and $g = r^4 \sin^2\theta > 0$, so that the invariant volume element equals $\sqrt{g}\, \mathrm{d}^3 x = r^2 \sin\theta\, \mathrm{d}r \mathrm{d}\theta \mathrm{d}\varphi$.

The divergence $A^\mu{}_{;\mu}$ of a vector A^μ is a scalar. With (2.42) and (2.33):

$$A^\mu{}_{;\mu} = A^\mu{}_{,\mu} + \Gamma^\mu{}_{\sigma\mu} A^\sigma = A^\mu{}_{,\mu} + \frac{(\sqrt{-g})_{,\sigma} A^\sigma}{\sqrt{-g}}, \qquad (3.26)$$

because $g_{,\sigma}/2g = (\sqrt{-g})_{,\sigma}/\sqrt{-g}$. We may write this as follows:

$$A^\mu{}_{;\mu} \sqrt{-g} = (A^\mu \sqrt{-g})_{,\mu}. \qquad (3.27)$$

Consequently,

$$\int A^\mu{}_{;\mu} \sqrt{-g}\, \mathrm{d}^4 x = \int (A^\mu \sqrt{-g})_{,\mu}\, \mathrm{d}^4 x \qquad (3.28)$$

is invariant. For the volume of integration we choose a 3-volume V times an infinitesimal $\mathrm{d}x^0$ that we subsequently eliminate again from the equation. Assuming now that $A^\mu{}_{;\mu} = 0$ we infer that

$$0 = \int_V (A^0 \sqrt{-g})_{,0}\, \mathrm{d}^3 x + \int_V (A^i \sqrt{-g})_{,i}\, \mathrm{d}^3 x, \qquad (3.29)$$

or, with Gauss's theorem

$$\left\{ \int_V A^0 \sqrt{-g}\, \mathrm{d}^3 x \right\}_{,0} = -\oint_{\partial V} A^i \sqrt{-g}\, \mathrm{d}\sigma_i. \qquad (3.30)$$

At this point we make a connection with physics by choosing $A^\mu = \rho u^\mu$ where $\rho = $ rest mass density and $u^\mu = $ 4-velocity. Exercise 3.6 invites the reader to show that for this A^μ the classical limit of (3.30) coincides with the *continuity equation* in integral form. On this ground we accept (3.30) with $A^\mu = \rho u^\mu$ as the integral form of the continuity equation in GR. The differential form is then

$$(\rho u^\mu)_{;\mu} = 0. \qquad (3.31)$$

Exercise 3.6: Show that with $A^\mu = \rho u^\mu$ (3.31) and (3.30) in the non-relativistic limit reduce to

$$\frac{\partial \rho}{\partial t} + \nabla \cdot \rho \boldsymbol{v} = 0\ ; \qquad \frac{\partial}{\partial t} \int_V \rho\, \mathrm{d}^3 x = -\oint_{\partial V} \rho \boldsymbol{v} \cdot \mathrm{d}\boldsymbol{\sigma}.$$

Hint: For weak fields and $\beta \ll 1$ we have (3.23); in this limit $\rho u^\mu \to (\rho, \rho v^i/c)$. Furthermore $\sqrt{-g} \simeq 1$.

Exercise 3.7: Prove the equation for the rate of change of the density ρ along the worldline $x^\mu(s)$:

$$d\rho/ds = -\rho u^\mu{}_{;\mu} . \tag{3.32}$$

Hint: (3.31) $\to \rho_{;\mu} u^\mu + \rho u^\mu{}_{;\mu} = 0$ and $\rho_{;\mu} = \rho_{,\mu}$.

3.4 The field equations

We now have to generalise the classical field equation $\nabla^2 \Phi = 4\pi G \rho$ to one that determines the metric tensor. The story how this is done has been told many times. The basic idea is that the local energy density fixes the local curvature of spacetime:

$$\text{curvature} \propto \text{energy density} . \tag{3.33}$$

The left hand side of (3.33) will involve the Riemann tensor as that determines the curvature. The Riemann tensor contains second and lower-order derivatives of the metric tensor $g_{\alpha\beta}$, which is attractive on general grounds. So it appears that we must relate $\nabla^2 \Phi$ to the Riemann tensor. The energy density on the right hand side could just be ρc^2, but as we shall see, things aren't that simple.

To find the relation between the potential Φ and the Riemann tensor, consider two test particles A and B moving along their respective geodesics $x^\mu(s, \lambda)$ and $x^\mu(s, \lambda + \delta\lambda)$. A is for example an observer on board the Space Station, who sees satellite B passing at some distance. This situation has been analysed in exercise 2.18. The vector ξ^μ connecting A and B satisfies equation (2.63) for the geodesic deviation:

$$\frac{D^2 \xi^\mu}{D s^2} = R^\mu{}_{\alpha\beta\nu} u^\alpha u^\beta \xi^\nu ; \qquad u^\alpha = \frac{\partial x^\alpha}{\partial s} . \tag{3.34}$$

We elaborate this tensor equation in the local rest-frame of A. In that frame $g_{\mu\nu} = \eta_{\mu\nu}$ according to (3.8), and $dt = d\tau$. This means that A promotes his clock to the master clock indicating co-ordinate time. Furthermore, all Γ's are zero (exercise 3.2), so that $D/Ds = c^{-1}D/D\tau = c^{-1}d/dt$. Moreover $x^\mu = (ct, 0, 0, 0)$ in this frame $\to u^\mu = (1, 0, 0, 0)$. We are left with

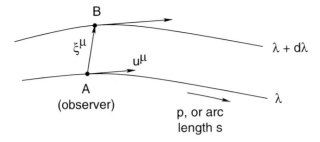

Fig. 3.3. The geodesic deviation as an aid in finding the vacuum equations.

$$\frac{d^2 \xi^\mu}{dt^2} = c^2 R^\mu{}_{00\nu}\, \xi^\nu . \qquad (3.35)$$

At this point A recalls that according to classical mechanics both he and B move in a stationary gravitational field: $\ddot{\boldsymbol{r}}_A = \boldsymbol{F}(\boldsymbol{r}_A)$ and $\ddot{\boldsymbol{r}}_B = \boldsymbol{F}(\boldsymbol{r}_B)$. Setting $\xi^i = r^i_B - r^i_A$ we have

$$\frac{d^2 \xi^i}{dt^2} = F^i(\boldsymbol{r}_A + \boldsymbol{\xi}) - F^i(\boldsymbol{r}_A) \simeq F^i{}_{,k}\, \xi^k = -\Phi_{,ik}\, \xi^k , \qquad (3.36)$$

since the force is related to the gradient of the gravitational potential as $F^i = -\Phi_{,i}$. By comparing (3.35) and (3.36) we find

$$\Phi_{,ik} = -c^2 R^i{}_{00k} . \qquad (3.37)$$

This is the relation between the second derivatives of Φ (that determine the tidal forces) and the Riemann tensor. The use of the indices is sloppy, but that's all right as long as we are in the exploratory stage. The classical field equation is $\Phi_{,ii} = -c^2 R^i{}_{00i} = 4\pi G \rho$. It follows quite generally from (2.50) that $R^0{}_{000} = 0$, and so we have found that

$$R_{00} = R^\alpha{}_{00\alpha} = -\frac{4\pi G \rho}{c^2} . \qquad (3.38)$$

However, this is not a tensor equation. The simplest guess would be that in an arbitrary reference frame we should use $R_{\mu\nu} = R^\alpha{}_{\mu\nu\alpha}$. In vacuum ($\rho = 0$) we would then get

$$\boxed{R_{\mu\nu} = 0} \qquad (3.39)$$

Although derived for weak fields, this is indeed the correct vacuum field equation, also for the strong fields in the vicinity of compact objects and black holes. An equivalent form is (see exercise):

$$\boxed{G^{\mu\nu} \equiv R^{\mu\nu} - \tfrac{1}{2} g^{\mu\nu} R = 0} \qquad (3.40)$$

Eq. (3.39) or (3.40) cannot be proven in the strict sense of the word. Their status of vacuum equations rests on the fact that predictions inferred from them are sofar in agreement with the observations. The deeper reason why this is so remains a mystery – apparently this is how Nature works. We refer to Pais (1982) for an account of Einstein's *Werdegang* to arrive at the correct field equation.

Nonzero energy density

What to do if the matter energy density $\rho c^2 \neq 0$? Evidently, we cannot simply replace the right hand side of (3.39) or (3.40) by a constant times ρ. To proceed we study how ρ behaves under Lorentz transformations.[5] Consider a volume V_0 at rest, containing point masses with number density n_0, rest mass m_0 and negligible random motion ('cold dust'). The mass density is $\rho_0 = n_0 m_0$. Observe V_0 from a frame moving with velocity v. The mass of each particle becomes $m = \gamma m_0$ with $\gamma = (1 - \beta^2)^{-1/2}$ and $\beta = v/c$. Lorentz contraction makes that volume and number density transform as $V = V_0/\gamma$ and $n = \gamma n_0$, respectively. Hence, the mass density transforms as $\rho = \gamma^2 \rho_0$, and is therefore not a scalar. Nor is it a component of a vector, as that produces one factor γ at most, according to (1.8). However, the transformation of a second rank tensor may yield a factor γ^2 since

$$T^{\alpha'\beta'} = L^{\alpha'}{}_{\mu} L^{\beta'}{}_{\nu} T^{\mu\nu}, \qquad (3.41)$$

and if we take ρ as the 00-element of a second rank tensor of which it is the only nonzero element in the local rest-frame, $T^{\mu\nu} = \rho \delta^{\mu}_0 \delta^{\nu}_0$, we obtain $T^{0'0'} = (L^{0'}{}_0)^2 \rho = \gamma^2 \rho$ in the moving frame. It seems therefore that ρ is also part of a second rank tensor. But which? The crucial step is to recognise that one should take $T^{\mu\nu} = \rho u^{\mu} u^{\nu}$ where ρ is now the *rest* mass density and u^{μ} the 4-velocity. Since $u^{\mu} = c^{-1} dx^{\mu}/d\tau \simeq (\gamma, v^i/c)$ we have $T^{00} = \gamma^2 \rho$.

It would seem now that (3.38) is to be replaced by $R_{\mu\nu} = -(4\pi G/c^2) T_{\mu\nu}$ or equivalently $R^{\mu\nu} = -(4\pi G/c^2) T^{\mu\nu}$. However, that leads to inconsistencies. The trouble is that $R^{\mu\nu}{}_{;\nu}$ is in general nonzero, so that $T^{\mu\nu}{}_{;\nu}$ would also be nonzero. And as explained in a moment, conservation of mass or geodesic motion would no longer be guaranteed. The proper continuation turns out to be to replace (3.40) by

$$\boxed{G^{\mu\nu} = R^{\mu\nu} - \tfrac{1}{2} g^{\mu\nu} R = -\frac{8\pi G}{c^2} T^{\mu\nu}} \qquad (3.42)$$

with $T^{\mu\nu} \equiv \rho u^{\mu} u^{\nu}$ = stress-energy tensor; ρ = rest mass density and u^{μ} = bulk 4-velocity of the cold dust. The value of the constant $-8\pi G/c^2$ will be derived in the next section by considering the classical limit.

---

[5] We follow Price (1982). This beautifully written article is highly recommended.

Equation (3.42) has several attractive features. Because $G^{\mu\nu}{}_{:\nu} = 0$ according to (2.61), we have $T^{\mu\nu}{}_{:\nu} = (\rho u^\mu u^\nu)_{:\nu} = 0$, or

$$u^\mu (\rho u^\nu)_{:\nu} + \rho u^\nu u^\mu{}_{:\nu} = 0 . \tag{3.43}$$

Multiply this by u_μ, use $u_\mu u^\mu = 1$ and $u_\mu u^\mu{}_{:\nu} = 0$ (exercise 3.10). We are left with $(\rho u^\nu)_{:\nu} = 0$, which is the continuity equation. A second consequence is that $u^\nu u^\mu{}_{:\nu} = 0$, i.e. the matter moves along geodesics according to (2.53). Thus, equation (3.42) with $T^{\mu\nu} = \rho u^\mu u^\nu$ describes the dynamics of a collection of particles with a rest mass density ρ in their own gravitational field ('cold dust'). Mass is conserved, and there is only gravitational interaction between the mass elements because each moves along a geodesic. For that reason, too, there are no collisions and the gas pressure is negligible. This simple form of matter corresponds to the current state of the universe, with the galaxies serving as the particles. Other forms of matter in which for example the pressure is important can be handled by adapting the stress-energy tensor $T^{\mu\nu}$ accordingly. We return to this issue in § 3.6.

This may be the right place to draw attention to the power of the principle of general covariance. The field equations have the same form in all reference frame, rotating, accelerating or other – it does not matter. The reader who has checked the derivation of the Schwarzschild and Robertson-Walker metric will have noticed that we make in fact a series of co-ordinate transformations. We make one whenever it comes in handy, and there is no penalty because the form of the field equations does not change. Whatever co-ordinates we choose, the field equations deliver a metric tensor so that $\mathrm{d}s^2 = g_{\alpha\beta}\, \mathrm{d}x^\alpha \mathrm{d}x^\beta$ is the correct metric in those co-ordinates. But the real advantage lies deeper: the formulation of GR and the field equations would be practically impossible without exploiting general covariance. Take for example the stress-energy tensor $T^{\mu\nu} = \rho u^\mu u^\nu$ of cold dust. It appears on stage by asking how the rest mass energy density ρc^2 transforms in SR, which suggests that it is the 00-element of a second rank tensor $\rho u^\mu u^\nu$. Next we declare this form valid in all reference frames. It follows that what appears as energy density ρc^2 in the local rest frame shows up partly as momentum fluxes in another frame. The conclusion that all elements of the stress-energy tensor $T^{\mu\nu}$ contribute to the curvature of spacetime is both inescapable and gratifying.

Exercise 3.8: Prove that (3.39) and (3.40) are equivalent.

Hint: Forward: $R_{\mu\nu} = 0$ hence $R^{\mu\nu} = 0$ and $R = R^\mu{}_\mu = 0$, i.e. $G^{\mu\nu} = 0$ ($G_{\mu\nu} = 0$ as well). Backward: $G^{\mu\nu} = 0 \to 0 = G^\mu{}_\mu = R^\mu{}_\mu - \frac{1}{2}g^\mu{}_\mu R = -R$. Therefore $R^{\mu\nu} = 0$ and $R_{\mu\nu} = 0$.

Exercise 3.9: Why is ξ^i in (3.35) and (3.36) the physical distance between A and B? In § 3.1 co-ordinate differences $\xi^i = r_B^i - r_A^i$ were said to be meaningless.

Hint: The physical distance is determined by (3.7), and what is g_{ij}?

Exercise 3.10: Show that $u_\mu u^\mu{}_{;\nu} = 0$ (not to be confused with the geodesic equation (2.53): $u^\nu u^\mu{}_{;\nu} = 0$).

Hint: $1 = u^\mu u_\mu = g_{\mu\nu} u^\mu u^\nu \rightarrow 0 = (g_{\mu\nu} u^\mu u^\nu)_{;\sigma} = g_{\mu\nu} (u^\mu{}_{;\sigma} u^\nu + u^\mu u^\nu{}_{;\sigma}) = 2 g_{\mu\nu} u^\mu u^\nu{}_{;\sigma} = 2 u_\nu u^\nu{}_{;\sigma}$. One may likewise prove that $u^\mu u_{\mu;\sigma} = 0$.

3.5 Weak fields (2)

This section is a little technical. We seek an expansion of the field equations in terms of the small parameter $\gamma_{\alpha\beta}$ for weak fields. We need that to be able to deal with the classical limit of the field equations, and later for handling gravitational waves. Once more we make the substitution [6]

$$g_{\alpha\beta} = \eta_{\alpha\beta} + \gamma_{\alpha\beta}, \qquad (3.44)$$

with $\gamma_{\alpha\beta}$ 'small'; $g_{\alpha\beta}$ and $\gamma_{\alpha\beta}$ may now depend on x_0. Take $\alpha = \sigma$ in (2.62) and substitute (3.44). The largest term in $R_{\mu\nu}$ turns out to be of the order of γ:

$$R_{\mu\nu} = \tfrac{1}{2} \eta^{\alpha\beta} (\gamma_{\alpha\beta,\mu\nu} - \gamma_{\mu\alpha,\beta\nu} - \gamma_{\beta\nu,\mu\alpha} + \gamma_{\mu\nu,\alpha\beta}) + O(\gamma^2). \qquad (3.45)$$

This can be written in the following form:

$$R_{\mu\nu} = \tfrac{1}{2} \Box \gamma_{\mu\nu} - \tfrac{1}{2}(\tau_{\mu,\nu} + \tau_{\nu,\mu}) + O(\gamma^2), \qquad (3.46)$$

where $\Box$ is the d'Alembert operator:

$$\Box \psi = \eta^{\alpha\beta} \psi_{,\alpha\beta} = \left(\frac{1}{c^2} \frac{\partial^2}{\partial t^2} - \nabla^2 \right) \psi \qquad (3.47)$$

and

---

[6] $\eta_{\alpha\beta}$ and therefore $\gamma_{\alpha\beta}$ is not a tensor in GR. The use of tensors is usually very convenient but there are exceptions, and is never a must. This is one such exception.

$$\tau_\mu = \eta^{\alpha\beta}(\gamma_{\mu\alpha,\beta} - \tfrac{1}{2}\gamma_{\alpha\beta,\mu}) \ . \tag{3.48}$$

Verification is a matter of substitution. The next step is a transformation of the co-ordinates. An exercise shows that there always exists a transformation $x^\mu \to \tilde{x}^\mu$ so that $\tilde{\tau}_\mu = O(\tilde{\gamma}^2)$. We work now in these new co-ordinates and omit all terms of second and higher order in $\tilde{\gamma}$. Then $\tilde{R}_{\mu\nu} = 0$ reduces to (we drop the ~ again):

$$R_{\mu\nu} \simeq \tfrac{1}{2}\Box\gamma_{\mu\nu} = 0 \ ; \qquad \tau_\mu \simeq 0 \ . \tag{3.49}$$

Here '$\simeq$' means accurate to first order in γ. These co-ordinates are called *harmonic co-ordinates*.[7] For stationary fields (3.49) leads to $\nabla^2\gamma_{00} = \gamma_{00,ii} = 0$. Hence $R_{\mu\nu} = 0$ implies (3.18) in these harmonic co-ordinates. But in (3.18) no special co-ordinates had been chosen. Exercise 3.11 shows that $\tilde{\gamma}_{00} = \gamma_{00}$ for stationary fields, so that (3.18) is always valid.

We also need the equivalent of (3.49) for the Einstein tensor $G^{\mu\nu}$. We suppress details and give only the result:

$$G_{\mu\nu} \simeq \tfrac{1}{2}\Box h_{\mu\nu} \ ; \qquad h^{\mu\nu}{}_{,\nu} \simeq 0 \ , \tag{3.50}$$

with

$$\left.\begin{array}{ll} h_{\mu\nu} = \gamma_{\mu\nu} - \tfrac{1}{2}\eta_{\mu\nu}\gamma \ ; & \gamma = \gamma^\sigma{}_\sigma \ ; \\ \gamma_{\mu\nu} = h_{\mu\nu} - \tfrac{1}{2}\eta_{\mu\nu}h \ ; & h = h^\sigma{}_\sigma \ . \end{array}\right\} \tag{3.51}$$

Since $R_{\mu\nu}$ and $G_{\mu\nu}$ are of order γ, we may raise and lower indices with $\eta^{\alpha\beta}$ which may be moved through $\Box$ in (3.49) and (3.50). Therefore we may move the indices up and down in these expressions as we please. The condition $h^{\mu\nu}{}_{,\nu} = 0$ is called the *Lorentz gauge* because of the strong analogy with the Lorentz gauge in electrodynamics ($A^\nu{}_{,\nu} = 0$ with A^ν = vector potential). The field equation for weak field now follows from (3.42) and (3.50):

$$\boxed{\Box h^{\mu\nu} = -\frac{16\pi G}{c^2}T^{\mu\nu}} \tag{3.52}$$

The one remaining issue is to show that the constant in eq. (3.42) has the value $-8\pi G/c^2$. From the definition of the Einstein tensor (2.60) we infer that $G^\mu{}_\mu = R - 2R = -R$. Substitution back into (2.60) produces $R^{\mu\nu} = G^{\mu\nu} - \tfrac{1}{2}g^{\mu\nu}G^\alpha{}_\alpha$. We now write (3.42) as $G^{\mu\nu} = kT^{\mu\nu}$ and compute k. In the classical limit G^{00} is by far the largest element of $G^{\mu\nu}$ since $u^\mu \simeq (1, v^i/c)$. Hence $G_{00} \simeq \eta_{0\alpha}\eta_{0\beta}G^{\alpha\beta} = G^{00} \simeq k\rho$, and $G^\mu{}_\mu = \eta_{\mu\alpha}G^{\mu\alpha} \simeq G^{00} \simeq G_{00}$. It

---

[7] Co-ordinates obeying the 4 restrictions $g^{\alpha\beta}\Gamma^\mu{}_{\alpha\beta} = 0$ are called harmonic. For weak fields this amounts to $\tau_\mu = 0$ to first order in γ, see Weinberg (1972), p. 161 ff.

follows that $R_{00} \simeq G_{00} - \frac{1}{2}\eta_{00}G^{\alpha}{}_{\alpha} \simeq \frac{1}{2}G_{00} \simeq \frac{1}{2}k\rho$. Comparison with (3.38) yields $k = -8\pi G/c^2$.

Exercise 3.11: Show that τ_μ can be made of order $O(\xi^2)$ by the transformation $x^\alpha \to \tilde{x}^\alpha = x^\alpha + \xi^\alpha(x)$ with ξ^α and its derivatives 'small'.

Hint: $\tilde{x}^\alpha{}_{,\mu} = \delta^\alpha_\mu + \xi^\alpha{}_{,\mu}$; $g_{\mu\nu}$ tensor $\to g_{\mu\nu} = \tilde{g}_{\alpha\beta}\tilde{x}^\alpha{}_{,\mu}\tilde{x}^\beta{}_{,\nu}$, from which $g_{\mu\nu} = \tilde{g}_{\mu\nu} + \xi_{\mu,\nu} + \xi_{\nu,\mu} + O(\xi^2)$. Hence $\tilde{\gamma}_{\mu\nu} = \gamma_{\mu\nu} - \xi_{\mu,\nu} - \xi_{\nu,\mu}$. Substitute in $\tilde{\tau}_\mu = \eta^{\alpha\beta}(\tilde{\gamma}_{\mu\alpha,\beta} - \frac{1}{2}\tilde{\gamma}_{\alpha\beta,\mu})$:

$$\tilde{\tau}_\mu = \tau_\mu - \Box\xi_\mu - \eta^{\sigma\rho}(\xi_{\sigma,\mu\rho} - \tfrac{1}{2}\xi_{\sigma,\rho\mu} - \tfrac{1}{2}\xi_{\rho,\sigma\mu}) + O(\xi^2) .$$

The term $\eta^{\sigma\rho}(...)$ is zero (interchange ρ and σ in the third term – why is that allowed?). Choose ξ_μ so that $\Box\xi_\mu = \tau_\mu$, then $\tilde{\tau}_\mu = O(\xi^2) = O(\tilde{\gamma}^2)$ and $\tilde{R}_{\mu\nu} = \frac{1}{2}\Box\tilde{\gamma}_{\mu\nu} + O(\tilde{\gamma}^2)$. There is still gauge freedom left because ξ_μ is determined up to an arbitrary solution of $\Box\xi_\mu = 0$.

Exercise 3.12: For weak stationary fields the metric in harmonic co-ordinates may be written as

$$g_{\alpha\beta} = \eta_{\alpha\beta} + \gamma_{\alpha\beta} = \eta_{\alpha\beta} + \frac{2\Phi(\boldsymbol{r})}{c^2} e_{\alpha\beta} . \tag{3.53}$$

$e_{\alpha\beta} = 1$ if $\alpha = \beta$, otherwise 0 (this is not a tensor, see exercise 2.1; because $\gamma_{\alpha\beta}$ is not a tensor it cannot be expressed in terms of known tensors).

Hint: Since $\eta^{\alpha\beta}$ is diagonal and $\gamma_{\alpha\beta,0} = 0$ we have $\eta^{\alpha\beta}\gamma_{\mu\alpha,\beta} = -\gamma_{\mu i,i}$ and $\frac{1}{2}\eta^{\alpha\beta}\gamma_{\alpha\beta,\mu} = \frac{1}{2}\gamma_{00,\mu} - \frac{1}{2}\gamma_{ii,\mu}$. From $\tau_\mu = 0$: $\gamma_{ii,\mu} - 2\gamma_{\mu i,i} = \gamma_{00,\mu}$. Try $\gamma_{ij} = a\delta_{ij}$ and other γ's zero, except γ_{00}. Result: $a_{,j} = \gamma_{00,j}$. Take $a = \gamma_{00} = 2\Phi/c^2$, according to (3.19).

Exercise 3.13: Show that $\Phi \sim -v^2$ for weak stationary field, so that $\gamma_{\alpha\beta}$ from (3.53) is of order β^2 (this is not to say that Φ depends on v, but that its value is of order $-v^2$ with v = characteristic velocity of a particle at that position.)

Hint: Planetary orbits: $\Phi = -GM_\odot/r$; circular orbit: $mv^2/r = GM_\odot m/r^2$.

Exercise 3.14: A general invariant definition of energy does not exist in GR. However, it does in case of a single test particle. From (3.23) we retrieve the SR relations $p^\mu = (E/c, p^i)$ and $p_\mu = (E/c, -p^i)$ since $g_{\alpha\beta} = \eta_{\alpha\beta}$. Hence two possibilities: $E = cp_0$ or $E = cp^0$. Show that

$$E = cp_0 = m_0 c^2 u_0 \qquad (3.54)$$

is the right choice because it has the correct classical limit, and because E is constant when the metric does not depend on time.

Hint: E = constant from (2.40). Consider the classical limit with (3.22) and (3.53): $(m_0 c)^2 = p_\alpha p^\alpha = g_{\alpha\beta} p^\alpha p^\beta = g_{00}(p^0)^2 + g_{11} p^2$ ($p^2 = p^i p^i$). Furthermore $g_{00}(p^0)^2 = g_{00}(g^{00} p_0)^2 = g_{00}(p_0/g_{00})^2 = p_0^2/g_{00} \simeq (1 - 2\Phi/c^2) p_0^2$. We now have $(m_0 c)^2 \simeq (1 - 2\Phi/c^2)(p_0^2 - p^2)$ or $\pi_0^2 - \pi^2 \simeq 1 + 2\Phi/c^2$, with $\pi_0 = p_0/m_0 c$ and $\pi = p/m_0 c$, or $\pi_0^2 = 1 + \pi^2 + 2\Phi/c^2$. Take the square root and use that π and Φ/c^2 are small: $\pi_0 \simeq 1 + \frac{1}{2}\pi^2 + \Phi/c^2$, from which $E = cp_0 \simeq m_0 c^2 + p^2/2m_0 + m_0 \Phi$. The three terms have an obvious classical interpretation.

Exercise 3.15: Invariant definition of the energy of a test particle. Consider a particle with 4-momentum p^α and an observer W with 4-velocity u^α. Show that from W's point of view, the energy of the particle is

$$E = cp_\alpha u^\alpha . \qquad (3.55)$$

Hint: $E = cp_\alpha u^\alpha = c\bar{p}_\alpha \bar{u}^\alpha = c\bar{p}_0$ = energy that W assigns to the particle according to (3.54) ($\bar{}$ = local rest-frame of W). Every W assigns the same function $c\bar{p}_0$ to the particle, but not the same value.

3.6 Discussion

In this section we deal with more general forms of the field equations. In the first place we investigate what the expression for the stress-energy tensor $T^{\mu\nu}$ should be when the gas pressure p is nonzero. It seems reasonable that $T^{\mu\nu}$ will be of the form $T^{\mu\nu} = \rho u^\mu u^\nu + p A^{\mu\nu}$. The only symmetric tensors that are available to build $A^{\mu\nu}$ are $g^{\mu\nu}$ and $u^\mu u^\nu$. Therefore try

$$T^{\mu\nu} = \rho u^\mu u^\nu + \frac{p}{c^2}\left(a\, u^\mu u^\nu + b\, g^{\mu\nu}\right) . \qquad (3.56)$$

Here, too, $T^{\mu\nu}{}_{;\nu} = 0$ must hold. For $p = 0$ that resulted in the continuity equation and the geodesic equation (i.e. the equation of motion), and something similar will also be the case now. To see what happens we work in the classical limit and work out $\{\rho u^\mu u^\nu + (p/c^2)(a u^\mu u^\nu + b\eta^{\mu\nu})\}_{,\nu} = 0$. The result turns out to be the continuity equation and the Navier-Stokes equation provided $a = 1$ and $b = -1$, see Foster and Nightingale (1989, p. 73) or Schutz (1985, Ch. 4). Thus we have found that

$$T^{\mu\nu} = \rho u^\mu u^\nu + \frac{p}{c^2}\left(u^\mu u^\nu - g^{\mu\nu}\right). \tag{3.57}$$

In the early universe and in neutron stars the gas pressure reaches values of the order of $p \sim \rho c^2$. Such high pressures determine, together with other forms of energy, the structure of spacetime because pressure is a form of potential energy. Pressure *gradients*, on the other hand, occur in the equations of motion $T^{\mu\nu}{}_{;\nu} = 0$, but they have no influence on the curvature of spacetime (§ 5.3).

The reasoning leading to (3.57) is an example of how the principle of general covariance is used in practice. We start with the SR form of $T^{\mu\nu}$ and look for a so-called mimimal generalisation, i.e. it is forbidden to add terms that are identically zero in SR, such as $\rho R^{\mu\nu}{}_{\alpha\beta} u^\alpha u^\beta$. Often it amounts to $, \to :$ and $\eta^{\mu\nu} \to g^{\mu\nu}$, but it remains a matter of trial and error.

The next issue is the cosmological constant. Einstein considered an extra term in the field equation (3.42):

$$\boxed{G^{\mu\nu} + \Lambda g^{\mu\nu} = -\frac{8\pi G}{c^2} T^{\mu\nu}} \tag{3.58}$$

The historical motivation for a nonzero cosmological constant Λ was that (1) it is a term that logically may appear in the equation as $g^{\mu\nu}{}_{;\nu} = 0$, so that $T^{\mu\nu}{}_{;\nu} = 0$ is left intact, and (2) it permitted the possibility of a static spherical universe. This solution turned out to be unstable, and when it was subsequently discovered that the universe actually expands, the cosmological constant was abandoned. But nowadays it is back with flying colours, see § 9.5. The magnitude of Λ is of the order of (size universe)$^{-2}$ which is so small that (3.42) remains valid for all local physics. A physical explanation of the cosmological constant is postponed to § 9.5.

Finally, we ask how (3.42) is to be extended to include other fields. These fields have their own stress-energy tensor and it seems obvious that

$$G^{\mu\nu} = -\frac{8\pi G}{c^2} \sum_i T^{\mu\nu}_{(\text{field } i)}. \tag{3.59}$$

The coupling between $G^{\mu\nu}$ and the electromagnetic field, for example, is important for the metric of a *charged* black hole. Here we shall only have the opportunity to consider coupling of $G^{\mu\nu}$ to a scalar boson field. Models of the universe based on such equations exhibit inflation, a brief period of extremely rapid expansion, and may provide a solution for some of the problems of the standard model of the Big Bang.

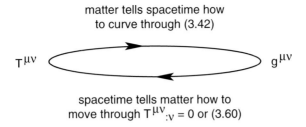

Fig. 3.4. Schematic structure of the equations of general relativity. The novel aspect is that both matter and spacetime have become active players in the dynamics of the world. Conceptually, one full round corresponds to one timestep in a numerical code. If $g_{\mu\nu}$ and $g_{\mu\nu,0}$ are given at $x^0 = 0$ and $T^{\mu\nu}$ is sufficiently well behaved, then $g_{\mu\nu}$ is determined for all x^0, up to 4 arbitrary functions. This freedom of choice corresponds to the freedom of choosing the co-ordinates (Wald, 1984, Ch. 10). After Rees, M. et al.: 1974, *Black Holes, Gravitational Waves and Cosmology*, Gordon and Breach, p. 2.

Structure of the equations

We draw attention to an important peculiarity of eq. (3.59), namely that it does not handle the various forms of energy on equal footing. All energies *other* than gravitational energy contribute to the geometry of spacetime through their $T^{\mu\nu}$ on the right hand side in (3.59). The gravitational field itself appears only in $G^{\mu\nu}$ on the left. Since $G^{\mu\nu}{}_{;\nu} = 0$ poses 4 extra restrictions, eq. (3.42) or (3.59) is a set of 6 nonlinear differential equations for the metric tensor. The nonlinearity is the mathematical expression of the fact that the energy density of the gravitational field acts as a source of gravity itself. The superposition principle of classical mechanics (gravitational field of two bodies is the sum of the individual fields) no longer applies in GR, except when the fields are weak, as in eq. (3.52).

The computational scheme of GR is shown in Fig. 3.4. The dynamics of the (matter) fields is fixed by $T^{\mu\nu}{}_{;\nu} = 0$ (for example the structure of a relativistic star, § 5.3). The motion of a test particle follows from a generalisation of (3.13):

$$\frac{Dp^\mu}{D\tau} = f^\mu . \tag{3.60}$$

The extra forces f^μ (e.g. the Lorentz force) push the particle off the geodesic, and the 4-momentum p^μ is no longer parallel transported along the orbit.[8]

---

[8] For information on numerical relativity see for instance Font, J.A., *Living Rev. Relativity* 6 (2003) 4 (http://www.livingreviews.org/lrr-2003-4).

But if there is only gravity, a particle experiences no acceleration in the parlance of GR.

Exercise 3.16: Show that (3.58) reduces to

$$\nabla^2 \Phi = 4\pi G\rho - \Lambda c^2 \qquad (3.61)$$

in the classical limit. A positive Λ is equivalent to a negative mass density, i.e. a repulsive force.

Hint: Write $k = -8\pi G/c^2$ for brevity. Combine (3.50) and (3.58): $\frac{1}{2}\Box h_{00} = -\Lambda + k\rho$. Combine (3.50) and (3.58) again, lower one index and contract: $\frac{1}{2}\Box h \simeq -4\Lambda + k\rho$ (T^{00} and G^{00} are the largest terms). From (3.51): $\Box \gamma_{00} = \Box(h_{00} - \frac{1}{2}h) = 2\Lambda + k\rho$. Stationary fields and (3.19): $\nabla^2 \Phi \simeq -\Box \Phi \simeq -\frac{1}{2}c^2 \Box \gamma_{00}$.

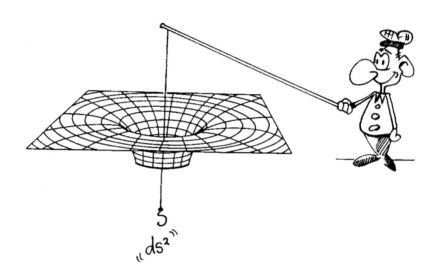

4

The Schwarzschild Metric

The stationary, spherically symmetric solution of the vacuum equations is of fundamental importance as it describes the field of a spherically symmetric body like a star – the simplest gravitational field one may think of. This solution was discovered by K. Schwarzschild in 1916, a few month after the publication of the vacuum equations. The field equations (3.39) are complicated nonlinear differential equations, but the spherical symmetry greatly simplifies the mathematics. The solution predicts small deviations from the Newtonian results for weak fields like that of the Sun, of which four have been confirmed experimentally (the perihelium precession, the deflection of light, the gravitational redshift, and the time delay of light signals). Together with the recent discoveries on the binary pulsar these are the most important quantitive verifications of GR. The solution also applies to the strong fields of compact objects such as neutron stars and black holes. In the latter case the solution predicts the existence of a singularity in spacetime at $r = 0$, which is, fortunately, unobservable for a distant observer.

4.1 Preliminary calculations

We employ spherical co-ordinates $x^0 = ct$, $x^1 = r$, $x^2 = \theta$ and $x^3 = \varphi$, and simplify the form of the metric (3.1) step by step:

1. Stationarity implies that $g_{\alpha\beta,0} = 0$.

2. ds^2 must be invariant under $dx^0 \to -dx^0$, so that terms $\propto dx^0 dx^i$ must be absent $\to g_{0i} = 0$.

3. Spherical symmetry implies that ds^2 is invariant under $d\theta \to -d\theta$ and $d\varphi \to -d\varphi$. It follows that $g_{r\theta} = g_{r\varphi} = g_{\theta\varphi} = 0$ because terms $\propto drd\theta$, $drd\varphi$ and $d\theta d\varphi$ must vanish.

4. At this point only the diagonal terms remain, and we write the metric as

$$ds^2 = Ac^2 dt^2 - Bdr^2 - Cr^2 d\theta^2 - Dr^2 \sin^2\theta\, d\varphi^2 \ . \quad (4.1)$$

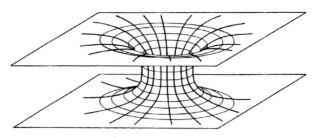

Fig. 4.1. A wormhole. Adapted from Misner et al. (1971).

The factors r^2 and $r^2 \sin^2\theta$ have been added for convenience.

5. According to (3.7), the subspace $x^0, r =$ constant has the metric of a sphere $dl^2 = r^2(C d\theta^2 + D \sin^2\theta \, d\varphi^2)$. An inhabitant of this space will not notice any effect of gravity (spherical symmetry). He concludes that his world is an ordinary spherical surface and takes $C = D$.

6. By choosing convenient units we can arrange that $C = 1$.

Following the literature we write (4.1) as

$$ds^2 = e^{2\nu} c^2 dt^2 - e^{2\lambda} dr^2 - r^2(d\theta^2 + \sin^2\theta \, d\varphi^2) , \qquad (4.2)$$

where $\nu = \nu(r)$ and $\lambda = \lambda(r)$.[1] Further simplification of the metric on the basis of symmetry considerations is not possible, except that we know that for $r \to \infty$ (4.2) must coincide with the Lorentz metric, so that $\lambda, \nu \to 0$ for $r \to \infty$. The elements of the metric tensor are:

$$\left.\begin{aligned} g_{00} &= e^{2\nu} ; & g_{11} &= -e^{2\lambda} ; \\ g_{22} &= -r^2 ; & g_{33} &= -r^2 \sin^2\theta , \end{aligned}\right\} \qquad (4.3)$$

and all other $g_{\alpha\beta}$ are zero. We have chosen the co-ordinates so that the 2-volume of each subspace $x^0, r =$ constant equals $4\pi r^2$, but that is not to say that r is the distance to the origin. There may not even be an origin. Suppose you move in the radial direction toward the origin, defined by insisting that the area of the sphere $r =$ constant decreases. But as you keep moving in that direction, the area may at some point begin to increase again so that $r = 0$ is never reached, Fig. 4.1. The message is that such topological constructions are in principle possible, and the field equations decide whether they are real or not. As it turns out, any particle entering a black hole must hit a singularity at $r = 0$. So a Schwarzschild *wormhole*, if it exists, is a different concept than Fig. 4.1 suggests, see Misner et al. (1971) and Wald (1984).

Of course one may use different co-ordinates. One possibility is to take

---

[1] ν and λ are in this chapter no longer available as indices.

4.1 Preliminary calculations

$B = C = D$ in (4.1), leading to isotropic co-ordinates, see Adler et al. (1965). Another possibility is Kruskal-Szekeres co-ordinates, § 6.4. Physical quantities are always independent of the choice of the co-ordinates. The calculations that follow are divided into three steps: first we need the Christoffel symbols. Then we use (2.57) to find $R_{\alpha\beta}$. Finally $\nu(r)$ and $\lambda(r)$ are solved from $R_{\alpha\beta} = 0$.

The Christoffel symbols

The easiest way is to find the geodesic equation with the help of variational calculus. We need this equation anyway, and once we have it we may just read the Γ's by comparing with (2.34), just like we did in § 2.5. We elaborate (2.36) with the help of (4.2):

$$\delta \int \left\{ e^{2\nu} (\dot{x}^0)^2 - e^{2\lambda} \dot{r}^2 - r^2 \dot{\theta}^2 - r^2 \sin^2\theta \, \dot{\varphi}^2 \right\} dp = 0 . \qquad (4.4)$$

Here $\dot{} = d/dp$, and $x^0(p), r(p), \theta(p), \varphi(p)$ is the parametric representation of a geodesic. Next we write down the Euler-Lagrange equations (2.37). Notation: $L = \{\cdots\} = $ integrand of (4.4):

(a). $\partial L/\partial x^0 = (\partial L/\partial \dot{x}^0)\dot{}$. Since $\partial L/\partial x^0 = 0$ we get

$$(2\dot{x}^0 e^{2\nu})\dot{} = 0 ; \qquad \dot{} = d/dp . \qquad (4.5)$$

Now use that $\dot{\nu} = d\nu/dp = (d\nu/dr)(dr/dp) = \nu' \dot{r}$ with $\nu' \equiv d\nu/dr$. Result:

$$\ddot{x}^0 + 2\nu' \dot{x}^0 \dot{r} = 0 ; \qquad ' = d/dr . \qquad (4.6)$$

Comparing with (2.34) we conclude that $\Gamma^0_{01} + \Gamma^0_{10} = 2\nu'$, and since $\Gamma^0_{01} = \Gamma^0_{10}$ we have

$$\Gamma^0_{10} = \nu' . \qquad (4.7)$$

The other $\Gamma^0_{\alpha\beta}$ are zero.

(b). $\partial L/\partial r = (\partial L/\partial \dot{r})\dot{}$, or after performing the differentiations:

$$2\nu' e^{2\nu} (\dot{x}^0)^2 - 2\lambda' e^{2\lambda} \dot{r}^2 - 2r(\dot{\theta}^2 + \sin^2\theta \, \dot{\varphi}^2) = (-2\dot{r} e^{2\lambda})\dot{} . \qquad (4.8)$$

After some cleaning up:

$$\ddot{r} + \nu' e^{2(\nu-\lambda)} (\dot{x}^0)^2 + \lambda' \dot{r}^2 - r(\dot{\theta}^2 + \sin^2\theta \, \dot{\varphi}^2) e^{-2\lambda} = 0 . \qquad (4.9)$$

Consequently:

$$\left. \begin{array}{ll} \Gamma^1_{00} = \nu' e^{2(\nu-\lambda)} ; & \Gamma^1_{11} = \lambda' ; \\[6pt] \Gamma^1_{22} = -r\, e^{-2\lambda} ; & \Gamma^1_{33} = -r \sin^2\theta \, e^{-2\lambda} , \end{array} \right\} \qquad (4.10)$$

and all other $\Gamma^1_{\alpha\beta}$ are zero. We get the remaining Christoffel symbols from the other two geodesic equations (see exercise):

$$\Gamma^2_{12} = \frac{1}{r} \; ; \quad \Gamma^2_{33} = -\sin\theta\cos\theta \; ; \tag{4.11}$$

$$\Gamma^3_{13} = \frac{1}{r} \; ; \quad \Gamma^3_{23} = \cot\theta \; . \tag{4.12}$$

At this point all nonzero Christoffel symbols have been found. Recall that $\Gamma^\mu_{\alpha\beta} = \Gamma^\mu_{\beta\alpha}$.

Computing $R_{\alpha\beta}$

The Γ's must now be inserted in (2.57). This requires a little perseverance. We illustrate that for R_{00}:

$$\begin{aligned} R_{00} &= \Gamma^\alpha_{0\alpha,0} - \Gamma^\alpha_{00,\alpha} - \Gamma^\alpha_{00}\Gamma^\beta_{\alpha\beta} + \Gamma^\alpha_{0\beta}\Gamma^\beta_{0\alpha} \\ &= -\Gamma^1_{00,1} - \Gamma^1_{00}\Gamma^\beta_{1\beta} + \Gamma^\alpha_{0\beta}\Gamma^\beta_{0\alpha} \\ &= -\Gamma^1_{00,1} - \Gamma^1_{00}\Gamma^\beta_{1\beta} + 2\Gamma^0_{10}\Gamma^1_{00} \; . \end{aligned} \tag{4.13}$$

$\Gamma^\alpha_{\mu\alpha}$ occurs twice in (2.57) and may be calculated with (2.33). But the gain is minimal because it is just as easy to find $\Gamma^\alpha_{\mu\alpha}$ through a summation:

$$\Gamma^\alpha_{0\alpha} = \Gamma^\alpha_{3\alpha} = 0 \; ; \quad \Gamma^\alpha_{1\alpha} = \nu' + \lambda' + \frac{2}{r} \; ; \quad \Gamma^\alpha_{2\alpha} = \cot\theta \; . \tag{4.14}$$

From (4.14) we now obtain

$$R_{00} = -\left\{\nu'' - \nu'\lambda' + (\nu')^2 + \frac{2\nu'}{r}\right\} e^{2(\nu-\lambda)} \; . \tag{4.15}$$

Without proof we mention the other nonzero components of $R_{\alpha\beta}$:

$$R_{11} = \nu'' - \nu'\lambda' + (\nu')^2 - \frac{2\lambda'}{r} \; ; \tag{4.16}$$

$$R_{22} = (1 - r\lambda' + r\nu')\,e^{-2\lambda} - 1 \; ; \tag{4.17}$$

$$R_{33} = R_{22}\sin^2\theta \; . \tag{4.18}$$

We may now compute the total curvature

$$\begin{aligned} R &= R^\alpha_{\;\alpha} = g^{\alpha\beta}R_{\alpha\beta} = \sum_\alpha R_{\alpha\alpha}/g_{\alpha\alpha} \\ &= -2\left\{\nu'' - \nu'\lambda' + (\nu')^2 + \frac{2\nu'}{r} - \frac{2\lambda'}{r} + \frac{1}{r^2}\right\}e^{-2\lambda} + \frac{2}{r^2} \; . \end{aligned} \tag{4.19}$$

In $\sum_\alpha \cdots$ the summation convention has been switched off momentarily. The Einstein tensor $G_{\alpha\beta} = R_{\alpha\beta} - \frac{1}{2} g_{\alpha\beta} R$ is also diagonal and we need only G_{00} and G_{11}:

$$G_{00} = -\frac{e^{2\nu}}{r^2} \frac{d}{dr} r(1 - e^{-2\lambda}) ; \qquad (4.20)$$

$$G_{11} = \frac{1}{r^2}(e^{2\lambda} - 1) - \frac{2\nu'}{r} . \qquad (4.21)$$

Exercise 4.1: Show that the two remaining geodesic equations for θ and φ are

$$(r^2 \dot{\theta})^{\cdot} = r^2 \sin\theta \cos\theta \, \dot{\varphi}^2 \quad \rightarrow \quad \ddot{\theta} + \frac{2}{r} \dot{r}\dot{\theta} - \sin\theta \cos\theta \, \dot{\varphi}^2 = 0 ; \qquad (4.22)$$

$$(r^2 \sin^2\theta \, \dot{\varphi})^{\cdot} = 0 \quad \rightarrow \quad \ddot{\varphi} + \frac{2}{r} \dot{r}\dot{\varphi} + 2\cot\theta \, \dot{\theta}\dot{\varphi} = 0 , \qquad (4.23)$$

and determine the corresponding Christoffel symbols (4.11) and (4.12).

4.2 The Schwarzschild metric

It is actually easier to solve $G_{\alpha\beta} = 0$ than $R_{\alpha\beta} = 0$. From (4.20) we find:

$$r(1 - e^{-2\lambda}) = b \quad \rightarrow \quad e^{2\lambda} = \frac{1}{1 - b/r} , \qquad (4.24)$$

with b constant. Substitute that in $G_{11} = 0$:

$$2\nu' = \frac{b/r^2}{1 - b/r} \quad \rightarrow \quad e^{2\nu} = A(1 - b/r) , \qquad (4.25)$$

since the expression on the left can be integrated to $2\nu = \log(1 - b/r) +$ const. The constant A must be 1 because (4.2) must be the Lorentz metric for $r \to \infty$. We insert these results in (4.2):

$$ds^2 = (1 - b/r) c^2 dt^2 - \frac{dr^2}{1 - b/r} - r^2(d\theta^2 + \sin^2\theta \, d\varphi^2) . \qquad (4.26)$$

To determine the constant b we note that the metric (4.26) can only describe the effect of a spherically symmetric mass M. In the classical limit the gravitational potential is $\Phi = -GM/r$, and according to (3.19) we have $g_{00} = 1 - 2GM/c^2 r$ for large r. Comparison with (4.26) shows that $b = 2GM/c^2$, or

$$ds^2 = (1 - r_s/r) c^2 dt^2 - \frac{dr^2}{1 - r_s/r} - r^2 (d\theta^2 + \sin^2\theta \, d\varphi^2) . \quad (4.27)$$

The quantity

$$r_s \equiv \frac{2GM}{c^2} \quad (4.28)$$

is called the *Schwarzschild radius*. The Sun has $r_s \simeq 3\,\text{km}$, see also Table 3.1. The components of the metric tensor, and λ and ν are now given by

$$\left. \begin{array}{ll} g_{00} = \mathrm{e}^{2\nu} = \mathrm{e}^{-2\lambda} = 1 - r_s/r \, ; & g_{11} = \dfrac{-1}{1 - r_s/r} \, ; \\[2mm] g_{22} = -r^2 \, ; & g_{33} = -r^2 \sin^2\theta \, . \end{array} \right\} \quad (4.29)$$

Relations (4.27) – (4.29) describe the standard form of the Schwarzschild metric. Birkhoff showed in 1923 that the Schwarzschild metric is the only spherically symmetric solution of the field equations exterior to a spherical, nonrotating, uncharged but not necessarily stationary mass distribution. This is known as *Birkhoff's theorem*.

The range of validity of the co-ordinates is as follows. Because of stationarity $-\infty < t < \infty$. Furthermore $0 \leq \theta \leq \pi$, $0 \leq \varphi < 2\pi$ because we have chosen θ and φ in the 'usual' way. For r there are two options:

1. There is a material surface at $r = R$ (and $R > r_s$). The object is a *(compact) star* and the metric is valid in the vacuum region outside the star, $r \geq R$. Inside the star the metric is different, see Ch. 5.

2. There is no material surface in the range $r > r_s$. In that case the object is a *black hole*. The metric is valid everywhere (no restrictions on r), but has two singularities at $r = 0$ and at $r = r_s$. Their physical nature will be dealt with in Ch. 6.

We briefly dwell on the question how the values of the co-ordinates may be measured. The co-ordinate time t may be determined for example by counting pulses from a laser. The laser emits pulses once every second in its own proper time, say. Now let the laser be at $r = \infty$ ('sufficiently far away') and have a proper frequency ν_0. At some finite r one measures a blueshifted frequency ν. We repeat the reasoning below (3.20): $\nu d\tau = \nu_0 dt$ ($d\tau = dt$ at $r = \infty$ because $g_{\alpha\beta} = \eta_{\alpha\beta}$). From (3.2) and (4.29):

4.2 The Schwarzschild metric

$$\frac{\Delta \nu}{\nu_0} = \frac{\nu - \nu_0}{\nu_0} = \frac{\mathrm{d}t}{\mathrm{d}\tau} - 1 = g_{00}^{-1/2} - 1$$

$$= \left(1 - \frac{r_s}{r}\right)^{-1/2} - 1 \simeq \frac{r_s}{2r} + \cdots . \qquad (4.30)$$

By measuring $\Delta \nu$ we may determine r/r_s everywhere. To fix r and r_s separately, we may either put a satellite in a circular orbit and measure its period Δt ($\rightarrow r$, see (4.46)), or we may construct a sphere with radius r, defined as the set of all spatial positions having the same frequency ν, and measure its area O. Then r is also known because r has been chosen so that a surface $r = \text{constant}$ has 2-volume $O = 4\pi r^2$. Finally, we may draw a grid of latitude and longitude circles on these spheres to determine θ and φ. This shows that measuring procedures to determine t, r, θ and φ do in principle exist.

No one will notice anything out of the ordinary as long as he or she stays on a shell $r = \text{constant}$, because the metric is classical: $\mathrm{d}l^2 = r^2(\mathrm{d}\theta^2 + \sin^2\theta\, \mathrm{d}\varphi^2)$. But strange things do emerge as one travels between shells: the 3-volume between two shells at $r = r_1$ and $r = r_2$ is larger than $\frac{4\pi}{3}(r_2^3 - r_1^3)$. And the distance between r_1 and r_2 is larger than the co-ordinate difference $r_2 - r_1$ (see the exercises below).

Exercise 4.2: Give a qualitative argument to illustrate that r_s is proportional to the mass M and not, for example, to some other power of M.

Hint: Require for example that the escape velocity is c at $R = r_s$. Another possibility is to say that $Mc^2 \sim$ potential gravitational energy $\sim (GM/r_s)M$.

Exercise 4.3: Calculate the total curvature R. Is spacetime curved or not?

Hint: A close look at (2.59) and (3.39) may spare you a surprise. What does 'curved' mean?

Exercise 4.4: Consider two spherical shells at $r = r_1$ and $r = r_2$. Calculate (1) the 3-volume between the shells; (2) the 2-volume (area) in the plane $\theta = \pi/2$ between r_1 and r_2; (3) the 1-volume (length) of a stick pointing radially to the star, end points at r_1 and r_2; (4) the 2-volume of spacetime enclosed by $t_1 \leq t \leq t_2$, $r_1 \leq r \leq r_2$ and θ, $\varphi = \text{constant}$.

Hint: Comes down to finding the invariant volume-element according to § 3.3. (1). Metric according to (3.7): $\mathrm{d}l^2 = (1 - r_s/r)^{-1}\mathrm{d}r^2 + r^2(\mathrm{d}\theta^2 + \sin^2\theta\, \mathrm{d}\varphi^2)$, hence $g \equiv \det\{g_{ik}\} = (1 - r_s/r)^{-1} \cdot r^2 \cdot r^2 \sin^2\theta$; 3-volume $= \int_{r_1}^{r_2} dr \int_0^\pi d\theta \cdot \int_0^{2\pi} d\varphi \sqrt{g} = 4\pi \int_{r_1}^{r_2} (1 - r_s/r)^{-1/2} r^2 \, dr > 4\pi \int_{r_1}^{r_2} r^2 dr = \frac{4\pi}{3}(r_2^3 - r_1^3)$.

(2). $\theta = \pi/2$, $d\theta = 0$ and $g = (1 - r_s/r)^{-1} \cdot r^2$; 2-volume $= \int_{r_1}^{r_2} dr \int_0^{2\pi} d\varphi \cdot \sqrt{g} > \pi(r_2^2 - r_1^2)$.

(3). $dl^2 = (1 - r_s/r)^{-1} dr^2$ because $d\theta = d\varphi = 0$. Hence $l = \int_{r_1}^{r_2} (1 - r_s/r)^{-1/2} \, dr > r_2 - r_1$.

(4). Metric: $ds^2 = (1 - r_s/r)c^2 dt^2 - (1 - r_s/r)^{-1} dr^2 \rightarrow g = -c^2$; 2-volume $= \int_{t_1}^{t_2} dt \int_{r_1}^{r_2} dr \sqrt{-g} = c(t_2 - t_1) \cdot (r_2 - r_1)$.

Exercise 4.5: To derive (4.30) we have used eq. (3.2), which is only valid for observers at rest. Consider an observer moving along r in the Schwarzschild metric at co-ordinate speed $v = dr/dt$. Prove the following relation between the co-ordinate time and the proper time of the observer:

$$\frac{d\tau}{dt} = \left\{ \left(1 - \frac{r_s}{r}\right) - \left(1 - \frac{r_s}{r}\right)^{-1} \left(\frac{v}{c}\right)^2 \right\}^{1/2}. \tag{4.31}$$

Hint: $d\theta = d\varphi = 0$ in (4.27), divide by $c^2 dt^2$. This result is a generalisation (1.6) of SR and also of (3.2). It serves as a warning that different kinds of redshift (here gravitational and Doppler) do not simply add!

4.3 Geodesics of the Schwarzschild metric

We could set out from the geodesic equations (2.34), but things become a lot easier if we utilise the constants of motion. These can be found in various ways. Because $g_{\alpha\beta,0} = 0$ we infer from (2.40) that $u_0 = g_{0\alpha} u^\alpha = g_{00} u^0 = (1 - r_s/r) c\dot{t} = $ constant:

$$(1 - r_s/r)\, c\dot{t} = \text{constant} \equiv e \, . \tag{4.32}$$

Recall that $t(p)$, $r(p)$, $\varphi(p)$ is the parametric representation of a geodesic and that $\dot{} = d/dp$ for a null geodesic and $\dot{} = d/ds = c^{-1} d/d\tau$ for a timelike geodesic. For a massive particle, (3.54) says that $e = E/m_0 c^2 = $ the total energy in units of its rest mass energy. Relation (4.32) is important because it fixes the rate of proper time τ of an object in geodesic motion with respect to co-ordinate time t. We may write it as

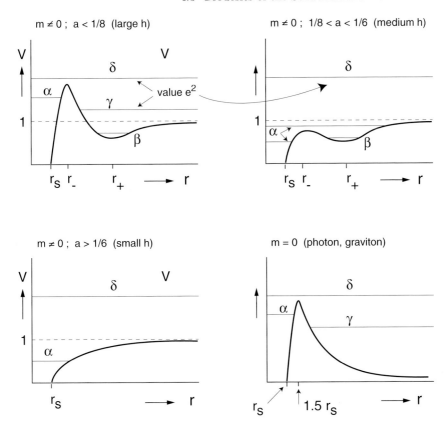

Fig. 4.2. Classification of the four different types of orbit in the Schwarzschild metric, here referred to as α, β, γ and δ-orbits (there is no generally accepted nomenclature). The figure shows $V(r)$ and e^2 as a function of r (not to scale). For massive particles with $\frac{1}{8} < a < \frac{1}{6}$ the potential has two extrema, but $V(r_-) < 1$. In this case there are two kinds of α-type orbits but no γ-type orbit. See also exercise 4.15.

$$\frac{d\tau}{dt} = \frac{1}{e}(1 - r_s/r) \, . \tag{4.33}$$

For a photon the meaning of e is not immediately clear, as $\dot{t} = dt/dp$ and p is an unspecified orbit parameter. This issue is elucidated in exercise 4.15. The metric does not depend on $x^3 = \varphi$ either $\to u_3 = g_{33}u^3 = g_{33}\dot\varphi = $ const. Due the spherical symmetry the geodesics must lie in a plane, and we may restrict ourselves to $\theta = \pi/2$, i.e. $g_{33} = -r^2$:

$$r^2\dot\varphi = \text{constant} \equiv h \, . \tag{4.34}$$

4 The Schwarzschild Metric

This is a generalisation of Kepler's second law – equal areas covered in equal times – which follows also directly from (4.23).

To find the equation for $\dot{r}$, it is actually easier to start from (4.27) than from (4.8). We 'divide' by dp^2:

$$(1 - r_s/r)\, c^2 \dot{t}^2 - \frac{\dot{r}^2}{1 - r_s/r} - r^2 \dot{\varphi}^2 = \left(\frac{ds}{dp}\right)^2 \equiv \kappa \,, \qquad (4.35)$$

with $\kappa = 0/1$ for a massless / massive particle, respectively. With the help of (4.32) and (4.34) we obtain

$$\frac{e^2}{1 - r_s/r} - \frac{\dot{r}^2}{1 - r_s/r} - \frac{h^2}{r^2} = \kappa \,, \qquad (4.36)$$

which we may write as

$$\dot{r}^2 = e^2 - V(r) \,; \quad \text{with} \quad V(r) = \left(1 - \frac{r_s}{r}\right)\left(\frac{h^2}{r^2} + \kappa\right). \qquad (4.37)$$

For $\kappa = 0$ (massless particles), V has a maximum at $1.5\, r_s$. For massive particles it is necessary to distinguish between a high and a low angular momentum h, measured by the parameter $a \equiv r_s^2/2h^2$. For $a > \frac{1}{6}$ (low angular momentum) $V(r)$ increases monotonously, and for $a < \frac{1}{6}$ (high angular momentum) $V(r)$ has two extrema at

$$\frac{r_\pm}{r_s} = \frac{1}{2a}\left(1 \pm \sqrt{1 - 6a}\right); \qquad a = \frac{r_s^2}{2h^2}. \qquad (4.38)$$

By using that $a < \frac{1}{6}$ it is easy to show that

$$1.5\, r_s \leq r_- \leq 3\, r_s \,; \qquad r_+ \geq 3\, r_s \,, \qquad (4.39)$$

while in the classical limit ($r_s \to 0$, i.e. $a \to 0$):

$$r_- \simeq 1.5\, r_s \,; \qquad r_+ \simeq r_s/a \,. \qquad (4.40)$$

A classification of the orbits is given in Fig. 4.2. The particle moves on a horizontal line $e^2 = $ constant in a region where $e^2 \geq V(r)$ to ensure that $\dot{r}^2 > 0$ in (4.37). Now $\dot{r}$ may only change sign where $e^2 = V(r)$, and the particle must reverse its radial direction there, as circular orbits at $e^2 = V(r)$ are unstable (see exercise). There are four different types of orbit. Assume for the sake of argument that we are dealing with a black hole, so that the metric is valid everywhere. A particle in an α-type orbit will be swallowed by the hole. A *massive* particle may be in a β-type orbit, performing an ellipse-like motion, but the orbit need not be *closed* – we only know that its r-range is restricted. Type-γ orbits are hyperbola-like, while particles in a δ-type orbit either fall into the hole or escape ($r \to \infty$). A radially moving photon has $V = 0$ (h and κ being zero), and is therefore always in an δ-type orbit. Massive particles can be in a stable circular orbit at $r = r_+$, the smallest of which is at $r = 3r_s$. Massless particles do not have stable circular orbits, see exercise.

The next step would be to determine $r(t)$ and $\varphi(t)$ by solving eqs. (4.32), (4.34) and (4.37). A simpler task (and sufficient for our purposes) is to derive the shape $r(\varphi)$ of the orbit. This may be done with *Binet's method* known from classical mechanics. We have $\dot r = dr/dp = (dr/d\varphi)(d\varphi/dp) = r_{,\varphi}\dot\varphi = hr_{,\varphi}/r^2$. Next, we introduce the variable $u = r_s/r$, and $u_{,\varphi} = -r_s r_{,\varphi}/r^2 = -u^2 r_{,\varphi}/r_s$. Result:

$$\left. \begin{array}{l} \dot r = hu^2 r_{,\varphi}/r_s^2 \\ r_{,\varphi} = -r_s u_{,\varphi}/u^2 \end{array} \right\} \quad \to \quad \dot r = -h u_{,\varphi}/r_s \ . \tag{4.41}$$

Substitute this in (4.37) and use $h^2/r_s^2 = 1/2a$:

$$(u_{,\varphi})^2 = 2ae^2 - (1-u)(u^2 + 2\kappa a) \ . \tag{4.42}$$

Take $d/d\varphi$ and rearrange terms:

$$u_{,\varphi}\left(2u_{,\varphi\varphi} + 2u - 2\kappa a - 3u^2\right) = 0 \ . \tag{4.43}$$

We discard the solution $u = $ constant. The other solution is

$$u_{,\varphi\varphi} + u = \kappa a + \tfrac{3}{2}u^2 \ ; \qquad u = r_s/r \ . \tag{4.44}$$

$\kappa = 0/1$ for photon / massive particle. All relativistic effects are hidden in the nonlinear term $\tfrac{3}{2}u^2$. It may be ignored in the classical limit (since for $r \gg r_s \to u \ll 1 \to u^2 \ll u$), and then (4.44) has the solution $u = \kappa a + C\cos(\varphi - \varphi_0)$ or $r(\varphi) \propto [\kappa a + C\cos(\varphi - \varphi_0)]^{-1}$, which is an ellipse or hyperbola for $\kappa = 1$, and a straight line for $\kappa = 0$. In the next section we shall use eq. (4.44) to derive the perihelium precession and the gravitational deflection of light.

Exercise 4.6: Prove (4.38) to (4.40).

Exercise 4.7: Show that $r_+ = r_s/a$ is equivalent to Kepler's third law.

Exercise 4.8: In principle, circular orbits are possible at all locations where $e^2 = V(r)$. Investigate the stability of these orbits.

Hint: In (4.44) the root $u_{,\varphi} = 0$ has been divided out, so it is safer to use

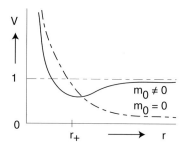

Fig. 4.3. The classical potential $V(r)$ may be obtained by taking the appropriate limit, see exercise. A massive particle can now only be in a β- or γ-type orbit, and a photon only in a γ-type orbit (a straight line); δ-orbits are only possible if $h = 0$ (head-on collision).

(4.37). Take $r = r_0 + \delta r$ and $e^2 - V(r_0) = 0$; result: $\delta \dot{r}^2 = -V_0' \delta r - \frac{1}{2} V_0'' \delta r^2$, where $\delta \dot{r}^2 \equiv (\delta \dot{r})^2$; $V_0' \equiv dV(r_0)/dr_0$, etc. Two possibilities:
1. $V_0' \neq 0 \;\to\; \delta \dot{r}^2 = -V_0' \delta r \;\to\; \delta \ddot{r} = -\frac{1}{2} V_0' \;\to\; \delta r$ always moves to a region where V is smaller, hence always unstable.
2. $V_0' = 0 \;\to\; \delta \dot{r}^2 = -\frac{1}{2} V_0'' \delta r^2 \;\to\; \delta \ddot{r} = -\frac{1}{2} V_0'' \delta r \;\to\;$ stable if $V_0'' > 0$ (in a minimum), otherwise unstable.

Conclusion: for $m_0 \neq 0$ only stable if $r = r_+$, i.e. at the bottom of the potential well; for $m_0 = 0$ *always* unstable.

Exercise. 4.9: We fire a bullet from $r = \infty$ towards a black hole with impact parameter d (= shortest distance between particle and hole if the orbit were a straight line). The bullet is in a γ-orbit and will miss the hole. Next we put the bullet in a δ-orbit by firing it at a higher velocity ($e \uparrow$), see Fig. 4.2. The bullet will now fall into the hole. Explain this paradox.

Hint: Fig. 4.2 is deceptive in that $V(r_-)$ changes as well. Show that $V(r_-) \simeq 2/(27a)$ for $a \ll 1$. Hence $V(r_-)/e^2 \propto (h/e)^2$. Calculate this constant with (4.32) and (4.34) by analysing the orbit at large r, where $\varphi \simeq d/r$. Show that $h/e \simeq v_\infty d/c$, so that $V(r_-)/e^2$ *increases* as we fire faster, and the bullet remains in a γ-orbit.

Exercise 4.10: Show that Fig. 4.2 reduces to Fig. 4.3 in the classical limit.

Hint: The transition to classical mechanics is obtained by letting $c \to \infty$. Show that $a, r_s, r_- \to 0$, $V(r_-) \to \infty$, while r_+ remains finite.

Exercise 4.11: Show that for a massive particle in a circular orbit with radius r

$$h = \left(\frac{rr_s/2}{1 - 3r_s/2r} \right)^{1/2} \quad ; \quad e = \frac{1 - r_s/r}{\sqrt{1 - 3r_s/2r}} . \qquad (4.45)$$

Hint: A circular orbit is only possible when $r = r_+$, hence $r/r_s = (1 + \sqrt{1 - 6a})/2a$. Solve for $a = r_s^2/2h^2$; e from $e^2 = V(r)$.

Exercise 4.12: Show that the period of a satellite in a circular orbit with radius r is given by (Δt = co-ordinate time, $\Delta \tau$ = proper time satellite):

$$\Delta t = \frac{2\pi r}{c} \left(\frac{2r}{r_s} \right)^{1/2} \quad ; \quad \Delta\tau = \frac{2\pi r}{c} \left\{ \frac{2r}{r_s} \left(1 - \frac{3r_s}{2r} \right) \right\}^{1/2}. \qquad (4.46)$$

Hint: (4.34) $\rightarrow 2\pi r^2/c\Delta\tau = h$; (4.33) $\rightarrow \Delta t/\Delta\tau = e/(1 - r_s/r)$; h and e from (4.45). Two points: (1). Apparently $d\tau/dt = (1 - 3r_s/2r)^{1/2} \neq \sqrt{g_{00}}$. Why is (3.2) not valid? (2). $(2\pi r/c)\sqrt{2r/r_s} = 2\pi r(GM/r)^{-1/2}$ is the classical expression. Does an observer at $r = \infty$ (who measures dt) notice any deviation from classical mechanics?

4.4 The classical tests of GR

The classical tests of GR, in order of their confirmation, are (1) the perihelium precession of Mercury, (2) the deflection of light in a gravitational field, (3) the redshift of light escaping from a gravitational well. Much later came (4) the delay in radar signals reflected by planets. A fifth experiment to measure the geodesic and Lense-Thirring precession of a gyroscope is now operational (Gravity Probe B, Ch. 8). For detailed information on these matters we refer to Will (1993).

In the middle of the 19th century it became apparent that the perihelium precession of Mercury had an unexplained difference of $43'' \pm 0.5''$ per century. The total precession is $5600''$ per century, of which $5025''$ is caused by the precession of the equinox (due to the precession of the Earth's rotation axis), and $532''$ by other planets, mostly Jupiter and Venus. Around that time Adams (1845) and Leverrier (1846) were able to predict the location of a then unknown planet (Neptune) from perturbations in the orbit of Uranus. By analogy it was assumed that the perihelium precession was caused by zodiacal dust or by an unknown planet (Vulcan) located inside the orbit of

Fig. 4.4. The *Einsteinturm* near Potsdam, an elegant design of Erich Mendelsohn, became operational in 1924, five years after the discovery of the gravitational deflection of light. It was originally built to measure the gravitational redshift in the solar spectrum. Photo: R. Arlt, AIP.

Mercury. However, the mass of the zodiacal dust was far too small and Vulcan has never been found.

Another possibility for a classical explanation of the perihelium precession is the fact that the mass distribution of the Sun has a nonzero quadrupole moment due to rotation. The gravitational potential has a small extra term: $\Phi(r) = -(GM/r)[1 + \frac{1}{2}J_2(R_\odot/r)^2]$ in the equatorial plane, where J_2 is the dimensionless quadrupole moment. This extra term $\propto r^{-3}$ causes a precession. Modern measurements place the value of J_2 in the range $10^{-6} - 10^{-7}$, in which case the quadrupole moment will contribute at most $0.1''$ per century.[2] The dependence of the precession rate on the semi-major axis ℓ is $\ell^{-7/2}$ for a quadrupole moment and $\ell^{-5/2}$ for GR. The observed perihelium precession of Venus and the Earth suggests an $\ell^{-5/2}$ dependence.[3]

The fact that light rays are deflected in a gravitational field was demonstrated during the solar eclipse of 1919, and in those days after World War I that achievement generated a media-hype *avant la lettre*. The observations are very difficult and later experiments resulted in values ranging from $1.4''$ to $2.7''$. In 1952 a value of $1.7'' \pm 0.1''$ was obtained. Nowadays Very Long Baseline Interferometry (VLBI) is used to measure the change in the position of a number of bright quasars in the ecliptic plane as the Sun passes by. These

---

[2] see Godier, S. and Rozelot, J.-P., *A & A 350* (1999) 310; Will (1993) § 7.3.
[3] see Adler et al. (1965) p. 202; Foster and Nightingale (1989) p. xiii.

observations have the advantage of being very accurate, and it is no longer necessary to wait for an eclipse. The result is $1.760'' \pm 0.016''$.[4]

Attempts to measure the gravitational redshift initially employed the solar spectrum. The effect is small: the solar spectral lines have a width of ~ 10 km s^{-1}, while the redshift is only 600 m s^{-1}. Moreover, there is a convective blueshift (rising gas being hotter than sinking gas) of about the same magnitude. Calibration is possible with the help of the Doppler shift induced by the known motion of the Earth with respect to the Sun. But due to lack of stability and other systematic effects the redshift could not be measured. It was only in 1962 that the gravitational redshift was convincingly detected in the solar spectrum. The Pound-Rebka experiment (1960) was actually the first quantitive measurement of the effect. In 1971 portable caesium clocks have been flown around the world on commercial jet flights, eastbound and westbound, and their readings were compared with a reference clock on the ground.[5] Such an experiment measures a mix of gravitational and special-relativistic redshifts. Gravity Probe A, a rocket experiment using a hydrogen maser clock measured the gravitational redshift with a precision of 10^{-4} in 1979.[6]

Shapiro proposed a fourth test in 1964: the delay of radar signals, and in retrospect it is amazing that this test had not been thought of earlier. The idea is that in the Schwarzschild metric the distance between r_1 and r_2 is larger than $r_2 - r_1$ (exercise 4.4), so that light needs more time to travel the distance between the points. This has been verified in radar reflection experiments on Mercury and Venus and by observations of the binary pulsars PSR 1534+12 and PSR 1855+09. Radio echo observations of VIKING attained a precision of 10^{-3}. Data from the Cassini spacecraft as it passed behind the Sun on its way to Saturn have recently improved the precision to 2×10^{-5}.[7]

Gravitational deflection of light

We now calculate the deflection angle of starlight from eq. (4.44), see Fig. 4.5. The classical photon orbit follows by omitting the nonlinear term: $u_{,\varphi\varphi} + u = 0 \to u = \text{const} \cdot \cos\varphi$. We write this zero-order solution as $u_0(\varphi) = \zeta \cos\varphi$, and $\zeta = r_s/r_0$ is the small parameter in the problem. This solution is a straight line $r\cos\varphi = r_0$. Now substitute $u = u_0 + \delta u$ in (4.44) and linearise:

$$\delta u_{,\varphi\varphi} + (1 - 3u_0)\delta u = q(\varphi); \quad q(\varphi) = \tfrac{3}{2}u_0^2. \qquad (4.47)$$

We need the solution of this equation with initial conditions $\delta u(0) = \delta u_{,\varphi}(0) = 0$, which we may find with the help of the method of variation of

---
[4] see Misner et al. (1971) p. 1104 and Will (1993) Ch. 1 and 7.
[5] Hafele, J.C. and Keating, R.E., *Science* **177** (1972), 166 and 168.
[6] Vessot, R.F.C. and Levine M.W., *General Relat. & Gravit.* **10** (1979) 181.
[7] Bertotti, B. et al., *Nature* **425** (1993) 374.

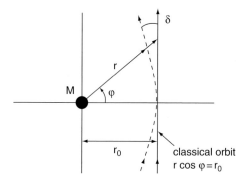

Fig. 4.5. Geometry of the deflection of light by the gravitational field of a central mass M. The deflection angle is 2δ.

constants (see exercise):

$$\delta u = s(\varphi)\int_0^\varphi c(\psi)q(\psi)\,d\psi - c(\varphi)\int_0^\varphi s(\psi)q(\psi)\,d\psi , \qquad (4.48)$$

where $s(\varphi)$ and $c(\varphi)$ are two independent solutions of the homogenous equation $f_{,\varphi\varphi}+(1-3\zeta\cos\varphi)f = 0$. These are the so-called *Mathieu functions*:

$$s(\varphi) = \sin\varphi + O(\zeta) ; \qquad c(\varphi) = \cos\varphi + O(\zeta) . \qquad (4.49)$$

The O-terms are actually series in $\sin n\varphi$ or $\cos n\varphi$ of order ζ or smaller, whose explicit expression we fortunately don't need. The whole solution is now

$$u = \zeta\cos\varphi + \frac{3\zeta^2}{2}\Big\{s(\varphi)\int_0^\varphi c(\psi)\cos^2\psi\,d\psi$$

$$-c(\varphi)\int_0^\varphi s(\psi)\cos^2\psi\,d\psi\Big\} . \qquad (4.50)$$

We demand that $u = 0$ ($r \to \infty$) for $\varphi = \pm(\pi/2+\delta)$. As we expect $\delta \sim \zeta \ll 1$, we expand up to first order in δ and ζ. Then $\cos(\pi/2+\delta) = -\delta$. Since there is already a factor ζ^2 in front of $\{\cdots\}$ we may take $\varphi = \pi/2$, $s = \sin$ and $c = \cos$ inside $\{\cdots\}$. This simplifies matters considerably:

$$0 \simeq -\zeta\delta + \frac{3\zeta^2}{2}\int_0^{\pi/2}\cos^3\psi\,d\psi = -\zeta\delta + \zeta^2 , \qquad (4.51)$$

or $\delta = \zeta = r_s/r_0$. The total deflection angle $\delta\psi$ is 2δ:

$$\delta\psi = 2r_s/r_0 \ . \tag{4.52}$$

This is twice as large as the result of a classical computation[8] that treats the photon as a massive particle with speed c. The deflection is therefore truly determined by the shape of null geodesics in a curved spacetime. For the eclipse geometry we have $\delta\psi = 2r_s/R_\odot = 2 \times 2.95 \text{ km}/6.96 \times 10^5 \text{ km} = 8.48 \times 10^{-6} \cong 1.75''$.

The HIPPARCOS satellite measured stellar positions with an accuray of $\sim 0.002''$. At this level of precision the deflection of light by the Sun can be detected over half of the sky! For let's suppose HIPPARCOS is looking perpendicularly to the Sun-Earth line. The deflection angle is then $\delta\psi = \delta = r_s/r_0$ where $r_0 = 1$ AU: $\delta\psi = 3/(1.5 \times 10^8) = 2 \times 10^{-8} \cong 0.004''$.

Binary pulsars

The derivation of the perihelium precession proceeds in a similar fashion as the deflection of light (see exercise). Much larger relativistic precessions have been measured in binary stellar systems consisting of a neutron star which is also a pulsar and another neutron star. These binary systems are laboratory test equipments on a cosmic scale that may be used to verify GR with greater accuracy and over a wider parameter range than is possible in solar system experiments. Six pulsars have now been found to be a member of a binary neutron star system. The most famous one is PSR 1913+16. The masses of the components are 1.441 $M_\odot$ (pulsar) and 1.387 $M_\odot$ (companion); $\epsilon = 0.617$; orbital period = 27907 s ($\simeq$ 7.8 hour); semi-major axis $\ell = 1.95 \times 10^6$ km (1.4 solar diameters); periastron precession = 4.22662° per year.[9] The recent discovery[10] of a new binary pulsar PSR J0737 3039 caused great excitement as the companion turned out to be a pulsar as well.[11] The system has an orbital period of only 2.45 hour and is much closer to us than PSR 1913+16, thus allowing even more precise tests of GR. The periastron precession is predicted to be 16.9° per year!

Exercise 4.13: Show that (4.48) is the required solution of (4.47).

Hint: After substitution in the equation it is found that the *Wronskian* $W \equiv c\,s_{,\varphi} - s\,c_{,\varphi}$ must be equal to 1. It follows from the homogenous equation that $W_{,\varphi} = 0$, hence $W = 1$ is only a normalisation.

---

[8] made by Von Soldner in 1801, see Will, C.M. *Am. J. Phys.* **56** (1988) 413.
[9] see Taylor, J.H. and Weisberg, J.M., *Ap. J.* *345* (1989) 434; Will (1993), Ch. 12 and p. 343.
[10] Burgay, M. et al., *Nature* **426** (2003) 531.
[11] Lyne, A.G. et al., *Science* *303* (2004) 1153.

Exercise 4.14: Show that the perihelium precession is given by

$$\omega_p = \frac{3(GM_\odot)^{3/2}}{c^2(1-\epsilon^2)\,\ell^{5/2}} \quad \text{rad s}^{-1}, \tag{4.53}$$

and that $\omega_p \simeq 43''$ per century for Mercury; ℓ = the semi-major axis and ϵ = the excentricity.

Hint: The classical orbit $u_0(\varphi)$ follows from (4.44): $u_{,\varphi\varphi} + u = a \to u = a(1 + \epsilon \cos\varphi) \equiv u_0(\varphi)$. The parameter $a \sim r_s/\ell \ll 1$ serves now as the small parameter. The excentricity ϵ need not be small. Insert $u = u_0 + \delta u$ in (4.44) and linearise $\to$ (4.47), except that $q(\varphi)$ and the O-terms in (4.49) are different functions of φ. The analysis proceeds quite analogously up to (4.50). We now need $du/d\varphi$:

$$\frac{du}{d\varphi} = -a\epsilon \sin\varphi + \frac{3a^2}{2}\left\{ \frac{ds}{d\varphi}\int_0^\varphi c(\psi)f(\psi)\,d\psi - \frac{dc}{d\varphi}\int_0^\varphi s(\psi)f(\psi)\,d\psi \right\},$$

with $f(\varphi) = (1 + \epsilon \cos\varphi)^2$; $du/d\varphi = 0$ for $\varphi = 0$, and we require it to be zero for $\varphi = 2\pi + \delta$ as well. Anticipate $\delta \sim a \ll 1$, and include terms up to first order in δ. Take $\varphi = 2\pi$, $c = \cos$, $s = \sin$ inside $\{\cdots\}$:

$$0 \simeq -a\epsilon\delta + \frac{3a^2}{2}\int_0^{2\pi}(1 + \epsilon\cos\psi)^2\cos\psi\,d\psi$$

$$= -a\epsilon\delta + 3a^2\epsilon\int_0^{2\pi}\cos^2\psi\,d\psi = -a\epsilon\delta + 3a^2\epsilon\pi\,.$$

Only the term $2\epsilon \cos^2\psi$ contributes to the integral. Now $\omega_p = \delta/P$ where P = period and $\delta = 3\pi a = 3\pi r_s^2/2h^2$. And $ch = r^2 d\varphi/d\tau \simeq r^2 d\varphi/dt = 2O/P$ with $O = \pi\ell^2\sqrt{1-\epsilon^2}$ = area of ellipse ($d\tau \to dt$ results only in higher order corrections). Kepler III: $\ell^3/P^2 = GM_\odot/4\pi^2 = c^2 r_s/8\pi^2 \to h^2 = \ell(1-\epsilon^2)\,r_s/2$. Mercury: $\ell = 0.387$ AU; $\epsilon = 0.206$; $P = 88$ days; $r_s = 2.95$ km; $\omega_p = 6.60 \times 10^{-14}$ rad s$^{-1} \cong 42.9''$ per century.

4.5 Gravitational lenses

The relativistic deflection of light causes a variety of wonderful effects. The gravitational field of a neutron star is so strong that it distorts and enlarges its own image to a considerable extent. Fig. 4.6 shows the image of a neutron star as it would look without relativistic effects, and the real image. Star

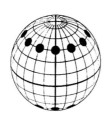
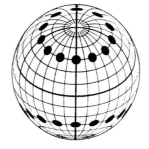

Fig. 4.6. The relativistic looks of a neutron star with radius $R = 2r_\mathrm{s}$. To the left, the image without relativistic effects. The real image with the light deflection by the strong gravity field included is shown on the right. The image magnification is computed in exercise 4.16. From Nollert, H.P., et al., *A. & A. 208* (1989) 153.

spots are much longer visible.

Einstein noted in 1936 that when two stars are positioned exactly behind each other, on the same line of sight, the light of the more distant star assumes the form of a ring around the nearby star, see Fig. 4.7. The chance of such a coincidence is very small. Chaffee (*Sci. Am.*, Nov. 1980) gives a fascinating account of the discovery of the first gravitational lens in 1979. It concerns two quasars Q0957+561 A and B with an angular separation of 6″ that have the same spectrum ($z = 1.41$). They turn out to be images of one and the same quasar whose light is deflected by an intervening galaxy at $z = 0.36$. It has been shown that gravitational lensing may be treated as a problem in geometrical optics in flat space with a refractive index $1 - 2\Phi(\mathbf{r})/c^2$, where Φ is to be gauged to zero at infinity. Since Φ is negative, the refractive index is larger than 1, which suggests that gravitational lenses may be modelled by a properly designed glass lens. According to the theory an odd number of images is formed, distorted and enlarged to different degree, but not all images may be visible. Later, *arcs* were discovered (images of a galaxy formed by a cluster located between the source and the Earth), and *radio rings* (Einstein ring image of compact radio source formed by intervening galaxy). At present of the order of 100 gravitational lenses are known.

Gravitational lenses are interesting for a number of reasons. In principle it is possible to determine the mass of the lens, including all dark matter. Another application is distance determination. The geometry of the object-lens-images system, see Fig. 4.7, can be derived from the angular distance between the images, the mass distribution of the lens, and the ratio of the distances of images and lens (= redshift ratio). The whole system may thus be drawn to scale. Since most quasars are variable, we may expect to observe

84   4 The Schwarzschild Metric

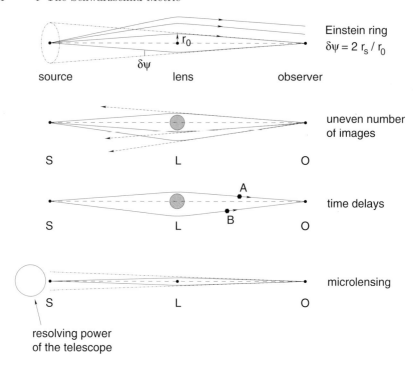

Fig. 4.7. Various gravitational lensing effects, see text.

the effect of path length differences, and this has now been seen in the double quasar Q0957+561 A and B. Intensity variations of Q0957+561 A lead those in Q0957+561 B by 417 ± 3 days.[12] This provides the missing absolute distance measure, so that now all distances are known. In principle it should be possible to measure the distance of the quasar in this way, independently of traditional astronomical methods. Such an independent measurement is of great importance for the determination of the Hubble constant H_0, which in turn sets the age and size of the universe.[13]

---

[12] Kundić, T. et al., *Ap. J. 482* (1997) 75.
[13] The literature on gravitational lenses is enormous. Some useful references are: Schneider, P., Ehlers, J. and Falco, E.E.: 1992, *Gravitational lenses*; Blandford, R.D. and Narayan, R.: 1992, *A.R.A.A. 30*, 311; Paczyński, B.: 1996, *A.R.A.A. 34*, 419; Wambsganss, J., *Living Rev. Relativity 1* (1998) 12 (http://www.livingreviews.org/lrr-1998-12); Mellier, Y.: 1999, *A.R.A.A. 37*, 127; Narayan, R. and Bartelmann, M.: 1999, in *Formation of Structure in the Universe*, A. Dekel and J.P. Ostriker (eds.), Cambridge U.P., p. 360.

Fig. 4.8. A gravitational lens of the Einstein cross type. The yellow-red galaxy in the centre of this Space Telescope image acts as a gravitational lens at $z = 0.81$, forming four visible images (blue dots) of a quasar at $z = 3.4$ that lies behind it and is invisible. The horizontal image size is $6.5''$ (see Ratnatunga, K.U. et al., *Ap. J.* *453* (1995) L5; Crampton, D. et al., *A & A 307* (1996) L53). Credit: NASA, HST, K. Ratnatunga and M. Im (JHU).

Microlensing

The lenses referred to above are *macrolenses*: the lensing is caused by the smooth mass distribution of the lens. *Microlensing* of compact (i.e. not extended) sources occurs when a point mass (a stellar-size object) crosses the light path of one of the images to the observer. For a brief time several subimages are formed, but their separation is so small that only a temporary change in brightness of the image can be observed. The duration of an event is hours to ~ 100 days and is, for a given lensing geometry, mainly determined by the lens mass. Microlensing was first discovered in Q2237+0305 (the 'Einstein cross'). The quasar images show uncorrelated brightness variations believed to be due to individual stars in the lensing galaxy crossing the line of sight. The most popular application is the search of *galactic* microlenses, which might reveal otherwise invisible dark objects. The idea is to use a CCD camera to monitor millions of stars in dense stellar fields in the Large Magellanic Cloud (LMC) or the galactic bulge, and to search for the characteristic brightness variations (symmetric time profile, independent of colour).

OGLE II saw about 100 events per year, in fields covering 11 square degrees on the galactic bulge ($\sim 2 \times 10^7$ stars). The event characteristics are consistent with the lenses being ordinary low mass stars, but the number of events is larger than expected. This has been interpreted to indicate that our

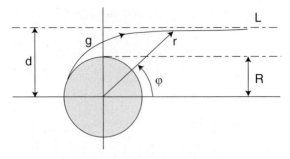

Fig. 4.9. Geometry of the magnification of a neutron star image in Fig. 4.6.

galaxy has a barred structure at its centre, protruding towards the Sun. The MACHO project has seen about 15 microlensing events in 5.7 year in fields covering the LMC containing 1.2×10^7 stars.[14] About 1/4 of these events can be explained as lensing by ordinary halo stars, and the remainder is reputed to lensing by MACHOs (= MAssive Compact Halo Object, such as brown dwarfs, neutron stars, old white dwarfs, black holes, etc.). About 20% of the expected galactic dark matter halo would be made of $0.3-0.7\,M_\odot$ MACHOs. But it cannot be excluded that the lenses are low mass stars in an LMC halo.

Exercise 4.15: Show that for photon orbits of the γ- and δ-type

$$e = h/d \ , \qquad (4.54)$$

d is the impact parameter, the shortest distance between photon and the origin if the orbit were a straight line. Both $e^2 = (h/d)^2$ and $V = (1 - r_{\rm s}/r)(h/r)^2$ are now $\propto h^2$, thus permitting a comparison of their relative values in Fig. 4.2.

Hint: Fig. 4.9 applies to ingoing as well as outgoing photons. A line through the origin parallel to the orbit at large r (where it is straight) determines d; d may have any value – at this point we are not interested in the orbit close to the central object. For $r \gg r_{\rm s}$ (small φ): $r \simeq d/\varphi$ or $u = r_{\rm s}/r \simeq r_{\rm s}\varphi/d$. Hence $u \to 0$ and $u_{,\varphi} \to r_{\rm s}/d$. Together with $\kappa = 0$ this fixes the value of the constant $2ae^2$ in (4.42): $2ae^2 = (r_{\rm s}/d)^2$; $2a = (r_{\rm s}/h)^2 \to e^2 = (h/d)^2$.

Exercise 4.16: With reference to Fig. 4.6 and 4.9, show that the gravitational field magnifies the image of a neutron star by a factor

$$d/R = (1 - r_{\rm s}/R)^{-1/2} \ , \qquad (4.55)$$

---

[14] Alcock, C. et al., *Ap.J. 542* (2000) 281.

4.5 Gravitational lenses

which amounts to 1.41 for $R = 2r_s$, the value corresponding to Fig. 4.6. Measure the diameters and verify if Nollert et al. did a proper job.

Hint: The image size is determined by the null geodesic g leaving the surface tangentially (why?) and approaching asymptote L for $r \to \infty$. Infer from the previous exercise that g obeys the equation

$$(u_{,\varphi})^2 + u^2 - u^3 = (r_s/d)^2 . \tag{4.56}$$

Evaluate (4.56) at the point where the ray leaves the surface. Tangential implies $u_{,\varphi} = 0$, while $u = r_s/R \to (r_s/R)^2 - (r_s/R)^3 = (r_s/d)^2$, etc.

5
Compact Stars

The Schwarzschild metric is only valid in vacuum, outside the star, but not in the stellar interior. Inside the star the metric is different, and in this chapter we shall investigate how relativistic effects influence the structure of a star. For main-sequence stars and even for *white dwarfs* the relativistic effects are small. In *neutron stars*, however, they play a dominant role. White dwarfs and neutron stars are two possible end products of stellar evolution. The third possibility is a *black hole*, an object that is smaller than its Schwarzschild radius.

5.1 End products of stellar evolution

The first equilibrium state in stellar evolution is the hydrogen burning phase. For this to happen, the star must have a minimum mass of $0.08\,M_\odot$.[1] The fusion of hydrogen produces helium and during this period the star is on the main sequence in the Hertzsprung-Russel diagram. When the stellar core runs out of hydrogen, it will contract and become hotter as it does so. This may be understood with the virial theorem:

$$E_{\rm t} = -\tfrac{1}{2} E_{\rm p}\,; \qquad E_{\rm tot} \equiv E_{\rm p} + E_{\rm t} = \tfrac{1}{2} E_{\rm p}\,, \qquad (5.1)$$

where $E_{\rm p}$, $E_{\rm t}$ and $E_{\rm tot}$ are the potential, the thermal and the total energy of the star, respectively ($E_{\rm p} < 0$; for definitions and proof see exercise). Now, $E_{\rm tot}$ will decrease, because the energy production by nuclear fusion diminishes while the radiative energy loss continues unabated. Hence $E_{\rm p} \downarrow$ and $E_{\rm t} \uparrow$. The stellar core contracts ($E_{\rm p} \downarrow$), and the density and temperature will rise (adiabatic compression; $E_{\rm t} \uparrow$). The star spends half of the liberated potential energy on radiative losses and the other half on compression (increase of $E_{\rm t}$).

---

[1] Stars lighter than $0.08\,M_\odot$ are called *brown dwarfs* – they undergo hardly any evolution.

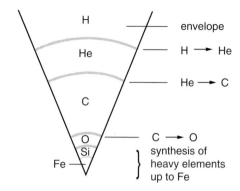

Fig. 5.1. A heavy evolved star consists of several shells, with fusion reactions in progress in the boundary layers (not to scale).

After the main sequence phase the core contracts and hydrogen proceeds in a shell. The outer layers expand, and the star becomes a red giant. The electron gas in the core becomes degenerate, and if $M \lesssim 0.5\,M_\odot$ the degeneracy pressure is able to withstand further contraction. In heavier stars the compression of the core continues until $T_c \sim 10^8$ K is reached, at which point fusion of helium sets in, through the triple-alpha reaction (3 ^{4}He $\to$ ^{12}C$+\gamma$; ^{16}O is also formed). Contraction follows once more when the helium in the core gets depleted. Stars heavier than about $6\,M_\odot$ attain a temperature $T_c \sim 8 \times 10^8$ K, which is sufficient to switch on carbon fusion. The next stage would be ^{16}O fusion, etc. Because all these reactions are strongly dependent on temperature, the star acquires a shell structure. The more massive the star, the more layers it will develop in due course of time, see Fig. 5.1.

Apart from rotation and magnetic fields, *mass loss* is a major complication when calculating stellar evolution. Mass loss occurs in the giant phase (by radiative pressure), but also due to instabilities during the shell burning stages. Stars with an initial mass below $\sim 6\,M_\odot$ are thought to lose enough mass to bring it below the Chandrasekhar limit of $1.4\,M_\odot$. The mass lost is often visible as a beautiful planetary nebula, see Fig. 5.3. These stars will end their life as a *white dwarf*, with a core of He, or C and O, depending on the initial mass. The remaining energy is radiated away, and the white dwarf just cools down progressively. The electron degeneracy pressure is almost independent of temperature and remains therefore in equilibrium with the gravitational force. Stars with $6 \lesssim M/M_\odot \lesssim 8$ have a degenerate core at the beginning of carbon fusion. The fusion switches on explosively because the pressure is independent of the temperature. This is known as the 'carbon flash'. These stars probably evolve into white dwarfs as well.

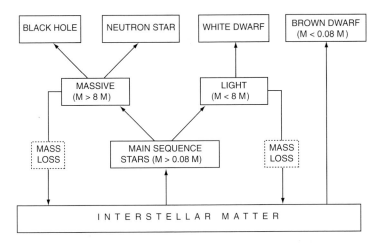

Fig. 5.2. Stars are born out of the interstellar medium and spend the largest fraction of their life on the main sequence with hydrogen fusion in the core. After the main sequence stage nuclear fusion proceeds in shells. The star is then a (super)giant, and loses mass through a strong stellar wind. Ultimately, light stars (progenitor mass $< 8\,M_\odot$) shed a planetary nebula and become a white dwarf. Heavy stars explode as a supernova leaving an expanding remnant and a neutron star or a black hole. Stars with progenitor mass $< 0.8\,M_\odot$ evolve so slowly that they are all still on the main sequence. Stellar evolution thus recycles and enriches the interstellar medium, and locks up matter in the four types of stellar remnant at the top of the diagram. After Bless (1995).

Heavy progenitors

The life of stars heavier than about $8 M_\odot$ is radically different and vaguely reminiscent of human tragedy – they carry their bulk with dignity until they can no longer cope and explode. The cores of these stars do not become degenerate and nuclear fusion continues until elements of the iron-group are formed. Further extraction of energy by fusion reactions is not possible, because nuclei of the iron-group have the largest binding energy per nucleon, Fig. 5.4. The star is now irrevocably on its way to total destruction. Due to the large mass of the star the compression of the core $(1 - 2\,M_\odot)$ continues until $T_c \sim 5 \times 10^9$ K. At that point much energy is lost through photo-desintegration of ^{56}Fe, an endothermic reaction ($^{56}\text{Fe} + \gamma \rightarrow 13\,^4\text{He} + 4n - 124$ MeV), and by the emission of neutrinos. The latter because the high Fermi energy of the electrons induces repeated inverse β-decay reactions of the type $e^- + (Z, A) \rightarrow (Z-1, A) + \nu_e \uparrow$. Nuclei of the type $(Z-k, A)$ are generally unstable and decay under emission of neutrons. The situation is now as follows. The electron density decreases and so does the associated electron pressure that sustained the core. At the same time free neutrons are formed in pro-

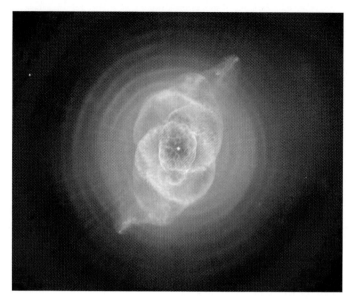

Fig. 5.3. End stages of stellar evolution (1). Hubble Space Telescope image of the planetary nebula NGC 6543, also known as the Cat's Eye nebula, at a distance of about 1 kpc. The nebula is ejected spasmodically from the central bright star as it develops into a white dwarf. Horizontal image size 1.2'. Credit: NASA, ESA, HEIC, and the Hubble Heritage Team.

gressively larger quantities. This runaway process seals the fate of the star. Eventually, the core contracts rapidly and collapses in about 0.1 s until nuclear densities are attained, $10^{14} - 10^{15}$ g cm^{-3}, and the *neutrons* become degenerate. The core now consists of a degenerate neutron gas, with a small amount of protons and electrons. The neutron degeneracy pressure is sufficient to halt further compression and a *neutron star* is formed. This is in fact a giant atomic nucleus held together by gravity rather than by the strong nuclear force. The gravitational energy ($\sim 10^{53}$ erg) is released in the form of neutrinos. These are exuded by the core in about 10 s, and escape almost all into space. The neutrino luminosity reaches therefore a brief but impressive peak of $\sim 10^{45}$ W. The collapsing outer layers bounce off the hard neutron core and a strong shock begins to propagate outwards. This shock prevents further collapse of the outer layers, aided by the capture of a small fraction of the escaping neutrinos. The collapse is reversed and a colossal explosion ensues, marking the birth of a supernova that radiates $\sim 10^{51}$ erg in the optical. The supernova remnant continues to expand, Fig. 5.5. In very heavy stars (for progenitor masses above $30\,M_\odot$) the outer layers are

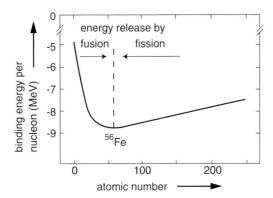

Fig. 5.4. Binding energy per nucleon as a function of atomic number. Quantum effects make that the curve is in reality not as smooth as indicated. Broadly speaking, energy is released in fusion reactions of nuclei lighter than iron. Beyond the iron-group the curve rises again due to the increasing Coulomb interaction of the protons. These nuclei release energy by fission.

only partially stopped. A *black hole* may form when the mass of the collapsed core exceeds the maximum mass of a neutron star (about $2\,M_\odot$). The progenitor mass required for black hole formation is not well known. Compact objects, finally, are quite numerous: about 5% of all objects of stellar-size mass in our galaxy is estimated to be a white dwarf, $\sim 0.5\%$ a neutron star and $(1-5) \times 10^{-4}$ are black holes.

Mass transfer in binaries

In a binary system the evolution of the components may be drastically altered by mass exchange. Matter accreting onto a white dwarf may cause various phenomena: *cataclysmic variables* (optical / UV emission of an accretion disc), thermonuclear fusion of the accreting matter, either steady (the *supersoft X-ray sources*), or in quasi-periodic explosions (a *nova*), or complete disruption of the star leaving nothing behind (*type Ia supernova*). Evolution of binary systems may produce quite exotic systems. Neutron star binaries, for example, such as the double pulsar PSR 1913+16, are believed to evolve from a binary system consisting of two ordinary massive stars. The more massive of the two evolves faster and explodes as a supernova, leaving behind a neutron star. After some time the lighter star becomes a red giant. The neutron star enters into the expanding envelope of its companion and begins to accrete matter. Tidal friction leads to the formation of a narrow binary system, and blows away the envelope. The system now consists of the red giant's helium core and the neutron star. If the mass of the helium star

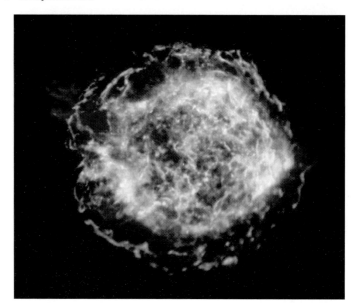

Fig. 5.5. End stages of stellar evolution (2). The supernova remnant Cassiopeia A, the relic of a massive star, as imaged by the Chandra X-ray Observatory. This supernova exploded around 1670, at a distance of about 3 kpc. It should have been almost as bright as Venus, but no 'new star' was reported. The green shell is synchrotron radiation from the outer shock wave. The ejected matter disperses relatively quickly into the interstellar medium. It is not clear whether the bright dot in the centre is the neutron star, because it does not show any periodicity. Red is Si emission (1.8 − 2.0 keV), green is 4.2 − 6.4 keV continuum, blue is Fe (6.5 − 7.0 keV). Exposure time 10^6 s. Horizontal image size 8′. See Hwang, U. et al., *Ap. J. 615* (2004) L117; Vink, J., *New Astr. Rev. 48* (2004) 61. Credit: U. Hwang, J.M. Lamming et al.

is above $2.5 M_\odot$ it eventually explodes as a supernova as well, and a narrow neutron star binary may result if the system is not disrupted. If the mass of the helium star is below $2.5 M_\odot$ it evolves into a white dwarf, and the end product may be a white dwarf-neutron star binary.

Neutron star observations

Our galaxy contains an estimated number of $\sim 10^8 - 10^9$ neutron stars, and most of these are invisible to us. There are basically three ways to observe neutron stars. *Radio pulsars* are rotating neutron stars equipped with a radio beacon that sweeps periodically over the Earth. There are about 1500 known pulsars, with periods ranging from 8.5 s down to 1.55 ms. Most are located in or close to the galactic disc. Precise timing of pulsars has yielded a wealth of information on neutron stars.[2] Next are the *X-ray binaries*, about 200 in our galaxy, in which a primary star loses mass that swirls around in an accretion disc, eventually falling onto the compact secondary. In doing so, the matter gets heated to X-ray temperatures. The secondary is usually a neutron star and in some cases a black hole.[3] *Low mass X-ray binaries* feature a low-mass solar-type primary with Roche lobe overflow. The X-ray emission is continuous, in most cases with occasional bursts. These so-called X-ray bursters are due to quasi-periodic runaway nuclear fusion of the accreting matter onto a neutron star companion. *High mass X-ray binaries* have a massive primary star with a strong wind, part of which accretes on a neutron star (these are all *X-ray pulsars*), in some cases on a black hole. Finally, X-ray emission of a few solitary nearby neutron stars has been detected. The nature of the X-ray emission is not understood – they could be young cooling neutron stars, or neutron stars accreting from the interstellar medium.[4]

5.2 The maximum mass M_c

The masses of white dwarfs and neutron stars are bounded by an upper limit which is a direct consequence of relativistic quantum statistics as shown by Landau in 1932. The star is a sphere of radius R containing A baryons that generate the mass and gravity, and $\sim A$ fermions providing the degeneracy pressure that balances gravity. White dwarfs: $\sim A/2$ protons, $\sim A/2$ neutrons and $\sim A/2$ electrons; neutron stars: A neutrons. The fermions are assumed to be free particles in a potential well with volume $V \sim R^3$. Every element $d^3r\, d^3p \sim \hbar^3$ of phase space may contain at most one fermion (we blissfully ignore spin). In a cold Fermi gas all states with $|\boldsymbol{p}| \leq p_f$ are occupied.[5] The total number of fermions is $A \sim (p_f R)^3/d^3r d^3p = (p_f R/\hbar)^3$, so that $p_f \sim \hbar A^{1/3}/R$. In case of relativistic degeneracy the Fermi energy is $E_f \sim p_f c \sim \hbar c A^{1/3}/R$. For non-relativistic degeneracy (that is, for $R > $ certain R_c at given A) we have $E_f \propto p_f^2 \propto R^{-2}$. The potential energy per baryon is $E_g \sim -GMm_b/R = -Gm_b^2 A/R$. The total energy per baryon is:[6]

---

[2] Lyne, A.G. and Graham-Smith, F.: 1998, *Pulsar Astronomy*, Cambridge U.P.
[3] See *Compact Stellar X-ray Sources*, W.H.G. Lewin, and M. van der Klis (eds.), Cambridge U.P., to appear.
[4] Treves, A. et al., *P.A.S.P.* **112** (2000) 297; Haberl, F., *Adv. Space Res.* **33** (2004) 638.
[5] The Fermi energy turns out to be so large that the energy distribution of the fermions is hardly affected by temperatures in the range $\lesssim 10^8$ K.
[6] In white dwarfs E_f refers to the electrons, that have negligible E_g, and E_g to the baryons, which have negligible E_f. The meaning of E is then the total energy of all particles divided by the number of baryons.

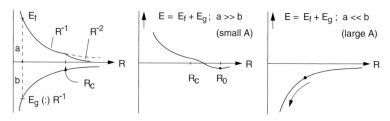

Fig. 5.6. Left: E_f and E_g as a function of R for given A. Center: the total energy E per baryon for small A, and for large A to the right.

$$E \simeq E_f + E_g \approx \begin{cases} a/R - b/R & R < R_c \; ; \\ aR_c/R^2 - b/R & R > R_c \; , \end{cases} \quad (5.2)$$

where $a = \hbar c A^{1/3}$ and $b = Gm_b^2 A$, see Fig. 5.6. It may be inferred from (5.2) that a relativistic star has a limiting mass of about $2\,M_\odot$, and that the radius of a white dwarf is about 5000 km and the radius of a neutron star about 3 km (see exercise). More detailed calculations obtain $M_c \simeq 1.4\,M_\odot$ for white dwarfs. This is referred to as the *Chandrasekhar limit*. The maximum mass of a neutron star depends on the equation of state $p(\rho)$, which is not well known for $\rho \sim 10^{14} - 10^{15}$ g cm^{-3}, see § 5.5.

Exercise 5.1: Prove the virial theorem (5.1) for non-degenerate stars.

Hint: If n and $\tfrac{3}{2}n\kappa T$ are the particle and energy density, then $E_t = \int_0^R (\tfrac{3}{2}n\kappa T) \cdot 4\pi r^2 dr$. Now $n\kappa T = p =$ pressure. Substitute and integrate partially: $E_t = -\tfrac{1}{2}\int_0^R (dp/dr)\,4\pi r^3\,dr$. Hydrostatic equilibrium: $dp/dr = -GM(r)\rho/r^2$ (for slow rotation!) $\to M(r) = \int_0^r 4\pi r^2 \rho\,dr$. Result: $E_t = \tfrac{1}{2}\int_0^R GM(r)\cdot dM(r)/r \equiv -\tfrac{1}{2}E_p$.

Exercise 5.2: Show that compact stars have a maximum mass M_c and a typical radius R_c which are entirely determined by fundamental constants:

$$M_c \sim \left(\frac{\hbar c}{Gm_b^2}\right)^{3/2} m_b \simeq 1.9 M_\odot \; , \quad (5.3)$$

$$R_c \sim \left(\frac{\hbar c}{Gm_b^2}\right)^{1/2} \frac{\hbar}{mc} \; . \quad (5.4)$$

$R_c \sim 3\,\mathrm{km}$ for $m = m_b$ (neutron star) and $5000\,\mathrm{km}$ for $m = m_e$ (white dwarf).

Hint: Take $R > R_c$ and $dE/dR = 0 \to R_0 = (2a/b)R_c \propto A^{-2/3}$; R_0 decreases as $A \uparrow$. But $R_0 \geq R_c \to a \geq b$ (ignore the factor of 2). For $a \leq b$ we get therefore Fig. 5.6, right; equilibrium is not possible when A is too large. Hence $A \leq A_c = (\hbar c/G)^{3/2} m_b^{-3}$ and $M_c = A_c m_b$. Take m_b = neutron mass. Since $R_0 \simeq R_c$, the value of R_c is roughly the one at which the degeneracy becomes relativistic: $E_f \sim \hbar c A_c^{1/3}/R_c \sim mc^2$. In retrospect we see that $M_c = (R_c/\lambda_c)^3 m_b$ with $\lambda_c = \hbar/mc$ = Compton wavelength. Interpretation?

5.3 The Tolman-Oppenheimer-Volkoff equation

We shall now derive the structure equations for spherically symmetric relativistic stars in hydrostatic equilibrium. We note that the interior metric of the star may still be written in the form (4.2) – (4.3), since these have been derived solely from symmetry arguments that apply here as well. Our task is therefore to find the new functions $\lambda(r)$ and $\nu(r)$ within the star with the help of the field equations. We begin by elaborating the stress-energy tensor $T_{\mu\nu} = \rho u_\mu u_\nu + (p/c^2)(u_\mu u_\nu - g_{\mu\nu})$ according to (3.57). Since the mass distribution is stationary we have $1 = u^\mu u_\mu = u^0 u_0 = g^{00}(u_0)^2 = (u_0)^2/g_{00} = \mathrm{e}^{-2\nu}(u_0)^2$ according to (4.3) ($g^{00} = 1/g_{00}$ because $g_{\alpha\beta}$ is diagonal). Consequently,

$$u_\mu = (\mathrm{e}^\nu, 0, 0, 0) . \tag{5.5}$$

With (4.3) we get

$$\left.\begin{array}{ll} T_{00} = \rho\,\mathrm{e}^{2\nu} ; & T_{11} = (p/c^2)\,\mathrm{e}^{2\lambda} ; \\ T_{22} = pr^2/c^2 ; & T_{33} = (pr^2/c^2)\sin^2\theta . \end{array}\right\} \tag{5.6}$$

It follows that

$$\left.\begin{array}{l} T^{00} = T_{00}/(g_{00})^2 = \rho\,\mathrm{e}^{-2\nu} ; \\ T^{11} = T_{11}/(g_{11})^2 = (p/c^2)\,\mathrm{e}^{-2\lambda} ; \\ T^{22} = p/(c^2 r^2) ; \\ T^{33} = p/(c^2 r^2 \sin^2\theta) . \end{array}\right\} \tag{5.7}$$

Both p and ρ are functions of r. Use has been made of $T^{00} = g^{0\mu}g^{0\nu}T_{\mu\nu} = (g^{00})^2 T_{00} = T_{00}/(g_{00})^2$, and likewise for T^{11}, etc. Because $G^{\mu\nu}{}_{;\nu} = 0$ we have

$T^{\mu\nu}{}_{:\nu} = 0$, and this equation determines the structure of the star (§ 3.6). An exercise invites the reader to show that this leads to

$$(\rho c^2 + p) \frac{\mathrm{d}\nu}{\mathrm{d}r} + \frac{\mathrm{d}p}{\mathrm{d}r} = 0 \,. \tag{5.8}$$

We use the covariant form of the field equations (3.42), $G_{\mu\nu} = -(8\pi G/c^2) T_{\mu\nu}$. Only $\mu = \nu = 0$ and $\mu = \nu = 1$ turn out to give an independent contribution, and we begin with $\mu = \nu = 0$. From (4.20) and (5.6):

$$\frac{\mathrm{d}}{\mathrm{d}r} r \left(1 - \mathrm{e}^{-2\lambda} \right) = \frac{8\pi G}{c^2} r^2 \rho \,. \tag{5.9}$$

Define the functions $m(r)$ and $M(r)$:

$$m(r) \equiv \frac{GM(r)}{c^2} \equiv \tfrac{1}{2} r \left(1 - \mathrm{e}^{-2\lambda} \right) , \tag{5.10}$$

and we may now write (5.9) as

$$\frac{\mathrm{d}M}{\mathrm{d}r} = 4\pi r^2 \rho \,; \qquad 0 \le r \le R \,. \tag{5.11}$$

R = stellar radius. Solve (5.10) for $\mathrm{e}^{-2\lambda}$ and use (4.3):

$$g_{11} = -\mathrm{e}^{2\lambda} = -\left\{ 1 - \frac{2m(r)}{r} \right\}^{-1} . \tag{5.12}$$

This amounts to a generalisation of g_{11} from (4.29). Continuity of the interior and exterior metric in $r = R$ requires $g_{11}^{\mathrm{int}} = g_{11}^{\mathrm{ext}}$, the latter given by (4.29), and leads to

$$\begin{aligned} M(R) &= M = \text{ mass of the star }; \\ 2m(R) &= r_{\mathrm{s}} = \text{ Schwarzschild radius }. \end{aligned} \tag{5.13}$$

Next comes $\mu = \nu = 1$. From (4.21) and (5.6):

$$\frac{1}{r^2} \left(\mathrm{e}^{2\lambda} - 1 \right) - \frac{2}{r} \frac{\mathrm{d}\nu}{\mathrm{d}r} = -\frac{8\pi G}{c^2} \frac{p}{c^2} \mathrm{e}^{2\lambda} \,. \tag{5.14}$$

Multiply (5.14) with $\mathrm{e}^{-2\lambda}$, then substitute $\mathrm{e}^{-2\lambda} = 1 - 2m(r)/r$, and solve for $\mathrm{d}\nu/\mathrm{d}r$:

$$\frac{\mathrm{d}\nu}{\mathrm{d}r} = \frac{m + (4\pi G/c^4) p r^3}{r(r - 2m)} \,. \tag{5.15}$$

Finally we eliminate $\mathrm{d}\nu/\mathrm{d}r$ with (5.8), and after some algebra we get:

5.3 The Tolman-Oppenheimer-Volkoff equation

$$\frac{dp}{dr} = -\frac{\{p + \rho c^2\}\{m + (4\pi G/c^4)pr^3\}}{r(r - 2m)}$$

$$= -\frac{G(\rho + p/c^2)(M + 4\pi r^3 p/c^2)}{r^2(1 - 2m/r)}. \quad (5.16)$$

This is the Tolman-Oppenheimer-Volkoff (TOV) equation. In the non-relativistic limit ($m \ll r$; $p \ll \rho c^2$) the classical equation $dp/dr = -GM(r)\rho/r^2$ for hydrostatic equilibrium re-emerges. Equations (5.11) and (5.16) determine the structure of a relativistic star in hydrostatic equilibrium. This is elaborated in the next sections. Outside the star the Schwarzschild metric applies. It turns out that GR-corrections are very small in white dwarfs, as may be anticipated from the value of Φ/c^2 (cf. Procyon B, Table 3.1). For that reason this chapter is mainly about neutron stars. However, GR-corrections are important for topics like the stability and the oscillation frequencies of white dwarfs.

An interesting point is the dual role of the pressure in (5.16). On the one hand the pressure gradient dp/dr delivers the force that prevents the star from collapsing. On the other hand p occurs in the stress-energy tensor $T_{\mu\nu}$ and acts therefore as a source of gravity, because pressure is a form of potential energy. These are the terms p/c^2 on the right side of (5.16). They increase $-dp/dr$ and therefore p. For a given density profile $\rho(r)$ the gradient $-dp/dr$ is always larger than in the case of classical gravity. The central pressure is therefore larger as well. The matter in a relativistic star has to withstand much larger internal forces to maintain hydrostatic equilibrium. The fact that neutron stars have a maximum mass is a direct consequence thereof.

Physical mass and bare mass

According to (5.11) and (5.13) the *total* or *gravitational mass* of the star is

$$M = \int_0^R 4\pi r^2 \rho \, dr. \quad (5.17)$$

This looks identical to the classical expression, but appearances are deceptive. The proper way to compute the mass of the star seems to be to multiply the density with the proper 3-volume element of space, $\sqrt{g}\, d^3 x = \{1 - 2m(r)/r\}^{-1/2} r^2 \sin\theta \, dr d\theta d\varphi$ (see exercise 4.4), and to sum up. This sum-of-all-mass-elements is called the *bare mass* M_b:

$$M_b = \int_V \rho \sqrt{g}\, d^3 x = \int_0^R \frac{4\pi r^2 \rho \, dr}{\sqrt{1 - 2m(r)/r}}$$

$$> \int_0^R 4\pi r^2 \rho \, dr = M \ . \tag{5.18}$$

$M_{\rm b}$ exceeds M because the 3-volume of a sphere with radius R is larger than $(4\pi/3)R^3$. But $M_{\rm b}$ cannot be measured. Whatever experiment we conduct *outside* the star to determine its mass (for example the period of an orbiting satellite), the result will always be M, because the metric there is the Schwarzschild metric of a central mass M. One may also say that $M_{\rm b}$ is larger than M due the *binding energy* of the star. A similar thing happens in the case of the mass defect of atomic nuclei. If the star is cut into small pieces, it takes an energy of $(M_{\rm b} - M)\,c^2$ to bring these to $r = \infty$, if their density is not altered. Almost the same amount of energy is released when a neutron star is formed.[7] We may estimate $M_{\rm b}$ from the factor $\{\cdots\}^{-1/2}$ in (5.18):

$$\left\{1 - \frac{2m(r)}{r}\right\}^{-1/2} \sim \left\{1 - \frac{2m(R)}{R}\right\}^{-1/2} = \left(1 - \frac{r_{\rm s}}{R}\right)^{-1/2}$$

$$\sim 1.2 \ , \tag{5.19}$$

for $r_{\rm s} = 3$ km, $R = 10$ km. It would follow that $(M_{\rm b} - M)\,c^2 \sim 0.2Mc^2 \sim 3 \times 10^{53}$ erg for a star of $M = 1\,M_\odot$. This colossal amount of energy is radiated in the form of neutrinos, and only a fraction $\sim 10^{-2} \cong 10^{51}$ erg in photons in the optical range of the spectrum (the visible supernova). We conclude that the birth of a neutron star in a gravitational collapse is accompanied by wholesale annihilation of mass.

Exercise 5.3: Prove (5.8) from $T^{\mu\nu}{}_{:\nu} = 0$.

Hint: Write out $T^{1\nu}{}_{:\nu} = 0$ using (2.51), (5.7), (4.10) and (4.14). $T^{i\nu}{}_{:\nu} = 0$ ($i = 0, 2, 3$) does not convey any extra information. Note that $T^{1\nu}{}_{:\nu}$ is *not* equal to $T^{11}{}_{:1}$, see exercise 2.12.

Exercise 5.4: Stellar evolution produces heavy elements up to iron, through fusion reactions. How are the elements heavier than iron formed?

---

[7] Actually it is less because the energy to compress the matter to $\rho \sim 10^{15}$ g cm^{-3} must still be subtracted, see Misner et al. (1971) p. 603 for details.

5.4 A simple neutron star model

The structure equations can be solved if we assume that ρ is constant. The fun of this simple and well known model is that it has some basic features in common with more realistic models. We put:

$$\rho = \begin{cases} \rho_0 & 0 \leq r \leq R \, ; \\ 0 & r > R \, . \end{cases} \quad (5.20)$$

From (5.11) we get immediately that:

$$M(r) = \frac{4\pi\rho_0}{3} r^3 \, , \quad (5.21)$$

while (5.10) tells us that

$$\frac{2m(r)}{r} = ar^2 \quad \text{with} \quad a = \frac{8\pi G \rho_0}{3c^2} \, . \quad (5.22)$$

Since $2m(R) = r_\text{s}$ we have

$$r_\text{s}/R = aR^2 \, . \quad (5.23)$$

Next we insert (5.22) into the TOV equation:

$$2\rho_0 c^2 \frac{dp}{dr} = -\frac{ar}{1 - ar^2} (p + \rho_0 c^2)(3p + \rho_0 c^2) \, , \quad (5.24)$$

from which $p(r)$ can be solved (see exercise):

$$p(r) = \rho_0 c^2 \frac{\sqrt{1 - ar^2} - \sqrt{1 - aR^2}}{3\sqrt{1 - aR^2} - \sqrt{1 - ar^2}} \, . \quad (5.25)$$

The central pressure in the star is

$$p(0) = \rho_0 c^2 \frac{1 - x}{3x - 1} \, ; \quad x = \sqrt{1 - aR^2} \, . \quad (5.26)$$

Apparently $p(0) \uparrow \infty$ for $x \downarrow \frac{1}{3}$, i.e. when $r_\text{s}/R = aR^2 \uparrow \frac{8}{9}$. In other words, when $R \downarrow \frac{9}{8} r_\text{s}$ a star with $\rho =$ constant collapses to become a black hole, because no material can support an infinite pressure. But constant-density stars, of course, do not exist. However, it has been proven that this result is generally valid, even when ρ is not constant: *A spherically symmetric star with radius $R < \frac{9}{8} r_\text{s}$ collapses to become a black hole.*[8]

We may reformulate this as follows. We have $M = (4\pi\rho_0/3)R^3 \geq (4\pi\rho_0/3) \cdot (9r_\text{s}/8)^3 = (4\pi\rho_0/3) \cdot (\frac{9}{8} \cdot 2GM/c^2)^3$. On solving for M we get:

---

[8] E.g. Wald (1984) p. 129; the only conditions are $\rho \geq 0$ and $d\rho/dr \leq 0$, but there is no requirement on the pressure p.

$$M \le M_c = \left(\frac{4c^2}{9G}\right)^{3/2}\left(\frac{3}{4\pi\rho_0}\right)^{1/2} = \frac{3.60\, M_\odot}{\sqrt{\rho_0/10^{15}}}\,. \tag{5.27}$$

Let us step back to discuss these results. Relation (5.27), on its own, says that if an object of density $\sim \rho_0$ is to have a radius larger than r_s ($\tfrac{9}{8}r_s$ to be precise), its mass cannot exceed M_c. But ρ_0 may have any value, and the results of this section cover normal stars and neutron stars. We have a look at normal stars first. In the classical limit, $aR^2 \ll 1$ and $x \simeq 1 - aR^2/2$, (5.26) says that $p(0) \simeq \rho_0 c^2 aR^2/4 \sim 2G\rho_0^2 R^2$. Equating that to the pressure $p = \rho_0\kappa T/m_b$ of a classical gas we find for the temperature $\kappa T \sim 2Gm_b\rho_0 R^2 \sim GMm_b/2R$. In other words, $\kappa T \sim$ potential energy per baryon at the surface. Inserting the solar mass M and radius R, the result is that $T \sim 10^7$ K – about the correct central temperature of the Sun.

Objects more compact than ordinary stars have a higher central pressure, and if the radius becomes of the order of r_s, (5.25) says that the pressure must be of order $\rho_0 c^2$, which for a classical gas implies that $\kappa T \gtrsim m_b c^2$ or $T \gtrsim 10^{13}$ K. But such high temperatures are unattainable due to very efficient cooling mechanisms (for example neutrino losses – a volume effect). Neutron stars 'solve' that by resorting to densities so high that degeneracy sets in, and the Pauli principle forces baryons to relativistic speeds regardless of temperature. The energy density and therefore the pressure may now attain values of $\rho_0 c^2$ and much more. Quantum statistics says that the pressure is $\sim \rho_0 c^2$ if there is one baryon per cubic Compton wavelength λ_c, i.e. $\rho_0 \sim m_b(\hbar/m_b c)^{-3} \sim 10^{17}$ g cm^{-3}. And this agrees in turn with (5.3) and (5.4) if we take $\rho_0 \sim M_c/R_c^3$.

Exercise 5.5: Prove (5.25).

Hint: Change to $y = p/\rho_0 c^2$ in (5.24) $\to \mathrm{d}y/\{(y+1)(3y+1)\} = -\tfrac{1}{2}ar\,\mathrm{d}r/(1-ar^2)$. Use $\{(y+1)(3y+1)\}^{-1} = \tfrac{1}{2}(y+\tfrac{1}{3})^{-1} - \tfrac{1}{2}(y+1)^{-1}$. Integrate from r to R.

Exercise 5.6: Prove for the $\rho = $ constant model that for $r \le R$:

$$g_{00} = \frac{1}{4}\left(3\sqrt{1-aR^2} - \sqrt{1-ar^2}\right)^2\,; \tag{5.28}$$

$$g_{11} = -(1-ar^2)^{-1}\,.$$

The metric of a constant-density star is now given by:

$$ds^2 = \frac{1}{4}\left(3\sqrt{1-aR^2} - \sqrt{1-ar^2}\right)^2 c^2 dt^2$$

$$-\frac{dr^2}{1-ar^2} - r^2(d\theta^2 + \sin^2\theta\, d\varphi^2), \qquad (5.29)$$

for $0 \le r \le R$. Verify that this metric has no singularities.

Hint: $g_{00} = e^{2\nu(r)}$; from (5.8): $d\nu = -(p+\rho_0 c^2)^{-1}dp$; integrate from r to R: $e^{\nu(r)} = e^{\nu(R)}/\{1 + p(r)/\rho_0 c^2\}$. Now use (5.25). Observe that $e^{\nu(R)}$ equals e^ν from (4.29) $\to e^{\nu(R)} = (1 - aR^2)^{1/2}$.

Exercise 5.7: Show that for a star with $\rho =$ constant

$$\frac{r_s}{R} = \frac{8}{9}\left(\frac{M}{M_c}\right)^{2/3}. \qquad (5.30)$$

Hint: r_s/R from (5.23), and R from (5.21); eliminate ρ_0 with (5.27).

Exercise 5.8: Consider a neutron star with constant $\rho = 10^{15}$ g cm^{-3} and $M = 1.8\,M_\odot$. Compute (a) the Schwarzschild radius; (b) the bare mass M_b, and (c) the rate of a clock at the centre of the star with respect to the clock rate at $r = \infty$.

Hint: (a): (5.27) $\to M = \frac{1}{2}M_c$; (5.30): $r_s/R = 0.560$ and $r_s = 1.8 \times 2.95$ km $= 5.3$ km, so that $R = 9.5$ km. Not bad for such a crude model.
(b): (5.18) $\to M_b = (4\pi\rho_0/a\sqrt{a})\int_0^{R\sqrt{a}} x^2(1-x^2)^{-1/2}dx$ (look up in a table). Result: $M_b = \frac{3}{2}M\{\arcsin y - y\sqrt{1-y^2}\}/y^3$ with $y = R\sqrt{a} = \sqrt{r_s/R} \to M_b = 1.25M$.
(c): From (3.2) and (5.28): $d\tau(0)/dt = \frac{1}{2}(3\sqrt{1-aR^2} - 1) = d\tau(0)/d\tau(\infty)$ because $d\tau(\infty)/dt = 1$ (why?) $\to d\tau(0)/d\tau(\infty) = 0.495$. A redshift while there is no gravity at the centre of the star - isn't that strange?

5.5 Realistic neutron star models

The structure equations may be integrated when the relativistic equation of state $p(\rho)$ is known. If the 'enclosed mass' M_i, the pressure p_i and the density ρ_i are known at radial position r_i, then we obtain M_{i+1} and p_{i+1} at the next level $r_{i+1} = r_i + \Delta r_i$ from eqs. (5.11) and (5.16). The equation of

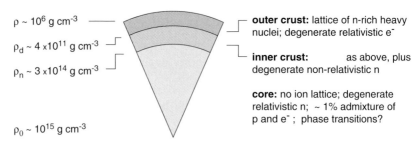

Fig. 5.7. Schematic structure of a neutron star (n = neutrons, p = protons, e^- = electrons). The figure is not to scale and the crust is in reality much thinner. The pressure in the crust is determined by the degenerate electrons, in the core by the degenerate neutrons.

state gives the corresponding density ρ_{i+1}. Starting values are enclosed mass $M_0 = 0$ and the central density ρ_0. During the integration $r > 2m(r)$ must hold everywhere. The stellar surface $r = R$ is defined by $p(R) = 0$. The g_{11} element of the metric tensor (4.3) and the function $\lambda(r)$ are fixed by relation (5.10). The element $g_{00} = e^{2\nu(r)}$ of the metric tensor (4.3) can be obtained by integrating (5.8) or (5.15). The value of ν is known at the outer boundary where the internal metric fits smoothly to the exterior Schwarzschild metric, see exercise 5.6. One may either integrate inwards, or start at $r = 0$ assuming an arbitrary value for $\nu(0)$ as the structure equations do not depend on ν. In the end a constant is added to all $\{\nu_i\}$ to reproduce the known value of $\nu(R)$.

The real problem is not the integration of the structure equations, but the equation of state (EOS). The properties of ultra-dense, cold matter in thermodynamic equilibrium may, broadly speaking, be divided in three regimes that also correspond to three different regions of the neutron star, see Fig. 5.7. As we already saw in § 5.1, the high Fermi energy of the electrons induces inverse β-decay ($e^- + p \rightarrow n + \nu_e$).[9] As a result the nuclei have more neutrons than normal. Above $\rho_d \simeq 4 \times 10^{11}$ g cm^{-3} the nuclei become oversaturated with neutrons, and free neutrons appear. This phenomenon is called *neutron drip*. For densities larger than ρ_n the ion lattice disintegrates, and, as a result, the structure of the core is deceptively simple: a degenerate relativistic neutron gas with a small admixture of protons and electrons. But there are major uncertainties as to the occurrence of phase transitions in the inner core. These include a hypothetical crystallisation of the neutrons, the formation of

---

[9] A neutron star is not a closed system to neutrinos, so complete equilibrium may not be achieved.

Table 5.1. Maximum mass M_c and other parameters for three EOS.[a]

EOS	M_c ($M_\odot$)	R_c (km)	ρ_0 [b] (10^{15} g cm^{-3})	surface redshift
soft (F) [c]	1.5 (1.7) [d]	7.9	5.1	0.49
medium (FPS)	1.8 (2.1)	9.3	3.4	0.53
stiff (L)	2.7 (3.3)	13.7	1.5	0.55

[a] Data from Cook, G.B., et al., *Ap. J.* **424** (1994) 823. All numbers are for nonrotating models.
[b] central density.
[c] The letter code refers to the EOS in Table 2 of Cook et al. (1994).
[d] In parenthesis the maximum mass for maximally rotating models.

a pion condensate or a transition to a quark-gluon plasma. The bottom line is that the EOS is reasonably well known in the crust ($\rho < \rho_n$), but not at all in the core ($\rho > \rho_n$), the region that largely fixes mass and radius of the star.

Rhoades and Ruffini[10] computed an upper limit to M_c by assuming that below a certain reference density ρ_ref of the order of ρ_n the so-called Harrison-Wheeler EOS applies, while for $\rho > \rho_\mathrm{ref}$ the EOS is required to obey causality ($p > 0$ and $dp/d\rho \leq c^2$)[11] but otherwise unspecified. They found that

$$M_c/M_\odot \leq 4.0 \cdot (\rho_n/\rho_\mathrm{ref})^{1/2} \ . \tag{5.31}$$

This upper limit is obtained for a maximally stiff EOS ($dp/d\rho = c^2$ for $\rho > \rho_\mathrm{ref}$). At the time, the importance of this result was the existence of a mass limit of about $4\,M_\odot$, independent of the details of the EOS. Since then, more realistic calculations have considerably brought down the value of M_c, see Table 5.1. An EOS is said to be *soft / hard* if it features a relatively low / high pressure at typical core densities. As the EOS shifts from soft to stiff, we see in Table 5.1 that M_c and the radius R_c both increase. Rotation adds 15 – 20% to M_c at most. Not in the table is the fact that the radius of the star increases as M decreases below M_c.

Constraining the EOS

Observations are on the verge of putting constraints on the EOS. Neutron star masses may be determined from orbital dynamics if the neutron star is a member of a binary system. In some X-ray binaries the so-called mass

---

[10] Rhoades, C.E. and Ruffini, R., *Phys. Rev. Lett.* **32** (1974) 324.
[11] $dp/d\rho$ equals the square of the signal speed of the medium.

function of the optical companion and of the X-ray emitting object could be determined. Six neutron star masses could thus be determined (Shapiro and Teukolski (1983), § 9.4). Precise timing of pulsars in binary systems has led to the determination of some 20 neutron star masses.[12] The masses range from 1 to 1.5 $M_\odot$, with a strong concentration near 1.35 $M_\odot$.

Neutron star radii may be estimated from the thermal emission of X-ray burst sources, for example. The idea is that the spectrum yields the temperature, and the radius follows from the observed flux density if we know or can somehow estimate the distance of the neutron star.[13] But due to many uncertainties precise radius measurements do not yet exist. The distances to some nearby neutron stars are known from a measurement of their parallax. Good determination of the radii of these neutron stars will be possible once the interpretation of their spectra is unambiguous. While an independent measurement of a neutron star's mass and radius seems asking for too much, a measurement of M/R is not. The XMM-Newton observatory has recently found indications for a redshift of $z = 0.35$ in the X-ray burst spectra of a neutron star.[14] Such a measurement would determine M/R, which in turn constrains the EOS, see exercise.

Lastly, we mention *quasi-periodic oscillations* as a possibility to constrain the EOS. Some low-mass X-ray binaries show quasi-periodic brightness oscillations (QPOs in the jargon of the X-ray community). There are often two well-defined frequencies in the range of 300 to 1200 Hz. These frequencies are so high that the oscillations are likely to be a byproduct of the accretion process very close to the neutron star. QPOs are therefore telling us something about the inner regions of the accretion disc, but there is no concensus about meaning of the message. One possibility is accretion onto a neutron star with a spin frequency of a few 100 Hz, and a moderately strong magnetic field ($10^7 - 10^{10}$ G), so that the accretion disc penetrates into the magnetosphere. The faster periodicity may result from clumpiness of matter due to instabilities near the so-called sonic point of the disc, where the radial accretion flow becomes supersonic. The QPO frequency would be the Keplerian frequency in the neighbourhood of the sonic point, and the lower frequency a beat phenomenon of the Keplerian frequency close to the sonic point and the spin frequency of the star. There many uncertainties and ramifications, but if these problems can be overcome QPOs may help to put contraints on the EOS.[15]

---

[12] Thorsett, S.E. et al., *Ap. J. 405* (1993) L29; Thorsett, S.E. and Chakrabarty, D., *Ap. J. 512* (1999) 288.
[13] Van Paradijs, J. and Lewin, W.H.G., *Class. Quantum Grav. 10* (1993) S117.
[14] Cottam, J. et al., *Nature 420* (2002) 51.
[15] See e.g. Miller, M.C. et al., *Ap. J. 508* (1998) 791.

Exercise 5.9: Show that the gravitational redshift z of a non-rotating spherical object is
$$1 + z = g_{00}(\text{emission site})^{-1/2} . \tag{5.32}$$

Hint: $1 + z = \lambda_0/\lambda = \nu/\nu_0$ (λ, λ_0 = emitted, observed wavelength); follow reasoning of (3.20) and $g_{00}(\infty) = 1$.

Exercise 5.10: Show that the gravitational redshift of a non-rotating object cannot exceed $z = 2$, and that the measurement of $z = 0.35$ implies that $M/R = 0.15\,M_\odot/\text{km}$. Suppose we know on astrophysical grounds that the star has a mass of $1.5\,M_\odot$. Which EOS in Table 5.1 are tenable?

Hint: g_{00}(emission site) has a minimum because $R > \frac{9}{8}r_s$ (§ 5.4); M must be below M_c (because if $M \downarrow$ then $R \uparrow$ and $z \downarrow$); $R = 10$ km $\rightarrow$ incompatible with EOS L and F, possibly compatible with FPS.

6
Black Holes

In the previous chapter we saw that a star may collapse completely. No force can prevent this, not even one we haven't discovered yet: any additional pressure would only generate more gravity than supporting force, and will accelerate the collapse. A complete collapse produces a *black hole*. Such a black hole would have a stellar-size mass, due to its formation history, but from a theoretical point of view there are no restrictions, and black holes of any mass may exist. Black holes were predicted by John Michell in 1784. He noticed that the escape velocity $(2GM/R)^{1/2}$ of a spherical mass may be greater than c if M/R is sufficiently large. He argued that such objects must be invisible because light cannot escape. In 1939 Oppenheimer and Snyder analysed the collapse and discovered that the collapsing matter cuts off all communication with the outside world. Gravitational collapse and black holes began to be seriously studied only after 1960. In 1963 Kerr found an axisymmetric solution of the vacuum equations, which was later realised to be the metric of a rotating black hole. In 1967 Wheeler coined the term 'black hole', and in 1975 Hawking discovered that a black hole emits black body radiation. Black holes may be thought of as lumps of pure gravity. They belong to the more advanced topics in GR, and we shall consider only a few elementary properties.

6.1 Introduction

Operationally, a black hole may be defined as an object that is smaller than its Schwarzschild radius. To an observer at $r = \infty$ a black hole appears as a hole in spacetime, that behaves like a black body with mass M and radius $r_s = 2GM/c^2$, and strong lensing effects near the edge. A black hole is entirely specified by 3 parameters: its mass M, angular momentum $\boldsymbol{L}$, and charge Q (theoretically, there is a fourth parameter: the magnetic monopole charge). All other information about the parent body is lost. A black hole may influence the outside world only through these parameters. This property led Wheeler to his famous aphorism 'black holes have no hair'. Magnetic field lines, incidentally,

Table 6.1. Massive dark objects in galactic nuclei [a]

system	type	Mass ($10^6 M_\odot$)	and radius (pc)	Reference
Milky way	S [b]	3.6 ± 0.3	0.005	Eisenhauer et al., *Ap. J.* **628** (2005) 246.
M31	S	170 ± 60	0.11	Bender et al., *Ap. J.* **631** (2005) 280.
M106	S	39 ± 3	0.13	Miyoshi et al., *Nature* **373** (1995) 127.
M32	E	3.4 ± 1.6	0.3	Van der Marel et al., *Nature* **385** (1997) 610.
M87	E	$(2.4 \pm 0.7) \cdot 10^3$	18	Ford et al., *Ap. J.* **435** (1994) L27.

[a] For reviews see Ferrarese, L. and Ford, H., *Space Sci. Rev.* **116** (2005) 523; Kormendy, J., in *Coevolution of Black Holes and Galaxies*, L.C. Ko (ed.), Cambridge U.P. (2004), p 1.
[b] S = spiral, E = elliptical galaxy.

do not count as hair: a charged rotating hole has a magnetic dipole moment $Q\boldsymbol{L}/Mc$ and an exterior magnetic field. The field is weak as the charge Q is expected to be very small. The Schwarzschild metric is the simplest black hole solution of the vacuum equations (mass M, nonrotating and uncharged). There are more general black hole solutions, for example the axisymmetric Kerr solution for a rotating uncharged hole with parameters M and L that we shall briefly consider in § 6.5. For the most general black hole characterised by M, L and Q see Wald (1984). Although GR allows black holes of any mass to exist, stellar and galactic evolutionary processes lead naturally to the formation of stellar-mass black holes and supermassive holes ($10^6 - 10^9 M_\odot$).

The mean density of a hole is $\bar\rho = M/(4\pi r_s^3/3) \sim 2 \times 10^{16}(M/M_\odot)^{-2}$ g cm^{-3}. For a supermassive hole of $\sim 10^8 M_\odot$ this is only ~ 1 (density of water). Black holes are therefore not necessarily associated with extremely dense matter. Having said that, the mean density of a hole of 10^{10} kg ($r_s \sim 10^{-15}$ cm!) is $\bar\rho \sim 7 \times 10^{56}$ g cm^{-3}. This is so high that these small holes are believed to form only during the Big Bang – if at all. But evidence for the existence of such primordial black holes is lacking.

6.2 Observations

Black holes can be observed only indirectly, when they interact with their environment. It is generally believed that the enormous luminosity $L \gtrsim 10^{47}$ erg s^{-1} of quasars and active galactic nuclei (AGNs) is caused by accretion of $1 - 100 M_\odot$ per year onto a massive black hole ($10^6 - 10^9 M_\odot$), for the following reasons. The emission is often variable on time scales t_v of days to hours, in some cases even 10^3s. Causality puts an upper limit to the source size of

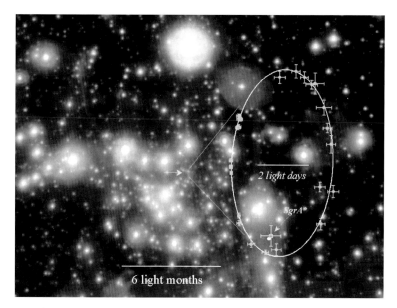

Fig. 6.1. Astrometric observations using adaptive optics techniques have shown that $S2$, a massive main-sequence star, is in a highly elliptic Keplerian orbit around the compact radio source Sgr A* at the galactic centre (large cross), with a period of about 15 yr. The combination of radial velocities (from the spectrum of S2) and proper motion data allows a precise determination of the orbital parameters *and* the distance of Sgr A*. The data strongly indicate that the gravitational potential is that of a point mass from 0.8 light days to 2 ly. The only compelling interpretation is a supermassive black hole. Horizontal image size: 15″. See Schödel, R. et al., *Nature* *419* (2002) 964; Eisenhauer, F. et al., *Ap. J.* *597* (2003) L121. The most recent mass determination of the hole is $3.6 \pm 0.3\,M_\odot$, and the distance to the galactic centre is 7.6 ± 0.3 kpc (Eisenhauer, F. et al., *Ap. J.* *628* (2005) 246). Credit: R. Genzel (private communication).

ct_v, which must therefore be small. An object radiating $L = 10^{47}$ erg s^{-1} will, by the Eddington limit argument,[1] have a mass of $\gtrsim 10^9 M_\odot$. Another way to estimate the mass of the central object is to say that if it converts mass into energy with an efficiency η to sustain its luminosity L for a time Δt, then it must have acquired a mass $M = L\Delta t/\eta c^2$. Accretion onto a black hole is the most efficient mechanism known for releasing gravitational energy. For a

---

[1] The radiation of the source exerts a radiation pressure on the infalling electrons through Thomson scattering which by charge neutrality is mediated to all accreting matter. The limiting (Eddington) luminosity L_Edd is attained when the radiation pressure equals the acceleration of gravity. For steady spherical accretion $L < L_\mathrm{Edd} \simeq 1.3 \times 10^{38} (M/M_\odot)$ erg s^{-1}, see Frank et al. (1992).

non-rotating hole $\eta = 0.057$ (see exercise) and for a maximally rotating hole η may be as large as 0.42 (while fusion of hydrogen to helium liberates only 0.007 of the rest mass energy). A reasonable estimate for accretion-powered sources is $\eta \sim 0.1$. Many AGNs (such as double radio lobe sources) must have existed at least 10^7 yr. Hence, $M \sim 10^8 M_\odot$. One way or another, we have a very large mass in a small volume, and such mass distributions, whatever their nature, are believed to develop quickly into a black hole.

Observations of stellar dynamics reveal that many nearby galaxies harbour 'heavy dark objects' within a very small radius around their centres (Table 6.1). Among these, the case for a massive black hole in the galactic centre at the location of the compact radio source Sgr A*, and in the nuclei of M31 and M106 (= NGC 4258) is very strong. Sgr A* is arguably the most convincing black hole candidate we have, see Fig. 6.1. Near-infrared and X-ray observations of Sgr A* reveal variability on a timescale of 10 minutes, indicating that the object cannot be larger than ~ 20 Schwarzschild radii (of a $3.6 \times 10^6 M_\odot$ hole). As VLBI techniques improve, they will eventually permit to resolve the Schwarzschild radius of the hole ($\sim 10 \mu$ arcsec). Perhaps we may one day observe the shadow cast by the event horizon of 'our' black hole in Sgr A*.[2]

The arguments for the existence of black holes are admittedly indirect, in the sense that they do not address the immediate vicinity of the hole. But that situation is beginning to change rapidly. For example, X-ray spectroscopy of the 6.4 keV iron line of MCG-6-30-15 (a Seyfert 1 galaxy) indicates that the emission comes from a hot disc around a spinning black hole. The inner radius of the emission appears to lie at about *one* Schwarzschild radius.[3]

Recently, a new class of X-ray sources has been discovered, the ultra-luminous X-ray sources, in star forming regions of nearby galaxies. They may point to the existence of intermediate-mass black holes of $10^2 - 10^4 M_\odot$.[4] Moving down the mass scale to stellar-mass black holes, evolution calculations indicate that they should be numerous: a fraction $\sim 10^{-4}$ of the stellar population. But only a few have been found, in bright X-ray binaries with $L_x \sim 10^{37} - 10^{38}$ erg s^{-1}. This points towards accretion onto a compact object of mass $M_x \sim 1 M_\odot$. A lower limit for the mass of the X-ray source can be inferred if the radial velocity profile of the companion star can be measured. If $M_x \gtrsim 3 M_\odot$ it has to be a black hole because the mass exceeds the maximum mass of a rotating white dwarf or neutron star ($\sim 3 M_\odot$). There

---

[2] On Sgr A* see Schödel, R. et al., *Nature* **419** (2002) 694; Genzel, R. et al., *Nature* **425** (2003) 934; Melia, F. and Falcke, H., *A.R.A.A.* **39** (2001) 309.
[3] Fabian, A.C., *Mon. Not. R. Astron. Soc.* **335** (2002) L1.
[4] E.g. Miller, M.C. and Colbert, E.J.M., *Intl. J. Mod. Phys. D* **13** (2004) 1.

are now 18 confirmed stellar mass black hole candidates.[5] Their companion is often a solar-type star, in three cases a massive O or B star.

6.3 Elementary properties

From the previous section it seems that the case for the existence of black holes, while formally still open, is tightening rapidly. We move on to review some of their properties. To this end we study the orbit of a test mass falling radially into the hole. The test mass moves along a geodesic in the Schwarzschild metric, with $h = 0$ because $\dot{\varphi} = 0$. From (4.37) we see that

$$\frac{1}{c^2}\left(\frac{\mathrm{d}r}{\mathrm{d}\tau}\right)^2 = e^2 - 1 + \frac{r_s}{r}, \qquad (6.1)$$

or, for $e = 1$:

$$\frac{\mathrm{d}r}{\mathrm{d}\tau} = -c\left(\frac{r_s}{r}\right)^{1/2}. \qquad (6.2)$$

Apparently, the choice $e = 1$ means that the mass has zero velocity at $r = \infty$. Equation (6.2) is of the type $\sqrt{r}\,\mathrm{d}r = \text{const} \cdot \mathrm{d}\tau$ and is readily integrated:

$$\tau = -\frac{2r_s}{3c}\left(\frac{r}{r_s}\right)^{3/2} + \text{const}. \qquad (6.3)$$

The singularity

It follows that the test mass traverses the distance between any finite value r_0 and $r = 0$ (where the collapsing matter has accumulated earlier) in a finite *proper time* $\Delta\tau$, Fig. 6.2 (left). This remains so if we do a more complete complete calculation with $e \neq 1$ to allow for nonzero velocity in $r = \infty$. We assumed that the vacuum metric is everywhere correct, but $r = 0$ is a singularity where the density becomes formally infinite. The test mass will be crushed by infinitely large forces as it arrives there. But before that happens quantummechanical effects take over, as classical GR loses its validity for length scales near the Planck length $L_p \simeq 1.6 \times 10^{-33}$ cm.[6] But nothing out of the ordinary happens when an observer crosses $r = r_s$. This is merely a co-ordinate singularity, a consequence of the way the co-ordinates are defined

---

[5] McClintock, J.E. and Remillard, R.A., in *Compact Stellar X-ray Sources*, W.H.G. Lewin and M. van der Klis (eds.), Cambridge U.P. (to appear), also astro-ph/0306213; see further Frank et al. (1992) § 6.7.

[6] The Planck mass and length are defined as the mass and Schwarzschild radius of a black hole whose Compton wavelength equals the Schwarzschild radius, see § 13.2.

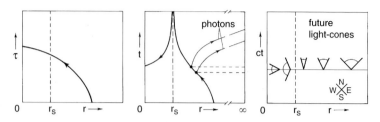

Fig. 6.2. The worldline of a test mass falling into a black hole. It reaches the origin $r = 0$ in a finite amount of proper time τ. To an observer at $r = \infty$ (whose clock runs synchonously with co-ordinate time t), the object turns dark and freezes to immobility just ouside the horizon $r = r_s$. To the right: the future light-cone as a function of r, see text.

in the Schwarzschild metric (4.27). By choosing different co-ordinates the singularity may be avoided (§ 6.4). Note that an extended observer will be torn to pieces by the tidal forces long before $r = r_s$ is reached (see exercises).

The event horizon

A completely different picture emerges if we analyse the situation as seen by an observer at $r = \infty$. He uses his own proper time to describe the fall, but that is identical to the co-ordinate time t.[7] With $e = 1$ and (4.32) we can transform proper time into co-ordinate time: $dr/d\tau = (dr/dt) \cdot (dt/d\tau) = (1 - r_s/r)^{-1} dr/dt$. Insert this in (6.2):

$$\frac{dr}{dt} = -c\left(1 - \frac{r_s}{r}\right)\left(\frac{r_s}{r}\right)^{1/2}, \qquad (6.4)$$

or, with $y \equiv r/r_s$:

$$\frac{c}{r_s} dt = -\frac{y\sqrt{y}}{y-1} dy, \qquad (6.5)$$

which can be integrated to

$$\frac{ct}{r_s} = -\left(\tfrac{2}{3} y\sqrt{y} + 2\sqrt{y} + \log \frac{\sqrt{y}-1}{\sqrt{y}+1}\right) + \text{const}. \qquad (6.6)$$

When $r \gg r_s$ only the first term contributes. For $r \simeq r_s$ we put $r = r_s + \delta$ or $y = 1 + \delta/r_s$. In this way we obtain the following approximation:

---

[7] This follows from (3.2): $d\tau(\infty)/dt = \sqrt{g_{00}(\infty)} = 1$.

$$\frac{ct}{r_{\rm s}} \simeq \begin{cases} -\dfrac{2}{3}\left(\dfrac{r}{r_{\rm s}}\right)^{3/2} + {\rm const.} & (r \gg r_{\rm s}) \ ; \\ -\log\delta + {\rm const.} & (r = r_{\rm s}+\delta) \ . \end{cases} \qquad (6.7)$$

For large r we thus recover (6.3), but not for $r \simeq r_{\rm s}$, as $t \uparrow \infty$ for $\delta \downarrow 0$. By inverting (6.7) we see that $\delta = {\rm const} \cdot \exp(-ct/r_{\rm s})$, or

$$\left. \begin{array}{l} r = r_{\rm s} + {\rm const} \cdot \exp(-t/t_{\rm c}) \ ; \\[2mm] t_{\rm c} = \dfrac{r_{\rm s}}{c} = \dfrac{2GM}{c^3} \simeq 10^{-5}\dfrac{M}{M_\odot} \ {\rm sec} \ . \end{array} \right\} \qquad (6.8)$$

According to an observer in $r = \infty$ the test mass slows down and hovers just outside $r = r_{\rm s}$, never actually reaching $r = r_{\rm s}$, Fig. 6.2 (middle panel). The time scale $t_{\rm c}$ for this to happen is very small, about $10\,\mu$s for $M = 1M_\odot$. Light emitted by the test mass will shift progressively to the red (see exercise). Measured in proper time, the number of photons emitted before crossing $r = r_{\rm s}$ is finite, and that number of photons also arrives at $r = \infty$, but spread out over time to $t = +\infty$. The object will therefore turn dark and vanish from sight. As we shall see, no signal from the interior region $r < r_{\rm s}$ will ever reach the exterior $r > r_{\rm s}$. For that reason $r = r_{\rm s}$ is called the *horizon*. There exist different kinds of horizon in GR, so we need to be more precise. The Schwarzschild metric has an *event horizon*: signals emitted by events inside the event horizon $r = r_{\rm s}$ will never be visible to external observers – however long they wait. The other main type of horizon is the *particle horizon* in cosmology. The particle horizon is the distance of particles beyond which an observer cannot see at this moment in time (but at a later moment he can), see § 11.2. In the professional jargon the term horizon is often employed without any type indication.

The future light-cone

Fig. 6.2 (middle panel) shows the worldline of the test mass in Schwarzschild co-ordinates r and t also inside the horizon. From (4.32): ${\rm d}t/{\rm d}\tau = e/(1 - r_{\rm s}/r)$ we see that ${\rm d}t < 0$ for ${\rm d}\tau > 0$ in $r < r_{\rm s}$: t appears to run backwards. No deep significance should be attached to this – it merely means that the Schwarzschild co-ordinates (event labels) t and r are awkward to use when $r \leq r_{\rm s}$. The right panel of Fig. 6.2 displays the future light-cone as a function of r. We may find that by putting ${\rm d}s^2 = {\rm d}\theta = {\rm d}\varphi = 0$ in (4.27):

$$\frac{c\,{\rm d}t}{{\rm d}r} = \pm \left| 1 - \frac{r_{\rm s}}{r} \right|^{-1} . \qquad (6.9)$$

Relation (6.9) divides the r,t plane in 4 sections N, S, E, W. It is intuitively clear that N is the future light-cone in the exterior region $r > r_{\rm s}$. The light-cone becomes progressively narrower close to $r = r_{\rm s}$. The *co-ordinate velocity*

116   6 Black Holes

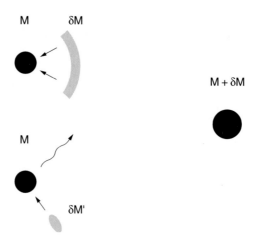

Fig. 6.3. Growth of a nonrotating black hole as seen by an external observer. Top left: a spherically symmetric shell collapses onto the hole. Bottom left: asymmetric collapse with emission of gravitational radiation. Right: the final product may be the same in both cases: a black hole of mass $M + \delta M$ ($\delta M' < \delta M$ on account of the energy lost by radiation).

dr/dt of a photon and of a particle becomes zero near the horizon, but the locally measured velocity does not, see exercise. But inside the hole the future light-cone is W (exercise). Even if a particle in $r < r_s$ emits a photon radially outwards (as seen in its own rest-frame), then $dr_{\text{photon}} < 0$, as the light-cone, by definition, contains all future worldlines. The exercise uses that $dr < 0$ somewhere, but we may take $dr > 0$ in $r < r_s$ as initial condition. Such particles/photons must come from $r = 0$ and they will fly through the horizon into the outside world (in their own perception – it takes until after $t = \infty$ before an observer in $r = \infty$ sees them). These are the time-reversed orbits, which do exist as a mathematical possibility, for example according to eq. (6.1). They are referred to as 'non-causal' because they depend on things happening in $r = 0$, about which we cannot say anything.

Growth of a black hole

How can a black hole ever grow if an observer at $r = \infty$ sees falling test masses 'freeze' on the horizon? The crux is that this is true for test masses, which do not affect the metric, but not for finite masses. Consider a spherically symmetric shell of matter falling into a black hole of mass M, Fig. 6.3. Because of the spherical symmetry an external observer finds himself in a Schwarzschild metric with mass $M + \delta M$ – provided the shell is sufficiently far from the hole so that the gravitational interaction between hole and shell is small. Birkhoff's theorem, § 4.2, says that the metric is stationary. This means that regardless of how the shell collapses, a black hole of mass $M + \delta M$ must form. The details of the collapse are quite complicated,[8] but Birkhoff's theorem allows us to infer the final result. It has further been proven that the surface $A = 4\pi r_s^2$ of the horizon cannot decrease, $dA \geq 0$, regardless of the (a)symmetry of the collapse, and that a near spherically symmetric collapse produces a black hole with the Kerr metric, see Schutz (1985) § 11.3 for more information.

Exercise 6.1: Consider the following derivation: $dr/d\tau = (dr/dt)(dt/d\tau) = (dr/dt) g_{00}^{-1/2}$, so that

$$\frac{dr}{dt} = \left(1 - \frac{r_s}{r}\right)^{1/2} \frac{dr}{d\tau} = -c\left(1 - \frac{r_s}{r}\right)^{1/2} \left(\frac{r_s}{r}\right)^{1/2},$$

which is different from (6.4). Which of the two is wrong and why?

Exercise 6.2: Throw a stone radially into a $1M_\odot$ black hole from $r = \infty$ at 30 km s^{-1}. How much proper time does the stone need to travel the interval $[10\, r_s, 0]$? Does the initial speed matter? How much proper time does it take a photon to traverse that distance?

Hint: From (6.1) and the initial condition: $e^2 - 1 = 10^{-8}$. Hence put $e = 1$! $\Delta\tau \simeq 210\,\mu$s. The photon is a catch.

Exercise 6.3: What speed does an observer at rest in $r = 1.1\, r_s$ measure for the stone and the photon as they rush by into the hole?

Hint: The locally measured speed $v \equiv $ d(locally measured distance)/d(locally measured time) $= \sqrt{-g_{rr}}\, dr/d(\tau\text{ observer}) = (\sqrt{-g_{rr}}\, dr)/(\sqrt{g_{00}}\, dt) = (1 - r_s/r)^{-1}(dr/dt)$, and dr/dt is known. Stone: (6.4) $\to v = -c\sqrt{r_s/r} \simeq -0.95\, c$. Photon: $ds^2 = 0$ in (4.27) $\to dr/dt = -c(1 - r_s/r) \to v = -c$.

Exercise 6.4: Consider stable circular orbits in the Schwarzschild metric. Prove that the difference in binding energy of an orbit at $r = \infty$ and smallest possible orbit $r = r_+ = 3r_s$ is given by:

$$\Delta E = \left(1 - \tfrac{2}{3}\sqrt{2}\right) m_0 c^2 \simeq 5.7 \times 10^{-2}\, m_0 c^2. \tag{6.10}$$

---

[8] See e.g. Shapiro and Teukolski (1983) § 17.5.

This implies that at most 5.7% of the rest mass energy is liberated as the mass is processed through an accretion disc. For a maximally rotating black hole this figure may rise to 42%.

Hint: $\Delta E = m_0 c^2 \Delta e$, see below (4.32); $e(\infty) - e(3r_s)$ from (4.45).

Exercise 6.5: Convince yourself about the future light-cones in Fig. 6.2.

Hint: Consider a timelike worldline (not necessarily a geodesic). From (4.27) and $ds^2 > 0 \;\to\; |cdt/dr| > |1 - r_s/r|^{-1}$ for $r > r_s$ (and $<$ for $r < r_s$). Together with $dt > 0$ (future light-cone) this leaves only N in the exterior region. Note that dr cannot be zero inside the hole; a particle *must* move there, whatever forces are applied $\to$ dr either positive or negative. For the subclass of geodesics this follows also from (6.1). On passing $r = r_s$ we know that $dr < 0 \;\to\; dr < 0$ everywhere $\to$ W remains (see text for $dr > 0$).

Exercise 6.6: The difference in gravitational acceleration over a length ℓ at a distance r from a black hole with mass M is

$$\delta g \sim \frac{M}{M_a} \frac{R_a^2 \ell}{r^3} g_a. \tag{6.11}$$

M_a, g_a = mass of Earth, acceleration at the Earth's surface. Use a classical estimate. How large is this tidal acceleration δg for $\ell = 1.8$ m, $M = 1\,M_\odot$, $r = 1000$ km?

Hint: $\delta g / g_a \simeq 24$.

Exercise 6.7: A laser with proper frequency ν_0 falls into a black hole on a radial geodesic $r_0(t_0)$ with $e = 1$, see Fig. 6.4, emitting photons to $r = \infty$, also radially. Prove that the frequency ν observed at $r = \infty$ equals

$$\nu/\nu_0 = 1 - \sqrt{r_s/r_0} \propto \exp(-t/2t_c). \tag{6.12}$$

Hint: Tricky, because we must allow for the extra redshift due to the photons escaping from an ever deepening gravitational well. If the laser would be at rest in r_0 we have $\nu_0 d\tau = \nu dt_0 \;\to\; \nu/\nu_0 = d\tau/dt_0 = (1 - r_s/r_0)^{1/2}$. But now $\nu_0 d\tau = \nu dt_1$:

$$\frac{\nu}{\nu_0} = \frac{d\tau}{dt_1} = \frac{d\tau}{dt_0}\frac{dt_0}{dt_1} = \left(1 - \frac{r_s}{r_0}\right)\frac{dt_0}{dt_1}, \tag{6.13}$$

see Fig. 6.4; $d\tau/dt_0$ from (4.32) with $e = 1$ and not from (3.2). The relation between t_0 and t_1 follows from the outgoing null geodesics: $dr/dt = c(1 - r_s/r)$

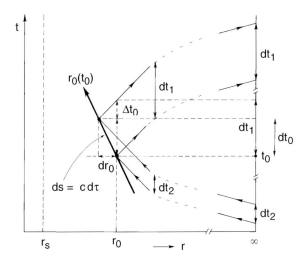

Fig. 6.4. This figure covers several situations, all involving masses falling into a black hole along a radial material geodesic $r_0(t_0)$: (1) a falling laser, emitting photons to $r = \infty$; (2) a falling detector observing photons arriving from $r = \infty$; (3) a falling mirror reflecting photons back to $r = \infty$.

((4.27) with $ds = d\theta = d\varphi = 0$) → all radial null geodesics are congruent and can be mapped onto each other by a vertical translation:

$$dt_1 = dt_0 + \Delta t_0 = dt_0 - dr_0 \cdot \left(\frac{dr}{dt}\right)^{-1}_{\text{null geodesic in } r_0, t_0}. \qquad (6.14)$$

dr_0 is negative and follows from (6.4). Result:

$$\frac{dt_0}{dt_1} = \left(1 + \sqrt{r_s/r_0}\right)^{-1}, \quad \text{and} \qquad (6.15)$$

$$\frac{dr_0}{dt_1} = \frac{dr_0}{dt_0}\frac{dt_0}{dt_1} = -c\left(1 - \sqrt{r_s/r_0}\right)\sqrt{r_s/r_0}. \qquad (6.16)$$

This proves the middle part of (6.12). Now let $r_0 = r_s + \delta$ with δ small, then $1 - \sqrt{r_s/r_0} \simeq \delta/2r_s$. From (6.16): $d\delta/dt_1 \simeq (-c/2r_s)\delta$ → $\delta \propto \exp(-t_1/2t_c)$.

Exercise 6.8: As you fall into a black hole along geodesic $r_0(t_0)$ with $e = 1$ you spend your last moments observing photons emitted by a laser at $r = \infty$ (proper frequency ν_0), see Fig. 6.4. Show that the observed frequency ν is:

6 Black Holes

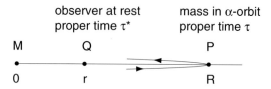

Fig. 6.5. Computing the gravitational acceleration in the Schwarzschild metric, see text and Fig. 4.2.

$$\nu/\nu_0 = \left(1 + \sqrt{r_s/r_0}\right)^{-1} \sim \tfrac{1}{2} \quad \text{as } r_0 \downarrow r_s. \tag{6.17}$$

Hint: Variant of the previous problem, using incoming instead of outgoing null geodesics, and now $\nu d\tau = \nu_0 dt_2$. The photons get blueshifted as they fall, but the redshift due to the observer's motion with respect to the laser is apparently stronger (for $e = 1$). You may also calculate the redshift $\nu/\nu_0 = dt_2/dt_1$ of light from $r = \infty$ reflected back to $r = \infty$ off a radially falling mirror. Observe that redshifts do not simply add!

Exercise 6.9: Prove that the gravitational acceleration g at the surface of a neutron star with mass M and radius R is equal to

$$g = \frac{-1}{\sqrt{1 - r_s/R}} \frac{GM}{R^2}. \tag{6.18}$$

Hint: More generally, the question is what acceleration a rocket must deliver to keep a mass in P at rest in the Schwarzschild metric, see Fig. 6.5. Strategy: put a test mass in a radial α-type orbit (§ 4.3). The test mass is dropped in P at zero velocity and an observer at rest in Q measures the acceleration of the mass as it flies by. There are three times in this problem: τ, τ^* and co-ordinate time t. The speed measured by an observer at rest in Q equals:

$$\frac{d\ell}{d\tau^*} = \frac{\sqrt{-g_{rr}}\,dr}{d\tau^*}\frac{dt}{d\tau} = \frac{1}{e}\frac{dr}{d\tau}, \tag{6.19}$$

since $d\tau^*/dt = \sqrt{g_{00}}$ and $dt/d\tau$ from (4.33). Differentiate once more:

$$\frac{d^2\ell}{d\tau^{*2}} = \frac{1}{e}\frac{d^2r}{d\tau^2}\frac{d\tau}{d\tau^*} = \frac{1}{e}\frac{d^2r}{d\tau^2}\frac{d\tau}{dt}\frac{dt}{d\tau^*}, \tag{6.20}$$

and $d\tau/dt$ and $dt/d\tau^*$ are known. Use (6.1) to show (1) that $e^2 = 1 - r_s/R$

(P is apex of the orbit $\to$ $dr = 0$ there), and (2) by differentiation: $(2/c^2)(dr/d\tau)(d^2r/d\tau^2) = -(r_s/r^2)(dr/d\tau)$ $\to$ $(d^2r/d\tau^2) = -(c^2 r_s/2r^2)$. Insert everything in (6.20) and let $r \to R$.

6.4 Kruskal-Szekeres co-ordinates

Schwarzschild co-ordinates are useful when $r \gg r_s$, but become inconvenient near r_s. The co-ordinate singularity at $r = r_s$ prevents one from stepping smoothly over the horizon. In 1960 Kruskal and Szekeres found a system of co-ordinates that does not suffer from this problem and is very expedient for use in the neighbourhood of $r = r_s$. The idea is to use a mesh of radial null geodesics as the co-ordinate lines of a new co-ordinate system. Since these are photon paths that actually cross the horizon we hope to eliminate in this way some of the odd behaviour of Schwarzschild co-ordinates. According to (4.27) radial null geodesics are given by $(1 - r_s/r)c^2 dt^2 - (1 - r_s/r)^{-1} dr^2 = 0$, which integrates to:

$$x_\pm \equiv ct \mp \{r + r_s \log(r/r_s - 1)\} = \text{constant}, \qquad (6.21)$$

and $x_+ = \text{constant}$ describes outgoing null geodesics, $x_- = \text{constant}$ the incoming null geodesics. For simplicity we restrict ourselves momentarily to $r > r_s$. The θ and ϕ co-ordinates remain unchanged and play no role. The next step is to introduce new co-ordinates u and v:

$$u + v = f(x_-); \qquad u - v = g(x_+), \qquad (6.22)$$

for arbitrary (well-behaved) functions f and g. We have now defined a class of co-ordinates in which $u \pm v = \text{constant}$ represents outgoing and incoming null geodesics. These null geodesics are therefore straight lines making an angle of $\pm 45°$ with the u and v co-ordinate axes – just like in the Minkowski space of SR. The final step is to choose appropriate functions f and g. Kruskal and Szekeres took $f(x) = 1/g(x) = \exp(x/2r_s)$, leading to

$$\left.\begin{aligned} u &= \left(\frac{r}{r_s} - 1\right)^{1/2} \exp(r/2r_s) \cosh \frac{ct}{2r_s} \\ v &= \left(\frac{r}{r_s} - 1\right)^{1/2} \exp(r/2r_s) \sinh \frac{ct}{2r_s} \end{aligned}\right\} \text{ for } r > r_s, \qquad (6.23)$$

$$\left.\begin{aligned} u &= \left(1 - \frac{r}{r_s}\right)^{1/2} \exp(r/2r_s) \sinh \frac{ct}{2r_s} \\ v &= \left(1 - \frac{r}{r_s}\right)^{1/2} \exp(r/2r_s) \cosh \frac{ct}{2r_s} \end{aligned}\right\} \text{ for } r < r_s. \qquad (6.24)$$

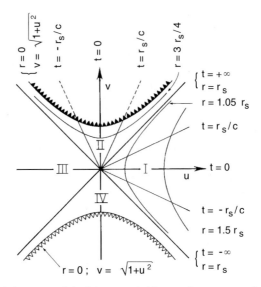

Fig. 6.6. Kruskal diagrams (1). Schwarzschild co-ordinates r and t as a function of the Kruskal-Szekeres co-ordinates u and v. The solid toothed line is the singularity $r = 0$.

The angles θ and φ remain unchanged. The inverse transformation is (exercise):

$$\left(\frac{r}{r_s} - 1\right) \exp(r/r_s) = u^2 - v^2 ; \tag{6.25}$$

$$\frac{ct}{r_s} = \log\left|\frac{u+v}{u-v}\right| . \tag{6.26}$$

The metric in these new co-ordinates is (see exercise):

$$\left.\begin{array}{l} ds^2 = \dfrac{4r_s^3}{r} \exp(-r/r_s)\,(dv^2 - du^2) - r^2\,d\Omega^2 ; \\[1em] d\Omega^2 = d\theta^2 + \sin^2\theta\,d\varphi^2 . \end{array}\right\} \tag{6.27}$$

From (6.25) we conclude that r/r_s is a function of $u^2 - v^2$.

Properties

Kruskal-Szekeres co-ordinates have a number of interesting properties. First of all, note that the u, v co-ordinates are a mix of the spatial co-ordinate r

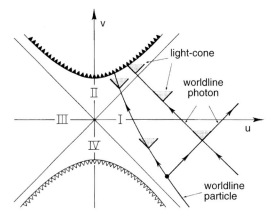

Fig. 6.7. Kruskal diagrams (2). Worldlines of photons and a massive particle. The light-cone has everywhere an opening angle of 45°.

and co-ordinate time t. The co-ordinates u and v have no obvious physical interpretation, and the reader is once more reminded of the fact that co-ordinates are merely event labels. Important is that the metric (6.27) is no longer singular at $r = r_\mathrm{s}$, but the singularity in $r = 0$ remains. It follows from (6.27) that radial null geodesics ($\mathrm{d}s = \mathrm{d}\theta = \mathrm{d}\varphi = 0$) are given by:

$$\mathrm{d}v = \pm \mathrm{d}u , \qquad (6.28)$$

so that they are indeed lines of $\pm 45°$ inclination in the u, v diagram ('Kruskal diagram'). From (6.25) we see that lines with $r = $ constant are hyperbolae, $u^2 - v^2 = $ constant. And (6.26) says that the lines $t = $ constant have $(u+v)/(u-v) = $ constant, i.e. $v = $ const $\cdot u$, that is, they are lines through the origin. The transformation is drawn in Fig. 6.6. Since $\cosh x + \sinh x = \mathrm{e}^x > 0$, we infer from (6.23) and (6.24) that $u + v > 0$: the Schwarzschild co-ordinates r, t are mapped onto region I ('our universe' $r > r_\mathrm{s}$) + region II (the black hole $r < r_\mathrm{s}$). According to (6.25) the singularity $r = 0$ is located at $v = +\sqrt{1+u^2}$. Fig. 6.7 shows the worldline of a particle falling into the hole (emitting a photon as an ultimate farewell message), and of an incoming photon. Clearly neither particles nor photons have the possibility to return to the exterior region I once they have entered II. All worldlines in II hit the singularity – there is no escape. The regions III and IV exist because we may also define (6.23) and (6.24) with an overall minus sign, the inverse transformation (6.25) and (6.26) being invariant for $(u, v) \to (-u, -v)$. Regions III + IV have no clear physical meaning – they contain the time-reversed orbits discussed below (6.9), and region IV is accordingly referred to as a *white hole* For more information see Misner et al. (1971) Ch. 31; Wald (1984) § 6.4.

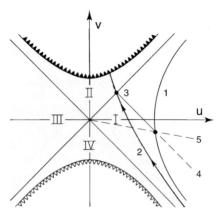

Fig. 6.8. Kruskal diagrams (3). Light reflected off a mirror on the surface of a collapsing star will, after a certain moment, no longer return to $r = \infty$. The grey part of the diagram has no physical meaning, see text.

Exercise 6.10: Prove the relations (6.25) – (6.27).

Hint: Use $\cosh^2 - \sinh^2 = 1$ and $\operatorname{arctgh} x = \frac{1}{2}\log\{(1+x)/(1-x)\}$ for $x^2 < 1$. For (6.27) differentiate (6.25): $(r/r_s^2)\exp(r/r_s)dr = 2(udu - vdv) \to dr = (2r_s^2/r)\exp(-r/r_s)(udu - vdv)$; (6.26): $cdt = 2r_s(u^2 - v^2)^{-1}(udv - vdu) = (2r_s^2/r)(1 - r_s/r)^{-1}\exp(-r/r_s)(udv - vdu)$. Substitute in (4.27).

Exercise 6.11: We send a light signal towards a mirror lying on the surface of a collapsing star, in an attempt to let a black hole reflect light. Show that the mirror will always see the beam and reflect the light, even inside the horizon. Nonetheless, light emitted after a certain time t_0 will never reach the outside world, whence the name 'a hole in spacetime'.

Hint: See Fig. 6.8: 1 = worldline external observer; 2 = worldline stellar surface; 3 = last possibility for reflected light to escape; 4 = corresponding null geodesic; 5 = definition of t_0.

Exercise 6.12: Let line 2 in Fig. 6.8 be the worldline of the collapsing stellar surface. Show that the grey part of the Kruskal diagram has no physical relevance.

Hint: Grey part of I and II contains worldlines of the stellar matter, but by shifting worldline 2 towards $t = -\infty$ regions I and II remain as a whole.

6.5 Rotating black holes: the Kerr metric

The Kerr metric is a stationary axisymmetric solution of the vacuum equation (3.39). We shall not actually solve the vacuum equations here, nor shall we engage in any detailed calculations. In the case of axial symmetry the metric tensor can no longer be made globally diagonal, and we intend to explain here one of the more spectacular consequences of the non-diagonality of the metric: the *frame-dragging* effect.

The metric tensor now depends on r and θ: $g_{\alpha\beta} = g_{\alpha\beta}(r,\theta)$. Furthermore $\mathrm{d}s^2$ should be invariant under the transformation $(\mathrm{d}t, \mathrm{d}\varphi) \to (-\mathrm{d}t, -\mathrm{d}\varphi)$ which implies $g_{t\theta} = g_{tr} = g_{\varphi\theta} = g_{\varphi r} = 0$. It would be incorrect to require $\mathrm{d}t \to -\mathrm{d}t$ and $\mathrm{d}\varphi \to -\mathrm{d}\varphi$ separately, because that does not correspond to the physical situation. Two cross terms remain: $g_{t\varphi}$ and $g_{r\theta}$. It turns out that $g_{r\theta}$ can also be made zero, but for $g_{t\varphi}$ this is not possible (without proof). The metric has the following form:[9]

$$\mathrm{d}s^2 = g_{tt}\mathrm{d}t^2 + 2g_{t\varphi}\mathrm{d}t\mathrm{d}\varphi + g_{\varphi\varphi}\mathrm{d}\varphi^2 + g_{rr}\mathrm{d}r^2 + g_{\theta\theta}\mathrm{d}\theta^2 \ . \tag{6.29}$$

The co-ordinates r and θ are no longer the same r and θ of the Schwarzschild metric; they coincide only in the limit $r \to \infty$. In that case (6.29) should be the Lorentz metric, that is: $g_{tt} \sim c^2$, $g_{rr} \sim -1$, $g_{\theta\theta} \sim -r^2$, $g_{\varphi\varphi} \sim -r^2 \sin^2\theta$, and furthermore $g_{t\varphi} \to 0$. Besides the mass M, the metric (6.29) contains a second parameter a, which occurs everywhere quadratically, except in $g_{t\varphi}$ which is linear in a. It turns out that $a = L/Mc$ where $L =$ total angular momentum (for the Sun $a = 0.092\,r_{\mathrm{s}} \simeq 0.28\,\mathrm{km}$). It follows that $\mathrm{d}s^2$ is invariant under $(a, \mathrm{d}t) \to (-a, -\mathrm{d}t)$ which is as it should be if (6.29) is the metric of a rotating black hole. Moreover (6.29) turns out to possess equatorial symmetry (invariance for $\theta \to \pi - \theta$).

Sofar we have always dealt with metric tensors that were diagonal, so that $g^{\alpha\alpha} = 1/g_{\alpha\alpha}$ (no summation), but here we encounter for the first time a nontrivial 2×2 submatrix:

$$\begin{pmatrix} g^{tt} & g^{t\varphi} \\ g^{t\varphi} & g^{\varphi\varphi} \end{pmatrix} = \begin{pmatrix} g_{tt} & g_{t\varphi} \\ g_{t\varphi} & g_{\varphi\varphi} \end{pmatrix}^{-1}$$

---

[9] See Shapiro and Teukolsky (1983) p. 357, Wald (1984) p. 312 ff and Schutz (1985) p. 297.

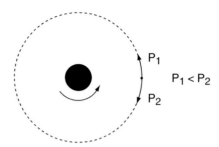

Fig. 6.9. A satellite in orbit around a rotating black hole. Due to frame-dragging the period P_1 of a prograde orbit is smaller than the period P_2 of a retrograde orbit. To first approximation $\Omega_1 - \Omega_2$ is given by (6.32). A *polar* orbit will show a precession of the orbital plane. In this way the LAGEOS geodetic satellites have measured the frame-dragging due the rotation of the Earth to a precision of 10% (Ciufolini, I. and Pavlis, E.C., *Nature 431* (2004) 958).

$$= \frac{1}{g_{tt}g_{\varphi\varphi} - g_{t\varphi}^2} \begin{pmatrix} g_{\varphi\varphi} & -g_{t\varphi} \\ -g_{t\varphi} & g_{tt} \end{pmatrix}, \qquad (6.30)$$

from which expressions for g^{tt}, $g^{\varphi\varphi}$ and $g^{t\varphi}$ follow. We now consider the constants of the motion. Because $g_{\alpha\beta,t} = g_{\alpha\beta,\varphi} = 0$ we know that u_t and u_φ are constant according to (2.40). Consider an ingoing particle with $u_\varphi = 0$. We have:

$$u^\varphi \equiv g^{\varphi\alpha}u_\alpha = g^{\varphi\varphi}u_\varphi + g^{\varphi t}u_t = g^{t\varphi}u_t ,$$
$$u^t \equiv g^{t\alpha}u_\alpha = g^{tt}u_t + g^{t\varphi}u_\varphi = g^{tt}u_t . \qquad (6.31)$$

With the help of these relations we calculate the rotation Ω of the particle at r, as measured by an observer at $r = \infty$:

$$\Omega \equiv \frac{d\varphi}{dt} = \frac{d\varphi/ds}{dt/ds} = \frac{u^\varphi}{u^t} = \frac{g^{t\varphi}}{g^{tt}} = -\frac{g_{t\varphi}}{g_{\varphi\varphi}} \simeq \frac{2GL}{c^2 r^3} , \qquad (6.32)$$

for $r \gg r_s$ and $\theta = \pi/2$. Use has been made of (6.31), and of (6.30) at the last = sign; the last expression in (6.32) is given without proof. At $r = \infty$ we have $\Omega = 0$ and $u^\varphi = 0$. The choice of $u_\varphi = 0$ implies that the particle begins its inward journey in the radial direction. But as it moves to finite r, the gravity of the rotating hole forces the particle to rotate with the hole. This is called *frame-dragging*. The hole 'drags space along' and this may be regarded as a manifestation of Mach's principle. One of the consequences of frame-dragging is that the period of a satellite moving in a prograde orbit is smaller than the period of a retrograde satellite, Fig. 6.9. In classical mechanics the gravitational field of a sphere is independent of its rotation. Not so in GR!

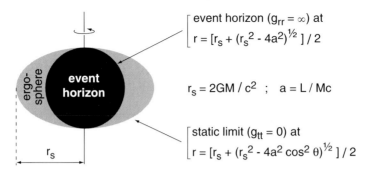

Fig. 6.10. A rotating black hole, with its event horizon and static limit. Rotation reduces the radius of the event horizon to a value between r_s and $r_s/2$. The space between the static limit and the horizon is called the ergosphere. Objects inside ergosphere may escape to $r = \infty$, but are forced to corotate with the hole. Black holes cannot spin arbitrarily fast: the angular momentum L is restricted by $a \leq r_s/2$. Holes with $a = r_s/2$ are said to be *maximally rotating*.

Next we have a look at circular photon orbits, $ds = dr = d\theta = 0$ in (6.29):

$$g_{tt}dt^2 + 2g_{t\varphi}dtd\varphi + g_{\varphi\varphi}d\varphi^2 = 0 \ . \tag{6.33}$$

These orbits are not null geodesics, so one would need some optical contraption like a set of mirrors to actually force the photon into a quasi-circular, polygon orbit. Dividing by dt^2 and solving for $d\varphi/dt$ produces

$$\frac{d\varphi}{dt} = \frac{1}{g_{\varphi\varphi}}\left(-g_{t\varphi} \pm \sqrt{g_{t\varphi}^2 - g_{tt}g_{\varphi\varphi}}\right) \ . \tag{6.34}$$

Suppose that $g_{tt} = 0$, then $d\varphi/dt = -2g_{t\varphi}/g_{\varphi\varphi} = 2\Omega$ or $d\varphi/dt = 0$. The former solution is a photon rotating with the hole, the latter is a retrograde photon. It is just able to beat the frame-dragging and is effectively at rest – for an observer at $r = \infty$, *not* for a local observer. And massive particles are forced to rotate when $g_{tt} = 0$. The surface $g_{tt} = 0$ is called the *static limit* and is located outside the *horizon*, which is defined by $g_{rr} = \infty$. The space between the horizon and the static limit is called the *ergosphere*. Rotation is compulsory in the ergosphere. Whatever force the rocket of a test particle in the ergosphere may exert, it cannot prevent the test particle from rotating with the hole (for an observer at $r = \infty$). But a particle may still escape from the ergosphere to $r = \infty$. In the Schwarzschild metric the surfaces $g_{tt} = 0$ and $g_{rr} = \infty$ coincide.

A discussion of the geodesics of the Kerr metric would be out of place here. We only mention that the smallest stable circular orbit of a test particle, at $3r_s$ in the Schwarzschild metric, moves inward (outward) for a prograde (retrograde) orbit in the equatorial plane. For example, for a hole rotating at 80% of its maximum rate ($a = 0.8 \cdot \frac{1}{2}r_s = 0.4r_s$) we have $r_{\text{progr}} = 1.45\,r_s$ and $r_{\text{retrogr}} = 4.21\,r_s$.

Exercise 6.13: We know that the exterior vacuum of a spherically symmetric neutron star has the Schwarzschild metric. Does the exterior vacuum of a rotating neutron star have the Kerr metric? Discuss the occurrence of frame-dragging in case your answer is negative.

6.6 Hawking radiation

In 1975 Hawking discovered that black holes should emit thermal radiation – in other words, that black holes are really black from a physical point of view. The effect is due to the fact that vacuum fluctuations, spontaneous creation and annihilation of particle anti-particle pairs occurring throughout space, develop an asymmetry near an event horizon.[10] The particle with negative energy may fall into the hole, and the other must then escape towards $r = \infty$ as a real particle. The reverse process is forbidden because a particle with negative energy cannot move as a real particle in the region outside the hole (but inside the horizon it can). A complete calculation requires quantum field theory near the horizon, see Wald (1984) p. 399 ff. Here we shall resort to an intuitive approach due to Schutz (1985).

Consider a photon pair created close to the horizon, at $r = r_s + \delta$. We analyse this process in the local rest-frame, i.e. a frame in radial free fall with zero velocity in $r = r_s + \delta$. Special relativity applies there, and the virtual particles have an energy $\pm\epsilon$. The observer is on a radial geodesic with $h = 0$ and from (4.37) we see that $0 = \dot{r}^2 = e^2 - \left(1 - r_s/(r_s + \delta)\right)$, or, using $\delta \ll r_s$: $e^2 \simeq \delta/r_s$. The observer reaches the horizon in a proper time interval $\Delta\tau$, which we may find from (4.37). We transform to the variable $x = r - r_s$:

---

[10] The effect occurs whenever there is an event horizon. For example an observer in Minkowski space subject to a constant acceleration a has an event horizon: signals from events in the region beyond the asymptote to his worldline will never be able to reach him. As a result he finds himself in a bath of thermal radiation with a temperature $T = \hbar a/2\pi\kappa c$. This is called the Unruh effect. The effect has been measured in electron storage rings, where it shows up in that it is impossible to achieve 100% polarisation. This is now understood to be the result of the thermal radiation that the accelerated electrons experience, see Bell, J.S. and Leinaas, J.M., *Nucl. Phys. B284* (1987) 488.

$$\left(\frac{\mathrm{d}x}{c\mathrm{d}\tau}\right)^2 = e^2 - \left(1 - \frac{r_\mathrm{s}}{r_\mathrm{s}+x}\right) \simeq \frac{\delta - x}{r_\mathrm{s}}, \qquad (6.35)$$

and $\Delta\tau$ follows by integration:

$$c\Delta\tau = -\sqrt{r_\mathrm{s}} \int_\delta^0 \frac{\mathrm{d}x}{\sqrt{\delta - x}} = 2\sqrt{r_\mathrm{s}\delta}. \qquad (6.36)$$

If $\Delta\tau \gg \hbar/\epsilon$ the pair will annihilate long before reaching the horizon, and when $\Delta\tau \ll \hbar/\epsilon$ they annihilate long after that (i.e. inside the hole). However, when $\Delta\tau \sim \hbar/\epsilon$ there is a chance that the photon with $\epsilon < 0$ stumbles into the hole while the other escapes towards $r = \infty$. This gives an approximate relation between ϵ and δ:

$$\epsilon \sim \frac{\hbar c}{2\sqrt{r_\mathrm{s}\delta}}. \qquad (6.37)$$

As the photon arrives in $r = \infty$, it is redshifted, and we compute its energy E there from (3.20) and (3.2): $E/\epsilon = \nu(\infty)/\nu(r_\mathrm{s}+\delta) = \mathrm{d}\tau(r_\mathrm{s}+\delta)/\mathrm{d}\tau(\infty) = \mathrm{d}\tau(r_\mathrm{s}+\delta)/\mathrm{d}t = \sqrt{g_{00}(r_\mathrm{s}+\delta)}$:

$$E \sim \frac{\hbar c}{2\sqrt{r_\mathrm{s}\delta}} \left(1 - \frac{r_\mathrm{s}}{r_\mathrm{s}+\delta}\right)^{1/2} \simeq \frac{\hbar c}{2r_\mathrm{s}}, \qquad (6.38)$$

which is independent of δ! Hawking's analysis showed that the photons have a Planck distribution corresponding to a temperature $\kappa T = \hbar c/4\pi r_\mathrm{s}$. A black hole emits thermal radiation with a temperature

$$T = \frac{\hbar c^3}{8\pi\kappa GM} \simeq 6.2 \times 10^{-8} \left(\frac{M}{M_\odot}\right)^{-1} \mathrm{K}. \qquad (6.39)$$

Physical consequences

Hawking radiation has two interesting consequences. The first is of a thermodynamic nature. Earlier on we said that the area of the horizon cannot decrease (this was without any consideration of quantum effects). We may cast this in a form reminiscent of a well-known thermodynamic relation. We have $\mathrm{d}A = 8\pi r_\mathrm{s}\mathrm{d}r_\mathrm{s} = 16\pi r_\mathrm{s}(G/c^2)\mathrm{d}M$, or:

$$\mathrm{d}Mc^2 = \frac{c^4}{16\pi G r_\mathrm{s}} \mathrm{d}A = T\mathrm{d}\left(\frac{\kappa c^3}{4\hbar G} A\right). \qquad (6.40)$$

This says that $\mathrm{d}E = T\mathrm{d}S$ with $E = Mc^2$ and the *entropy* S of the hole would then be equal to

$$S = \frac{\kappa c^3}{4\hbar G} A = \frac{\pi \kappa c^3}{\hbar G} r_\mathrm{s}^2, \qquad (6.41)$$

apart from a constant. We may now argue that the hole's entropy cannot decrease because A cannot. However, A may decrease when quantum effects are taken into account, but in that case it is no longer correct to regard the hole as an isolated system. The idea is that the total entropy of the hole and the emitted radiation cannot decrease.

A second consequence is that a black hole will evaporate, because it loses energy by emission of radiation. The mass of the hole must decrease according to $\mathrm{d}Mc^2/\mathrm{d}t = -4\pi r_s^2 \sigma T^4$, or with $\sigma = \pi^2 \kappa^4/60\hbar^3 c^2$:

$$\left.\begin{aligned} \frac{\mathrm{d}M}{\mathrm{d}t} &= -\frac{a}{M^2}, \quad \text{with} \\ a &= \frac{1}{2^{10} \cdot 15\pi} \cdot \frac{\hbar c^4}{G^2} = 4.0 \times 10^{24} \text{ g}^3 \text{ s}^{-1}. \end{aligned}\right\} \quad (6.42)$$

The evaporation rate is initially slow, but accelerates towards the end and the last stages proceed explosively. All kinds of particles are emitted, not only photons, but emission of particles with rest mass m_0 becomes important only when $\kappa T \gtrsim m_0 c^2$. From the exercise below we see that of all primordial black holes that may have formed during the Big Bang, those with $M \lesssim 2 \times 10^{14}$ g have evaporated by now – provided they did not accrete mass. According to (6.39) these holes have an initial temperature of $\gtrsim 6 \times 10^{11}$ K.

Exploding microscopic black holes behave not unlike elementary particles, with a characteristic emission spectrum of particles and photons. At present there is no evidence for their existence. They may perhaps be found in cosmic rays. An intriguing possibility is that microscopic black holes (or something resembling it) might be created in future particle accelerators, and be detected through their decay products.

Exercise 6.14: Show that the characteristic wavelength of the Hawking radiation at $r = \infty$ is r_s.

Exercise 6.15: Prove that the lifetime of a non-accreting black hole in vacuum is given by

$$t = \frac{M^3}{3a} \simeq 14 \times 10^9 \left(\frac{M}{1.7 \times 10^{14} \text{ g}}\right)^3 \text{ yr}. \quad (6.43)$$

7
Gravitational waves

Periodic solutions of the vacuum field equations correspond to periodic variations in the geometry of spacetime. Because the equations are nonlinear in $g_{\alpha\beta}$ analytic solutions can only be found in a few special cases. The physical origin of the nonlinearity is that the energy and momentum density of the gravitational field act in turn as a source of gravity. The situation is therefore much more complicated than in the case of electromagnetic waves in vacuum, which is a *linear* problem. However, we expect that the waves are very weak, and then we may use the linearized theory of § 3.5. According to (3.49) we have

$$\Box \gamma_{\mu\nu} = \left(\frac{1}{c^2} \frac{\partial^2}{\partial t^2} - \nabla^2 \right) \gamma_{\mu\nu} = 0 , \qquad (7.1)$$

with $g_{\mu\nu} = \eta_{\mu\nu} + \gamma_{\mu\nu}$, showing that there must exist waves in the metric that propagate at the speed of light. From (2.62) we infer that to first order in γ:

$$R^{\alpha}{}_{\mu\nu\sigma} = \tfrac{1}{2}\eta^{\alpha\beta} \left(\gamma_{\beta\sigma,\mu\nu} - \gamma_{\mu\sigma,\beta\nu} - \gamma_{\beta\nu,\mu\sigma} + \gamma_{\mu\nu,\beta\sigma} \right) . \qquad (7.2)$$

The waves show up in the Riemann tensor as well, so that we are really dealing with fluctuations in the structure (the 'geometry') of spacetime, and not with fluctuations in the definition of the co-ordinate system, for example. These waves have recently been detected, albeit indirectly, in the binary pulsar PSR 1913+16. Here we review their most important properties and detection techniques.

7.1 Small amplitude waves

We use the linearized theory of § 3.5, where we wrote $g_{\alpha\beta} = \eta_{\alpha\beta} + \gamma_{\alpha\beta}$ and $h_{\mu\nu} \equiv \gamma_{\mu\nu} - \tfrac{1}{2}\eta_{\mu\nu}\gamma^{\sigma}{}_{\sigma}$. For $h_{\mu\nu}$ the following equations were obtained:

$$\Box h^{\mu\nu} = 0 ; \qquad h^{\mu\nu}{}_{,\nu} = 0 . \qquad (7.3)$$

It is recalled that the theory is accurate to first order in γ, that we may raise and lower indices with $\eta_{\mu\nu}$ and that $\eta_{\mu\nu}$ commutes with $\Box$. We seek a plane wave solution:

$$h^{\mu\nu} = a^{\mu\nu} \exp(ik_\alpha x^\alpha) \quad \text{with} \quad k^\mu = (\Omega/c, \boldsymbol{k}) \,. \tag{7.4}$$

The constants $a^{\mu\nu}$ obey $a^{\mu\nu} = a^{\nu\mu}$, so that there are in total 10 independent numbers. Furthermore, $k_\alpha x^\alpha = k_0 x^0 + k_i x^i = k^0 x^0 - k^i x^i = \Omega t - \boldsymbol{k} \cdot \boldsymbol{r}$. Insert that in (7.3):

$$0 = \Box h^{\mu\nu} \equiv \eta^{\alpha\beta} h^{\mu\nu}{}_{,\alpha\beta} = a^{\mu\nu} \eta^{\alpha\beta} \{\exp(ik_\sigma x^\sigma)\}_{,\alpha\beta} \,. \tag{7.5}$$

Now $\{\exp(\cdot)\}_{,\alpha\beta} = -k_\alpha k_\beta \exp(\cdot)$, or

$$0 = -a^{\mu\nu} \eta^{\alpha\beta} k_\alpha k_\beta \exp(ik_\sigma x^\sigma) \,, \tag{7.6}$$

and we conclude that $0 = \eta^{\alpha\beta} k_\alpha k_\beta = \eta_{\alpha\beta} k^\alpha k^\beta = (k^0)^2 - \boldsymbol{k}^2$, or, with (7.4):

$$\Omega^2 = (kc)^2 \,. \tag{7.7}$$

A gravitational wave has the same dispersion relation as a plane electromagnetic wave in vacuum. From $h^{\mu\nu}{}_{,\nu} = 0$ we find

$$a^{\mu\nu} k_\nu = 0 \,. \tag{7.8}$$

These are 4 restrictions on the 10 constants $a^{\mu\nu}$. But we haven't chosen a co-ordinate frame yet, and that yields four more restrictions. These take the following simple form (see exercise):

$$a^\sigma{}_\sigma = 0 \quad \text{and} \quad a_{\mu\nu} t^\nu = 0 \,, \tag{7.9}$$

where t^ν is an arbitrary 4-vector obeying $k_\mu t^\mu \neq 0$. Relation (7.8) reveals that $k^\mu a_{\mu\nu} = a_{\nu\mu} k^\mu = \eta_{\nu\sigma} a^{\sigma\mu} k_\mu = 0$, so that $k^\mu a_{\mu\nu} t^\nu$ is already zero. Hence $a_{\mu\nu} t^\nu = 0$ gives only 3 independent restrictions. From $a^\sigma{}_\sigma = 0$ it follows that $h^\sigma{}_\sigma = 0$, so that $\gamma = h = 0$ according to (3.51). The distinction between $h_{\mu\nu}$ and $\gamma_{\mu\nu}$ has vanished:

$$h_{\mu\nu} = \gamma_{\mu\nu} \,. \tag{7.10}$$

All gauge freedom has now been exhausted and from the 10 free constants $a_{\mu\nu}$ only two are left. To proceed we take $t^\nu = (1,0,0,0)$. Then from (7.9): $a_{\mu 0} = 0$. In particular $a_{00} = 0$ and $a^0{}_0 = \eta^{0\nu} a_{\nu 0} = 0$. It then follows from $a^\sigma{}_\sigma = 0$ that $a^i{}_i = 0 \to a_{ii} = \eta_{i\nu} a^\nu{}_i = -a^i{}_i = 0$. Taking $\boldsymbol{k}$ along the x^3 axis, we find with (7.7) that $k^\mu = (\Omega/c)(1,0,0,1)$. Finally, we have from (7.8) that $0 = a_{\mu\nu} k^\nu = a_{\mu 0} k^0 + a_{\mu 3} k^3 = (\Omega/c) a_{\mu 3}$. In summary, $a_{\mu 0} = a_{\mu 3} = a_{ii} = 0$, so that $a_{\mu\nu}$ has the following format:

$$a_{\mu\nu} = \begin{pmatrix} 0 & 0 & 0 & 0 \\ 0 & a_{xx} & a_{xy} & 0 \\ 0 & a_{xy} & -a_{xx} & 0 \\ 0 & 0 & 0 & 0 \end{pmatrix} \,. \tag{7.11}$$

Only $h_{xx} = \gamma_{xx}$ and $h_{xy} = \gamma_{xy}$ are $\neq 0$. This choice of $a_{\mu\nu}$ is called the *transverse traceless gauge*, or *TT-gauge*. In the literature it is often denoted as $\bar{a}_{\mu\nu}$, $\bar{h}_{\mu\nu}, \bar{\gamma}_{\mu\nu},..$ to indicate that it refers to a special choice of the co-ordinates. Only the spatial components of $a_{\mu\nu}$ perpendicular to the direction of propagation are nonzero, that is, the wave is transverse. And 'traceless' obviously refers to $a^\sigma{}_\sigma = 0$. There are two independent wave modes, corresponding to the constants a_{xx} and a_{xy} in (7.11).

For a weak gravitational wave propagating along the x^3-axis (z-axis) in the TT-gauge we may summarize our results as follows, using $k_\alpha x^\alpha = \Omega(t - z/c)$ and λ = wavelength:

$$g_{\mu\nu} = \eta_{\mu\nu} + \gamma_{\mu\nu} \; ; \qquad \gamma_{\mu\nu} = a_{\mu\nu} \exp\{i\Omega(t - z/c)\} \; , \qquad (7.12)$$
$$\text{with} \qquad \lambda = 2\pi/k = 2\pi c/\Omega \; ,$$

and $a_{\mu\nu}$ is given by (7.11). The explicit form of the metric is

$$\mathrm{d}s^2 = c^2 \mathrm{d}t^2 - (1 - \gamma_{xx})\mathrm{d}x^2 - (1 + \gamma_{xx})\mathrm{d}y^2 + 2\gamma_{xy}\mathrm{d}x\mathrm{d}y - \mathrm{d}z^2 \; . \qquad (7.13)$$

Exercise 7.1: Prove that one may impose the restrictions (7.9).

Hint: In exercise 3.11 a transformation was used to obtain the linearized theory. However, there was still some gauge freedom left. We make once more a transformation $x^\alpha \to \bar{x}^\alpha + \xi^\alpha(x)$ for which then $\Box \xi_\alpha = 0$ must hold. From the hint in exercise 3.11 we see that $\bar{\gamma}_{\mu\nu} = \gamma_{\mu\nu} - \xi_{\mu,\nu} - \xi_{\nu,\mu}$, so that $\bar{\gamma} = \bar{\gamma}^\rho{}_\rho = \eta^{\rho\alpha}\bar{\gamma}_{\alpha\rho} = \eta^{\rho\alpha}(\gamma_{\alpha\rho} - \xi_{\alpha,\rho} - \xi_{\rho,\alpha}) = \gamma^\rho{}_\rho - 2\xi^\rho{}_{,\rho} = \gamma - 2\xi^\rho{}_{,\rho}$. Hence $\bar{\gamma} = \gamma - 2\xi^\rho{}_{,\rho}$. From (3.51): $\bar{h}_{\mu\nu} = \bar{\gamma}_{\mu\nu} - \frac{1}{2}\eta_{\mu\nu}\bar{\gamma}$. Show that this leads to:

$$\bar{h}_{\mu\nu} = h_{\mu\nu} - \xi_{\mu,\nu} - \xi_{\nu,\mu} + \eta_{\mu\nu}\xi^\rho{}_{,\rho} \; . \qquad (7.14)$$

Now take $\xi_\mu = b_\mu \exp(ik_\alpha x^\alpha)$ with k^μ from (7.4). This choice obeys $\Box \xi_\mu = 0$. We must now show that there exist a b_μ so that (7.9) holds. Write $h_{\mu\nu} = a_{\mu\nu} \exp(ik_\alpha x^\alpha)$ and $\bar{h}_{\mu\nu} = \bar{a}_{\mu\nu} \exp(\cdot)$, in accordance with (7.4), and substitute in (7.14):

$$\bar{a}_{\mu\nu} = a_{\mu\nu} - ib_\mu k_\nu - ik_\mu b_\nu + i\eta_{\mu\nu}b^\rho k_\rho \; , \qquad (7.15)$$

from which $\bar{a}^\mu{}_\mu = a^\mu{}_\mu - ib^\mu k_\mu - ik^\mu b_\mu + i\eta^\mu{}_\mu b^\rho k_\rho \to \bar{a}^\mu{}_\mu = a^\mu{}_\mu + 2ib^\mu k_\mu$ (since $\eta^\mu{}_\mu = \delta^\mu{}_\mu = 4$). Require $\bar{a}^\mu{}_\mu = 0 \to ib^\mu k_\mu = -\frac{1}{2}a^\mu{}_\mu$. Substitute in (7.15):

$$\bar{a}_{\mu\nu} = a_{\mu\nu} - ib_\mu k_\nu - ik_\mu b_\nu - \tfrac{1}{2}\eta_{\mu\nu}a^\rho{}_\rho \; . \qquad (7.16)$$

Require next that $\bar{a}_{\mu\nu}t^\nu = 0$ for a given t^ν:

$$ib_\mu(k_\nu t^\nu) = a_{\mu\nu}t^\nu - ik_\mu(b_\nu t^\nu) - \tfrac{1}{2}t_\mu a^\rho{}_\rho \ . \tag{7.17}$$

We are done if we can eliminate $b_\nu t^\nu$ on the right hand side. Multiply (7.17) with t^μ. The result is an equation from which $b_\nu t^\nu$ may be solved, if $k_\nu t^\nu \neq 0$. Substitute that again in the right hand side of (7.17). The final expression for b_μ is not important – what matters is that it exists.

7.2 The effect of a gravitational wave on test masses

We consider the dynamics of a free test mass in a gravitational wave. Its worldline is a timelike geodesic, determined by (2.34):

$$\frac{\mathrm{d}u^\alpha}{\mathrm{d}s} + \Gamma^\alpha{}_{\mu\nu}u^\mu u^\nu = 0 \ ; \qquad u^\alpha = \frac{\mathrm{d}x^\alpha}{\mathrm{d}s} \ . \tag{7.18}$$

We elaborate this in the TT-gauge, to first order in $\gamma_{\alpha\beta}$. From (2.24) we see that $\Gamma^\alpha{}_{\mu\nu} = O(\gamma)$. Therefore it suffices to expand u^μ and u^ν in the second term in (7.18) to zeroth order, see (3.23): $u^\mu \simeq (1, v^i/c) \simeq (1, 0, 0, 0)$. The equation for the test mass motion reads:

$$\frac{\mathrm{d}u^\alpha}{\mathrm{d}\tau} + c\Gamma^\alpha{}_{00} = 0 \ . \tag{7.19}$$

We conclude from (3.16) that $\Gamma^\alpha{}_{00} = \tfrac{1}{2}\eta^{\alpha\lambda}(2\gamma_{\lambda0,0} - \gamma_{00,\lambda}) = 0$ because $\gamma_{\mu\nu} = a_{\mu\nu}\exp(ik_\alpha x^\alpha)$ in the TT-gauge, and $\gamma_{\lambda0} = 0$ because $a_{\lambda0} = 0$. It follows that $u^\alpha(\tau) = u^\alpha(0)$, and if the test mass is at rest at $\tau = 0$, it remains at rest as the wave passes by.[1] Superficially, it seems that the test mass does not move. However, in the TT-gauge we are using very special co-ordinates. It turns out that the co-ordinates have been chosen so that they move along with the particle. Let us look at the behaviour of test masses on a circle in the $x^3 = 0$ plane, orthogonal to the direction of wave propagation, Fig. 7.1. The co-ordinates of P are $x^1 = l_0 \cos\theta$, $x^2 = l_0 \sin\theta$ and $x^3 = z = 0$. Because $g_{0i} = \gamma_{0i} = 0$ we can find the physical distance between the origin O and P by integrating (3.7) along OP. The integration is trivial because g_{ij} does not depend on x^1 and x^2. The distance l between O and P becomes ($i = 1$ or 2):

$$\begin{aligned}l^2 &= -g_{ij}x^i x^j \\ &= (1-\gamma_{xx})l_0^2 \cos^2\theta + (1+\gamma_{xx})l_0^2 \sin^2\theta - 2\gamma_{xy}l_0^2 \sin\theta\cos\theta \\ &= l_0^2(1 - \gamma_{xx}\cos 2\theta - \gamma_{xy}\sin 2\theta) \ , \end{aligned} \tag{7.20}$$

so that

7.2 The effect of a gravitational wave on test masses   137

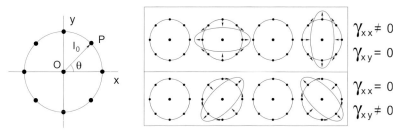

Fig. 7.1. Left: A ring of test particles perpendicular to a gravitational wave is periodically deformed as shown to the right. Each test mass moves along a geodesic and senses no acceleration. However, if O and P are materially connected, they experience a tidal acceleration $\ddot{l}$, see § 7.4.

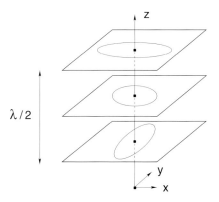

Fig. 7.2. A $\gamma_{xx} \neq 0$ gravitational wave propagating in space. The whole pattern moves with the speed of light along the direction of propagation (the z-axis), and γ_{xx} (and γ_{xy}) are independent of position in planes perpendicular to the direction of propagation. Since the expected wave frequencies are less than a few kHz, the wavelengths λ are large, at least 100 km.

$$l \simeq l_0\left(1 - \tfrac{1}{2}\gamma_{xx}\cos 2\theta - \tfrac{1}{2}\gamma_{xy}\sin 2\theta\right) . \qquad (7.21)$$

Since $\gamma_{xx}, \gamma_{xy} \propto \exp(ik_\alpha x^\alpha) = \exp\left(i\Omega(t - x^3/c)\right) = \exp(i\Omega t)$ we see that the ring of test masses is deformed periodically as in Fig. 7.1. There are two independent linearly polarized waves. The directions of polarization differ by an angle of 45°. From these two waves one may construct circularly polarized waves, as usual. In such a circularly polarized wave, the test particles of Fig. 7.1 describe small circles around their unperturbed position.

---

[1] This is no longer the case if we would work to second order in γ.

Exercise 7.2: Show that the distance between two test masses on the z-axis does not change. The wave is therefore transverse at least up to order γ.

Hint: Locate the particles in $x^1 = x^2 = 0$, and $x^3 = 0$ and $x^3 = \epsilon \rightarrow l^2 = -g_{33}\epsilon^2 = -\eta_{33}\epsilon^2 = \epsilon^2$ because $\gamma_{zz} = 0$.

Exercise 7.3: Estimate the acceleration experienced by an extended body due to the passage of a gravitational wave.

Hint: (7.12): the action of the wave (i.e. γ_{xx} and γ_{xy}) is independent of position in planes $\perp z$-axis, but different in planes at different z. A 'pencil' along the z-axis will not feel the wave (exercise 7.2). Pencil $\perp z$-axis: (7.21) $\rightarrow$ $a = \ddot{l} = -\frac{1}{2}l_0\ddot{\gamma}_{xx} \sim l_0\gamma\Omega^2$ assuming $\gamma_{xy} = 0$, $\gamma_{xx} = \gamma\cos\Omega t$ and $\cos\theta = 0$. The wave causes a tidal acceleration $\propto$ size of object. Take the Space Station ($l_0 = 100\,\text{m}$) and $\gamma = 10^{-6}$, $\Omega/2\pi = 5\,\text{Hz} \rightarrow a \sim 0.1\,\text{m s}^{-2} \rightarrow$ Station is periodically stretched and compressed with a force equivalent to $0.01\,g$.

7.3 Generation of gravitational radiation

The amplitude of gravitational waves is expected to be extremely small, $\gamma_{xx}, \gamma_{xy} \sim 10^{-20}$, and the reasons are twofold: the enormous distance of potential sources, and the fact that gravitational radiation is inherently weak because there is no dipole radiation. To illustrate this, consider electromagnetic radiation of a source of size $2R$. The radiation consists of the sum of the various multipole contributions, the dipole radiation usually being the strongest. At large distances from the source ($r \gg R$), the vector potential in the Lorentz gauge is given by:

$$\boldsymbol{A}^{\text{rad}}(\boldsymbol{r}, t) = \frac{1}{cr}\dot{\boldsymbol{d}}(t - r/c) + \frac{1}{cr}\sum \text{multipoles}, \qquad (7.22)$$

where $\boldsymbol{d} = \Sigma e_i \boldsymbol{r}_i$ is the electric dipole moment of the source and $\dot{} = \partial/\partial t$. The power emitted in electric dipole radiation is proportional to $\ddot{\boldsymbol{d}} \cdot \ddot{\boldsymbol{d}}$. The next terms in (7.22) are those of the magnetic dipole moment $\Sigma e_i(\boldsymbol{r} \times \boldsymbol{v})_i$ and the electric quadrupole moment $\Sigma e_i(3\boldsymbol{r}\boldsymbol{r} - r^2\boldsymbol{I})_i$ of the source. The power emitted in electric quadrupole and magnetic dipole radiation is a factor of $(kR)^2 \sim (R/\lambda)^2$ smaller than that in electic dipole radiation. In the case of gravitational radiation, the (mechanical) dipole moment equals $\boldsymbol{d} = \Sigma m_i \boldsymbol{r}_i$. However, $\dot{\boldsymbol{d}} = (\Sigma m_i \dot{\boldsymbol{r}}_i)^{\cdot} = (\boldsymbol{P}_{\text{tot}})^{\cdot} = 0$. There is no dipole radiation because

7.3 Generation of gravitational radiation

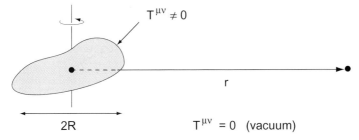

Fig. 7.3. A source of characteristic size R and Schwarzschild radius r_s radiates gravitational waves that are detected at a large distance r. The amplitude of the waves is given by (7.25).

the total momentum $\boldsymbol{P}_{\text{tot}}$ of the system is constant. And the analogon of magnetic dipole radiation is absent because the angular momentum is conserved. The first non-vanishing contribution is generated by a variable quadrupole moment.[2]

Assuming that the deviations from the Lorentz metric in the source are small, the generation of gravitational radiation is described by eq. (3.52):[3]

$$\Box h^{\mu\nu} = -\frac{16\pi G}{c^2} T^{\mu\nu} \; . \tag{7.23}$$

We shall now estimate the order of magnitude of $h^{\mu\nu}$ far from the source. The radiation field there consists of a superposition of spherical waves of different frequencies of the type:

$$h^{\mu\nu} = \frac{H^{\mu\nu}}{r} \exp\{i(\Omega t - kr)\} \; , \tag{7.24}$$

with $\Omega^2 = (kc)^2$, the dispersion relation (7.7), see exercise. We neglect the θ, φ dependence of $H^{\mu\nu}$ because all we are interested in is an order of magnitude. Directly exterior to the source, in $r \sim R$ (see Fig. 7.3), we have $h^{\mu\nu} \sim H^{\mu\nu}/R$. Next we estimate in (7.23) for $r \simeq R$: $\Box \sim R^{-2}$ and $T^{\mu\nu} \sim \rho u^\mu u^\nu \sim \rho v^2/c^2$, so that

---

[2] For sources of gravitational radiation see e.g. Schutz, B.F., *Class. Quantum Grav.* **13** (1996) A219; **16** (1999) A131.

[3] Strictly speaking $T^{\mu\nu}$ in (7.23) describes only motion due to other forces than gravity. Radiation from two compact binary stars whose motion is determined by gravity should actually be found by solving $\Box h^{\mu\nu} = 0$ with two Schwarzschild singularities in $\boldsymbol{r}_1(t)$ and $\boldsymbol{r}_2(t)$ as a boundary condition. However, it can be shown that the result coincides with the solution of (7.23) up to $O(\gamma)$ if one uses in $T^{\mu\nu}$ the velocities following from classical mechanics.

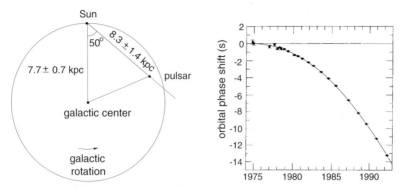

Fig. 7.4. The location in the galactic plane of the compact binary system of which PSR 1913+16 is a member (left), and the cumulative shift of the periastron passage since the discovery of the system (right). Adapted from Taylor, J.H., *Class. Quantum Grav.* **10** (1993) S167, and Damour, T. and Taylor, J.H., *Ap. J.* **366** (1991) 501.

$$\frac{h^{\mu\nu}}{R^2} \sim \frac{G}{c^2}\frac{\rho v^2}{c^2} \ .$$

One might object that $T^{\mu\nu} \sim \rho$ when $\mu = \nu = 0$, but according to (7.11) $h^{0\alpha}$ does not contribute. With the help of $M \sim \rho R^3$ we find that near the source $h^{\mu\nu} \sim r_s(v/c)^2 R^{-1}$, and this should also be equal to $H^{\mu\nu}/R$, or $H^{\mu\nu} \sim r_s(v/c)^2$. At the observer we have $h^{\mu\nu} \sim H^{\mu\nu}/r$, and we arrive at

$$\gamma^{\mu\nu} = h^{\mu\nu} \sim \left(\frac{v}{c}\right)^2 \frac{r_s}{r} \sim \begin{cases} \left(\dfrac{\omega R}{c}\right)^2 \dfrac{r_s}{r} & \text{for } v = \omega R \ , \\ \dfrac{r_s^2}{Rr} & \text{for } v^2 = GM/R \ . \end{cases} \quad (7.25)$$

Here we distinguish two archetypical cases: a bar rotating at a given angular frequency ω, and a binary system where v can be estimated by the classical circular orbit speed. This estimate (7.25) is valid if the source is far removed from spherical symmetry, and $v \ll c$. Without proof we mention that the average energy flux density F of a gravitational wave is given by (see e.g. Kenyon (1990)):

$$F = \frac{c^3}{16\pi G} \langle \dot\gamma_{xx}^2 + \dot\gamma_{xy}^2 \rangle \ ; \qquad \dot{} = \partial/\partial t \ . \quad (7.26)$$

The existence of gravitational waves has been demonstrated indirectly

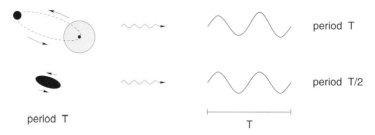

Fig. 7.5. Top: gravitational radiation of a rotating object has, theoretically, a period T because the source needs T seconds to return to the same configuration. Bottom: the radiation is emitted by the equivalent quadrupole, whose time dependence determines the spectrum. For small ellipticity the quadrupole rotates almost uniformly and the radiation is practically monochromatic with frequency $2/T$, since the quadrupole needs $T/2$ seconds to return to a physically identical configuration. Not all periodic sources emit at twice the fundamental frequency: a rotating bar does, but a harmonically oscillating bar emits at the fundamental frequency. For binaries in a highly elliptic orbit the radiation takes the form of a series of pulses separated by the orbital period T. The spectrum features emission of higher harmonics because these are now present in the time dependence of the quadrupole.

but convincingly in the compact binary system of which PSR 1913+16 is a member. The system loses energy in the form of gravitational radiation, and this shows up as a slowly decreasing orbital period P_b which is now 27906 s or 7.75 hr, Fig. 7.4. Observations carried out over the past 25 years have shown that $\dot{P}_b = -(2.422 \pm 0.006) \times 10^{-12}$, which agrees within the measurement error (0.3%) with the prediction of GR: $(\dot{P}_b^{\mathrm{obs}} - \dot{P}_b^{\mathrm{gal}})/\dot{P}_b^{\mathrm{GR}} = 1.0032 \pm 0.0035$. The term $\dot{P}_b^{\mathrm{gal}}$ is due to a small relative acceleration between the binary pulsar and the solar system. This is because the binary pulsar is closer to the galactic centre than the Sun, and is gradually overtaking us on account of its larger galactic orbital velocity. This causes a measurable correction $\dot{P}_b^{\mathrm{gal}} = -(0.012 \pm 0.006) \times 10^{-12}$. This high precision can be achieved because the pulsar is a very accurate clock, and because the system is clean.[4] The orbital shrinking due to emission of gravitational waves of the newly discovered binary pulsar J0737-3039A/B is expected to be detected soon, and should permit an even more accurate test.

Exercise 7.4: Prove that (7.24) is a solution of $\Box h^{\mu\nu} = 0$.

---

[4] Taylor, J.H. and Weisberg, J.M., *Ap. J.* **345** (1989) 434; Damour, T., *Class. Quantum Gravity* **10** (1993) S59; Taylor, J.H., *Class. Quantum Gravity* **10** (1993) S167.

Hint: $\Box = c^{-2}\partial_t^2 - \nabla^2 = c^{-2}\partial_t^2 - r^{-2}\partial_r r^2 \partial_r$, because we neglect the dependence on θ and ϕ.

Exercise 7.5: Estimate the order of magnitude and the time depencence of $\gamma^{\mu\nu}$ of the following sources: (a) asymmetric collapse of supernova 1987a in the Large Magellanic Cloud ($r = 52$ kpc); take $r_s \sim 4$ km ($\sim 1.4 M_\odot$) and $R \sim 10\,r_s$. (b) close encounter of two $1 M_\odot$ black holes in the centre of our galaxy ($r = 8$ kpc); take $r_s = 6$ km, and $R = 10^4$ km, for example. (c) the compact binary system containing PSR 1913+16 ($r \sim 8$ kpc); take $r_s \sim 8$ km ($2 \times 1.4 M_\odot$), $R =$ semi-major axis of relative orbit $= 2 \times 10^6$ km. (d) a rotating egg-shaped neutron star (due a strong magnetic field). Take $r = 2$ kpc (Crab pulsar), $r_s \sim 4$ km and $R \sim 2 r_s$.

Hint: (a) $\gamma^{\mu\nu} \lesssim 2.5 \times 10^{-19}$. On account of (6.8) we expect a brief radiation pulse of $\sim 10\,\mu$s. Unfortunately no detector was operational at the time of the event, in contrast to neutrino detectors. (b) $\gamma^{\mu\nu} \lesssim 10^{-20}$. The radiation is a pulse lasting $R/v \sim (R/r_s)^{1/2}(R/c) \sim 1$ s. (c) $\gamma^{\mu\nu} \lesssim 10^{-22}$. The radiation is periodic at $\frac{1}{2} \times$ orbital period $= 3.88$ hours, see Fig. 7.5. (d). $\gamma^{\mu\nu} \lesssim 3 \times 10^{-17}$! But the shape of the star will be almost spherically symmetric, hence $\gamma^{\mu\nu}$ considerably smaller.

Exercise 7.6: Compute $\dot{P}_b$ of PSR 1913+16 from Fig. 7.4, right.

Hint: Expand the period $P(t) = P_b + \dot{P}_b t + \cdots$. The number of periods n in a certain time interval equals $n = \int dt/P$, and $n_0 = \int dt/P_b$ if the period were constant. The cumulative shift Δt of the periastron passage is $\Delta t \simeq (n_0 - n) P_b$, or

$$\Delta t \simeq P_b \int \left(\frac{1}{P_b} - \frac{1}{P}\right) dt \simeq \int \left(1 - \frac{P_b}{P_b + \dot{P}_b t}\right) dt$$

$$\simeq \int \frac{\dot{P}_b t}{P_b} dt = \frac{\dot{P}_b t^2}{2 P_b} , \qquad (7.27)$$

and $P_b = 27906$ s, and from Fig. 7.4 we see that $\Delta t = -14$ s in $t = 18$ years.

Exercise 7.7: Show that the flux density of a weak gravitational wave with $\gamma \simeq 10^{-22}$ and a frequency of $\Omega/2\pi = 1$ kHz is about equal to the optical flux density of the full moon (~ 3 erg cm^{-2} s^{-1} at the Earth). In this sense sources of gravitational waves shine very brightly in the sky! Explain this paradox.

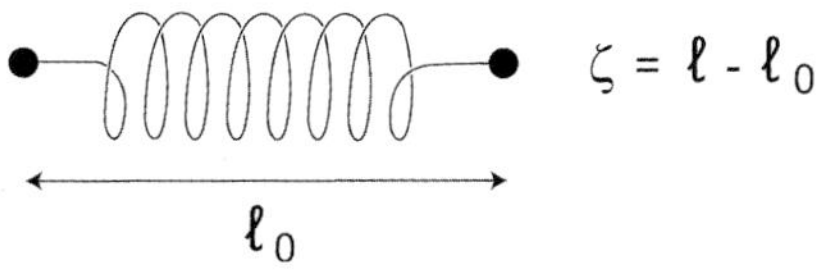

Fig. 7.6. An idealised detector for gravitational waves consisting of two masses connected by a spring.

Hint: Estimate $\dot{\gamma}_{xx}^2 + \dot{\gamma}_{xy}^2 \sim \Omega^2 \gamma^2$ in (7.26). The energy flux in gravitational waves is large, but the relative amplitude γ is small. The stiffer the medium, the smaller the amplitude of a wave at a given energy flux. Spacetime behaves as a very stiff medium. Sources of gravitational waves radiate in general considerable amounts of energy, but the waves pass through everything without leaving hardly any physical effect.

7.4 Bar detectors

Detection of gravitational waves is very difficult because the expected amplitudes are so small. Fourty years ago Weber experimented with aluminium bars that were isolated from the environment as much as possible. We may model such a detector as two masses connected by a spring, i.e. as a damped harmonic oscillator with frequency $\omega_0/2\pi$, see Fig. 7.6. The equation for the distance ζ of the masses is: $\ddot{\zeta} = -2\epsilon\dot{\zeta} - \omega_0^2 \zeta$. The effect of a weak gravitational wave can be described by adding the acceleration $\ddot{l}$ due to the wave on the right side.[5] Let $\gamma_{xy} = 0$ and $\theta = 0$ in (7.21), i.e. we consider one wave and a detector aligned along the x-axis of Fig. 7.2, so that $\ddot{l} = -\frac{1}{2} l_0 \ddot{\gamma}_{xx}$. The equation for ζ is then

$$\ddot{\zeta} + 2\epsilon \dot{\zeta} + \omega_0^2 \zeta = -\frac{1}{2} l_0 \ddot{\gamma}_{xx} . \qquad (7.28)$$

Since γ_{xx} is independent of position along the detector we may put $\gamma_{xx} = \gamma \cos \Omega t$. The maximum amplitude equals (see exercise):

$$\zeta_{max} = \tfrac{1}{2} l_0 \gamma Q ; \qquad Q = \frac{\omega_0}{2\epsilon} = \text{quality factor} . \qquad (7.29)$$

For $Q = 10^5$, $\gamma = 10^{-20}$ and $l_0 = 2$ m we have $\zeta_{max} \sim 10^{-13}$ cm, about the size of an atom, which nicely illustrates the detection problem. An additional

---

[5] See e.g. Misner et al. (1971) p. 1004 ff; Schutz (1985) p. 222.

Fig. 7.7. Close up of the MiniGRAIL detector under development in Leiden. It consists of a CuAl sphere of 68 cm diameter suspended by a thin rod. The resonance frequency is 2.9 kHz, the bandwidth 230 Hz. The sphere carries several transducers that mechanically amplify and detect the vibration. The theoretical sensitivity of this 20 mK cryogenic detector is $\sim 4 \times 10^{-21}$. A spherical detector can determine the direction $\boldsymbol{n}$ of the incoming wave (up to a $\pm \boldsymbol{n}$ uncertainty) because the relative excitation levels of the quadrupole modes of the sphere depends on $\boldsymbol{n}$. Image credit: A. de Waard and G. Frossati. See http://www.minigrail.nl/

complication is that of the order of Q waves are needed to excite a resonant detector to its full amplitude $\zeta_{\max}$, which renders detection of bursts of radiation more difficult. And $\zeta_{\max}$ is independent of l_0 since $\omega_0 \propto$ sound speed $/\ l_0$. Bar detectors are sensitive in a narrow frequency interval $\Delta\Omega \sim \omega_0/Q$ around ω_0, and seem therefore more suited for detection of quasi-periodic radiation, as emitted by narrow binary systems.

Noise is a problem of overwhelming importance. At room temperature the amplitude ζ of thermally excited oscillations is also about 10^{-13} cm. Weber had two detectors operating in coincidence at room temperature, at a frequency of $\omega_0/2\pi = 1660$ Hz. Coincidence measurements by independent detectors at different locations are essential to eliminate chance detections that are actually large noise peaks. There are still a few bar detectors operating at room temperature and they attain a sensitivity of $\gamma \sim 10^{-16}$. By cooling to liquid helium temperatures (around 4 K) the NIOBE, EXPLORER and ALLEGRO bar detectors reached a sensitivity of $\gamma \sim 6 \times 10^{-19}$. This development took place during the eighties and nineties of the previous century. To detect the bar vibrations they are amplified, usually by a resonant transducer that is read out by a squid. In the near future detectors of the third generation NAUTILUS and AURIGA will become operational. These will be cooled to ~ 0.1 K. The MiniGRAIL project develops a spherical cryogenic (20 mK) detector in the Netherlands, and a similar detector is being built in São Paulo.[6]

Exercise 7.8: Prove (7.29).

Hint: Take $\gamma_{xx} = \gamma \exp(i\Omega t)$ and $\zeta = \hat{\zeta} \exp(i\Omega t)$ in (7.28) $\to (-\Omega^2 + 2i\epsilon\Omega + \omega_0^2)\hat{\zeta} = \frac{1}{2} l_0 \gamma \Omega^2$. The solution is $\zeta = \text{Re}\{\hat{\zeta} \exp(i\Omega t)\} = \text{Re}\{|\hat{\zeta}| \exp(i\phi) \cdot \exp(i\Omega t)\} = |\hat{\zeta}| \cos(\Omega t + \phi)$ for certain ϕ. Ergo $\zeta_{\max} = \max_\Omega |\hat{\zeta}|$. A good detector has $\epsilon \ll \omega_0$ ($Q \gg 1$), and then the maximum is located practically at $\Omega = \omega_0$.

7.5 Interferometer detectors

An alternative detection technique is based on Michelson interferometers. These are more expensive but offer two advantages: the sensitivity can be higher and they cover a broad frequency band. We analyse the operation of such a detector, see Figs. 7.8 and 7.9. The laser beam enters the arms through a beam splitter. The beams then travel back and forth between two mirrors on each arm that are suspended so that they can move freely in the direction of the beam. We assume an ideal orientation: the gravitational wave propagates perpendicularly to the plane defined by the arms, that are aligned along the x and y-axis as in Fig. 7.2. The wave induces a frequency shift[7] $\delta\nu/\nu_0 = (\nu_2 - \nu_0)/\nu_0 = dt_0/dt_2 - 1$ in the returning beams with respect to the laser, see Fig. 7.10. The induced phase differences in the two arms have opposite

---

[6] For information on existing and planned bar detectors see Blair (1991); Saulson (1994); Ricci, F. and Brillet, A, *Annu. Rev. Nucl. Part. Sci.* 47 (1997) 111, and Ju, L. et al., *Rep. Prog. Phys.* 63 (2000) 1317.
[7] Actually $d\tau_0/d\tau_2 - 1$, but $g_{00} = 1$.

146   7 Gravitational waves

Fig. 7.8. Areal view of the LIGO interferometer at Hanford (WA), showing the central housing and the two arms of 4 km length. The other LIGO interferometer is located 3000 km away in Livingstone (LA). Courtesy of California Institute of Technology.

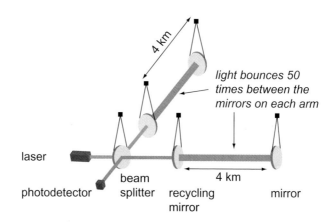

Fig. 7.9. Principle of the LIGO Michelson interferometer.

sign, and show up as intensity variations upon interference on the detector (a photodiode). We take once more $\gamma_{xy} = 0$, $\gamma_{xx} = \gamma \cos \Omega t$, and focus attention on the x-beam. Then (7.13) reduces to $c^2 dt^2 = (1 - \gamma \cos \Omega t) dx^2$:

$$dx = \pm c (1 - \gamma \cos \Omega t)^{-1/2} dt \simeq \pm c (1 + \tfrac{1}{2}\gamma \cos \Omega t) dt . \qquad (7.30)$$

7.5 Interferometer detectors

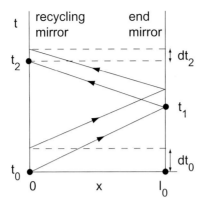

Fig. 7.10. Null geodesics of photons propagating between the mirrors along the x-arm of the interferometer. The geodesics of subsequent wave crests are not congruent because the metric depends on time. The mirrors have fixed spatial co-ordinates which we take to be $x = 0$ and $x = l_0$.

$+, -$ for beams propagating to the right and left, respectively. Since the motion of the mirrors in the direction of the beam is free, their co-ordinates $x = 0$ and $x = l_0$, according to § 7.2, do not change when a gravitational wave passes. Therefore we may integrate (7.30), for a beam propagating to the right in Fig. 7.10:

$$\frac{l_0}{c} = \int_{t_0}^{t_1} (1 + \tfrac{1}{2}\gamma \cos \Omega t)\, dt$$

$$= t_1 - t_0 + \frac{\gamma}{2\Omega}(\sin \Omega t_1 - \sin \Omega t_0). \qquad (7.31)$$

For the returning beam after reflection we take the $-$ sign in (7.30), and an extra $-$ sign because we integrate over x from l_0 to 0. As a result, the expression for a beam propagating to the left emerges by substituting $t_0 \to t_1$, $t_1 \to t_2$. Adding these two gives:

$$t_2 - t_0 = \frac{2l_0}{c} - \frac{\gamma}{2\Omega}(\sin \Omega t_2 - \sin \Omega t_0). \qquad (7.32)$$

To zeroth order $t_2 = t_0 + 2l_0/c$, which we use to eliminate t_2 in the first order term on the right:

$$t_2 - t_0 = \frac{2l_0}{c} - \frac{\gamma}{2\Omega}\{\sin \Omega(t_0 + 2l_0/c) - \sin \Omega t_0\}$$

$$= \frac{2l_0}{c} - \frac{\gamma}{\Omega}\sin\left(\frac{\Omega l_0}{c}\right)\cos(\Omega t_0 + \text{const}). \qquad (7.33)$$

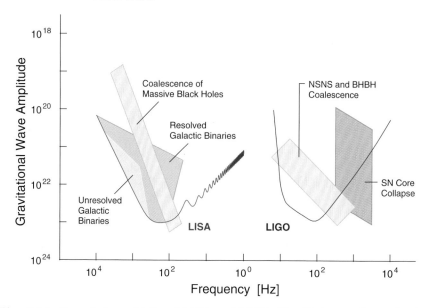

Fig. 7.11. Expected sensitivity of LISA and LIGO. The U-shape reflects the factor $\sin(\Omega L/c)$ in (7.35). From LISA System and Technology Study Report ESA-SCI(2000)11.

In reality the beam travels back and forth n times between the mirrors, and it is easy to see that the same relation holds with $l_0 \to L = nl_0 =$ effective arm length. Differentiation of (7.33) produces $dt_2 - dt_0 = \gamma \sin(\Omega L/c) \sin(\Omega t_0 + \text{const}) \cdot dt_0$, or:

$$\frac{\delta \nu}{\nu_0} = \frac{dt_0}{dt_2} - 1 \simeq -\gamma \sin\left(\frac{\Omega L}{c}\right) \sin(\Omega t + \text{const}) . \quad (7.34)$$

We have dropped the index 0 on t_0 on the right. The frequency shift is far too small to be measurable, but the phase difference $\delta\psi$ is not:

$$\delta\psi = 2\pi \int \delta\nu \, dt = \frac{\gamma \omega_0}{\Omega} \sin\left(\frac{\Omega L}{c}\right) \cos(\Omega t + \text{const}) , \quad (7.35)$$

with $\nu_0 = \omega_0/2\pi =$ laser frequency, $\Omega/2\pi =$ frequency gravitational wave. The factor $\sin(\Omega L/c)$ in (7.35) determines a broad frequency range where the detector is sensitive, centered on $\Omega L/c = \pi/2$ or $\Omega/2\pi = c/4L$. LIGO has an effective arm length $L \sim 500$ km and a laser frequency of $\nu_0 = 3 \times 10^{14}$ Hz ($\lambda = 1\,\mu$). The maximum sensitivity lies around $\Omega/2\pi \sim 150$ Hz, see Fig. 7.11, and the expected phase shift is very small: $\delta\psi \simeq \gamma \omega_0/\Omega = 2 \times 10^{12} \gamma$. The phase shift $\delta\psi$ of the y-beam has the opposite sign.

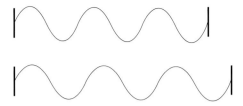

Fig. 7.12. A gravitational wave will stretch and compress the wavelength of the laser beam and the arm length of the interferometer in equal proportion. On this account no phase difference would develop, see text.

The physics of interferometer detectors

In view of the interest these interferometer detectors will draw in the coming decades we analyse their operation in some detail. Fig. 7.12 raises a basic question. A gravitational wave stretches the arm of the interferometer and the wavelength of the laser beam proportionally. Hence there are no phase differences and the detector will not work. Where is the catch? The argument is correct in the limit of small L. In that case (7.35) says that $\delta\psi \to 0$. But when L is so large that the travel time of the laser beam is of the order of the period of the gravitational wave, then the laser beam is no longer a standing wave but a travelling wave. The wave train becomes a local entity travelling with speed c with respect to the local track as it is alternatingly being stretched and compressed. And then phase differences do develop.

Consider a beam propagating to the right, assuming $\cos \Omega t > 0$. Then (7.30) tells us that $dx > c dt$. The *co-ordinate speed* of light is larger than c, and it is straightforward to see from (7.13) that the co-ordinate speed in the y-arm is smaller than c. This generates a time difference and hence a phase difference between the two beams as they interfere on the detector, see Fig. 7.13, top panels. However, we may also write (7.30) as

$$dl \equiv (1 - \tfrac{1}{2}\gamma \cos \Omega t)\, dx = c\, dt, \tag{7.36}$$

where dl is the physical length corresponding to the co-ordinate distance dx according to (7.21). In other words, $dl = c dt$ and that holds for forward and backward propagating beams. This says that the photons behave as cyclists moving at speed c with respect to the local track as it is periodically stretching and shrinking, see Fig. 7.13, lower panels. For $\cos \Omega t > 0$ the physical length of the x-track is reduced, that of the y-track increased by an amount $\delta L \sim \gamma L$ ($L = nl_0$). Two things happen now. The wave trains are slightly compressed and stretched (blue or redshifted), just like the track, but that is too small to be observable. In the second place there is a difference in arrival time $\delta t \sim \gamma L/c$, corresponding to a phase difference $\delta\psi \sim \gamma \omega_0 L/c$,

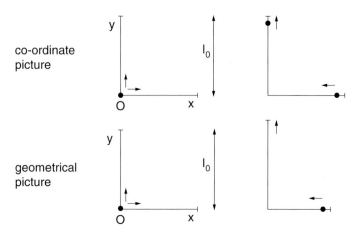

Fig. 7.13. An interferometer detector as a dual race track for photons. The gravitational wave propagates along the z-axis. A laser beam may be thought of as a series of wave crests that follow null geodesics as in Fig. 7.10. Here we follow only one wave crest. The top two panels are co-ordinate pictures. Two laser wave trains start in O at $t = 0$. As long as $\cos \Omega t > 0$, the co-ordinate speed of light on the x-track, dx/dt, is larger than c, but smaller than c on the y-track. However, the co-ordinate length of the track is constant. The top right panel shows the positions after a time $\delta t = l_0/c$. The wave trains arrive in O with a time difference (in reality the beams bounce back and forth many times). The lower two panels show the equivalent geometrical pictures, see text, and Fig. 2.1.

which is essentially (7.35) for $\Omega L/c \ll 1$. When $\Omega L/c \sim 1$ the computation of δt requires an integration and yields (7.35). Optimal operation (maximal δt) occurs when the duration c/L of the race comprises a quarter of the gravitational wave period. If the race takes longer (larger L) the relative stretching and compressing of the tracks reverses and the net δt becomes smaller. If $\Omega L/c = \pi$ the gain δt accumulated during the first quarter of the wave period is undone during the second quarter, and the net gain δt becomes zero, see (7.35). This conceptual picture of photons as cyclists on a shrinking or stretching road is also useful for understanding the shape of our past light-cone in cosmology, see § 11.2.

Detector signal

In order to give the reader some idea of the problems involved in the interferometric detection of gravitational waves, we close this chapter with a (much simplified) estimate of the flux on the detector. Denoting the unperturbed phase as $\psi_0 = \omega_0 t$, and time averaging as $\langle \cdot \rangle$, the photodiode measures an intensity

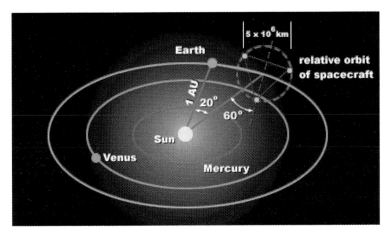

Fig. 7.14. The Laser Interferometer Space Antenna (LISA), a joint ESA-NASA project, to be launched around 2015. From LISA System and Technology Study Report ESA-SCI(2000)11.

$$I_{\text{out}} = \langle [A\cos(\psi_0 + \delta\psi + \alpha) + A\cos(\psi_0 - \delta\psi)]^2 \rangle$$
$$\simeq \tfrac{1}{2} I_0 \left[1 + \cos(2\delta\psi + \alpha)\right]. \tag{7.37}$$

The phase difference α between the beams is a matter of fine tuning the arm length. For zero phase difference the detector sees the full laser power I_0 so, ignoring optical losses, $I_0 = I_{\text{out}} = \langle (2A\cos\psi_0)^2 \rangle = 2A^2$. Relation (7.37) says that $I_{\text{out}} = 0$ for $\alpha = \pi$ and $\delta\psi = 0$, but optical imperfections will prevent complete nulling and we should expect rather a dark signal $I_{\text{out}} = \epsilon I_0$. So

$$I_{\text{out}} \simeq \tfrac{1}{2} I_0 \left[1 + \epsilon + \cos(2\delta\psi + \alpha)\right] \tag{7.38}$$

is more realistic. For example, an imbalance δA in the beam amplitudes can be shown to imply $\epsilon = \tfrac{1}{2}(\delta A/A)^2$ for $\alpha = \pi$. The interferometer should operate close to $\alpha = \pi$ because otherwise the detector sees a large fraction of I_0 and the associated laser noise, from which the small superposed signal can no longer be extracted. But at $\alpha = \pi$ we have $I_{\text{out}} \simeq I_0 [\epsilon + 2(\delta\psi)^2]$ which is even worse since, as derived above, $\delta\psi \simeq 2 \times 10^{12} \gamma \simeq 2 \times 10^{-9}$ for $\gamma = 10^{-21}$. The signal is distorted, $\propto (\delta\psi)^2$, and so small that it would drown in the dark current. The solution is rapid phase modulation around $\alpha = \pi$. Phase modulators between the beamsplitter and the first mirrors (not shown in Fig. 7.9) add a phase $\phi \sin\omega_m t$ to one beam and $-\phi \sin\omega_m t$ to the other. We take $\alpha = \pi$, and since $\delta\psi \ll \phi \ll 1$ we expand to first order in $\delta\psi$ and to second order in ϕ:

$$I_{\text{out}} \simeq \tfrac{1}{2} I_0 \left[1 + \epsilon + \cos(2\delta\psi + 2\phi\sin\omega_m t + \pi)\right]$$

$$\simeq \tfrac{1}{2} I_0 \left[\epsilon + \phi^2 - \phi^2 \cos 2\omega_\mathrm{m} t + 4\phi\,\delta\psi \sin \omega_\mathrm{m} t \right] . \qquad (7.39)$$

To see that this a much better arrangement, let's take $\epsilon \sim \phi^2 \sim 10^{-6}$. The dark signal is $\sim 10^{-6} I_0$ and has a zero and a double frequency component. The signal $\delta\psi$ is now encoded as the *amplitude* of a periodic signal at the modulation frequency (which is in the MHz range). This is a great advantage. The modulation depth is $4\phi\,\delta\psi/(\epsilon + \phi^2) \sim 2\delta\psi/\phi \sim 4 \times 10^{15}\gamma \sim 4 \times 10^{-6}$ for $\gamma = 10^{-21}$, which is small but not impossible. The different frequency dependence allows easy separation of the various components. The phase modulation has an important extra bonus in that it is very effective in suppressing certain types of noise.

Suppose we want to keep the phase difference α constant at the 10^{-3} radian level. Since the laser wavelength λ is 1 μm, that corresponds to a distance of only $10^{-3}\lambda/2\pi \sim 0.1$ nm over an arm length of 4 km! It follows that an active phase locking system is indispensable, as the seismic perturbations are much larger. The question is how that can be done without disturbing the measurements. The trick is, briefly, to reset the phase at a rate that is outside the measuring bandwidth (Fig. 7.11). For more information on these issues and many other experimental finesses and complications we refer to Blair (1991) and Saulson (1994).

Exercise 7.9: Check the details of the derivation of (7.37) and (7.39).

Hint: Take $a = \psi_0 + \delta\psi + \alpha$ and $b = \psi_0 - \delta\psi$ and use $\cos a + \cos b = 2\cos[(a+b)/2]\cos[(a-b)/2]$; $(a+b)/2$ is a fast variable and $\langle \cos^2[(a+b)/2] \rangle = \tfrac{1}{2}$ → $I_\mathrm{out} = 2A^2 \cos^2[(a-b)/2]$. Use $2\cos^2 x = 1 + \cos 2x$ → $I_\mathrm{out} = A^2[1 + \cos(a-b)]$. For (7.39) write $a = 2\delta\psi$ and $b = 2\phi \sin\omega_\mathrm{m} t$. Then $\cos(a+b+\pi) = -\cos(a+b) \simeq -[1 - \tfrac{1}{2}(a+b)^2] \simeq -1 + \tfrac{1}{2}(2ab + b^2)$, since $a \ll b \ll 1$. Finally $2\sin^2 x = 1 - \cos 2x$.

Projects under development

Interferometers for the detection of gravitational radiation are in an advanced state of development. The two most important are the LIGO project (USA), Fig. 7.8,[8] and the Italian/French Virgo project, a single 3 km interferometer

---

[8] Abramovici, A. et al., *Science* **256** (1992) 325; Barish, B.C. and Weiss, R., *Physics Today*, October 1999, 44; and http://www.ligo-wa.caltech.edu.

7.5 Interferometer detectors

under construction at Cascina near Pisa.[9] These projects should be taking science data on a regular basis within a few years. Two smaller projects are the British-German GEO-600, and the Japanese TAMA-300 (both operational).

The seismic background renders measurements below ~ 10 Hz impossible on Earth. Detection of low frequency gravitational waves must be done from space. Through Doppler tracking of the ULYSSES and GALILEO spacecraft an upper limit of $\gamma \leq 10^{-15}$ has been set in the range $0.1 - 10$ mHz. ESA and NASA are studying the ambitious LISA project (Laser Interferometer Space Antenna),[10] see Fig. 7.14.

---

[9] Ricci, F. and Brillet, A., *Annu. Rev. Nucl. Part. Sci.* **47** (1997) 111, and http://www.virgo.infn.it/
[10] LISA System and Technology Study Report, ESA-SCI(2000)11, July 2000; websites: http://sci.esa.int/categories/futureprojects/ and http://lisa.jpl.nasa.gov.

8
Fermi-Walker Transport

In § 2.4 we investigated parallel transport of a vector along an arbitrary worldline $x^\mu(s)$. The motivation was that we should be able to compare, at different places along the orbit, the vectors associated with a point mass, such as the speed or the spin. The vectors are supposed to be known along the orbit, and we compare the vector $\boldsymbol{A}$ with $\boldsymbol{A}'$, obtained by parallel transport, see Fig. 2.4. If these two do not coincide we say that the vector has intrinsically changed due to influences other than gravity. The actual change of the vector $\boldsymbol{A}$ along the worldline is a matter of studying the dynamics. We know that the 4-velocity u^μ is by definition tangent vector and $u^\mu u_\mu = 1$, but the change of the spin vector for example depends on the applied torque. Here we analyse a seemingly innocuous question: a spinning top moves along a worldline that is not a geodesic, i.e. the top experiences an acceleration, but there are no external torques. How does the spin axis behave? The result will be used to derive the *Thomas precession* of the electron and the *geodesic precession* of a gyroscope.

8.1 Transport of accelerated vectors

A test mass moves along its worldline W due to gravity and other forces, and $x^\mu(s)$ is determined by eq. (3.60), see Fig. 8.1. Now imagine that the test mass carries orthonormal unit vectors, the 4-velocity u^μ and n_i^μ ($i = 1, 2, 3$). In the local rest-frame $u^\mu = (1, 0, 0, 0)$. The $n_i^\mu = (0, \boldsymbol{n}_i)$ are spacelike, $n_i^\mu n_{j\mu} = -\delta_{ij}$ and $u^\mu n_{i\mu} = 0$. The unit vectors $\boldsymbol{n}_i$ may be thought of as defined by the spin axes of ideal precession-free gyroscopes (no external torques). Having defined the physical situation in the rest-frame, we now seek a mathematical description of the change or 'transport' of u^μ and n_i^μ, or rather of A^μ (a linear combination of u^μ and the n_i^μ) along $x^\mu(s)$ in an arbitrary reference frame. We surmise that the transport law is a generalisation of parallel transport, and try to achieve our goal with an extra term in (2.28). Accordingly, we define the following operator on $x^\mu(s)$:

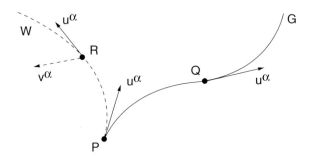

Fig. 8.1. Introducing Fermi-Walker transport. If there is only gravity, a test mass with initial 4-velocity u^α in P moves on a unique geodesic G, but in the presence of additional non-gravitational forces it moves on a non-geodesic worldline W. The 4-velocity $u^\alpha = \mathrm{d}x^\alpha/\mathrm{d}s$ is always tangent to G and to W, and $u^\alpha u_\alpha = 1$ (as always). Parallel transport $\mathrm{D}A^\alpha/\mathrm{D}s = 0$ along G carries $u^\alpha(P)$ over into $u^\alpha(Q)$ because G is a geodesic. But parallel transport along W produces some $v^\alpha(R) \neq u^\alpha(R)$. We seek a generalised (Fermi-Walker) transport law $\delta A^\alpha/\delta s = 0$ that carries u^α over into itself and preserves the value of the inner product $A^\alpha B_\alpha$ of two vectors along an arbitrary worldline.

$$\frac{\delta A^\mu}{\delta s} \equiv \frac{\mathrm{D}A^\mu}{\mathrm{D}s} - K^\mu{}_\alpha A^\alpha \ . \tag{8.1}$$

$\mathrm{D}/\mathrm{D}s$ is the operator (2.26) for parallel transport. We lower the index on the right hand side of (8.1) by multiplying with $g_{\nu\mu}$. The result is $\mathrm{D}A_\nu/\mathrm{D}s - K_\nu{}^\alpha A_\alpha$ (see exercise), and thus we define for covariant vectors

$$\frac{\delta A_\nu}{\delta s} \equiv \frac{\mathrm{D}A_\nu}{\mathrm{D}s} - K_\nu{}^\alpha A_\alpha \ , \tag{8.2}$$

where $\mathrm{D}A_\nu/\mathrm{D}s$ is now given by (2.27). The transport law would then be

$$\frac{\delta A^\mu}{\delta s} = 0 \quad \text{or} \quad \frac{\delta A_\nu}{\delta s} = 0 \ , \tag{8.3}$$

for contravariant and covariant vectors, respectively. With the help of (8.1) and (2.26) we obtain

$$\frac{\delta A^\mu}{\delta s} \equiv \frac{\mathrm{d}A^\mu}{\mathrm{d}s} - \left(K^\mu{}_\nu - \Gamma^\mu{}_{\nu\sigma} u^\sigma \right) A^\nu = 0 \ . \tag{8.4}$$

This is the explicit form of the so called *Fermi-Walker transport law* for a contravariant vector. In order to be able to handle tensors of higher rank we define for two vectors X and Y, conform relation (2.44):

$$\frac{\delta}{\delta s} XY = \frac{\delta X}{\delta s} Y + X \frac{\delta Y}{\delta s} \ . \tag{8.5}$$

We now proceed to determine the tensor $K^{\mu\nu}$. The inner product $A^\mu B_\mu$ of two vectors A^μ and B^μ (i.e. two linear combinations of u^μ and the

8.1 Transport of accelerated vectors

n_i^μ) is constant in the local rest-frame. But $A^\mu B_\mu$ is scalar and therefore one and the same constant in all frames. This implies according to (2.47) that $DA^\mu B_\mu/Ds = dA^\mu B_\mu/ds = 0$, though DA^μ/Ds and DB^μ/Ds in general do not vanish since they are not parallel-transported. We elaborate $0 = \delta(A^\mu B_\mu)/\delta s \equiv (\delta A^\mu/\delta s)B_\mu + A^\mu(\delta B_\mu/\delta s)$:

$$0 = A^\mu \frac{DB_\mu}{Ds} + B_\mu \frac{DA^\mu}{Ds} - A^\mu K_\mu{}^\alpha B_\alpha - B_\mu K^\mu{}_\alpha A^\alpha$$

$$= \frac{D}{Ds}(A^\mu B_\mu) - K^{\mu\alpha} A_\mu B_\alpha - K^{\mu\alpha} A_\alpha B_\mu$$

$$= -(K^{\mu\alpha} + K^{\alpha\mu}) A_\mu B_\alpha . \qquad (8.6)$$

It follows that $K^{\mu\nu}$ must be antisymmetric, $K^{\mu\alpha} = -K^{\alpha\mu}$. It seems natural to expect that $K^{\mu\alpha}$ depends on the 4-velocity, and therefore we try

$$K^{\mu\nu} = a^\mu u^\nu - u^\mu a^\nu , \qquad (8.7)$$

for a certain vector a^μ. A component of a^μ parallel to u^μ does not contribute to (8.7), so we may impose without restriction that

$$a^\mu u_\mu = 0 , \qquad (8.8)$$

and then we also have that

$$K^{\mu\nu} u_\nu = a^\mu . \qquad (8.9)$$

The unknown vector a^μ may be found by requiring that u^μ obey the transport law $\delta u^\mu/\delta s = 0$. With the help of (8.1), (8.8) and (8.12) we get:

$$0 = \frac{Du^\mu}{Ds} - (a^\mu u_\alpha - u^\mu a_\alpha) u^\alpha = \frac{Du^\mu}{Ds} - a^\mu , \qquad (8.10)$$

because of (8.8) and $u_\alpha u^\alpha = 1$. Consequently:

$$a^\mu = \frac{Du^\mu}{Ds} . \qquad (8.11)$$

By comparing with (3.60) we see that a^ν is equal to the non-inertial acceleration f^μ of P divided by $m_0 c^2$.

One might object that expression (8.7) is not the most general choice, and that

$$K^{\mu\nu} = a^\mu u^\nu - u^\mu a^\nu + H^{\mu\nu} \qquad (8.12)$$

with antisymmetric $H^{\mu\nu}$ would also satisfy the requirements. We now show that $H^{\mu\nu} = 0$ implies the absence of any rotation of spatial vectors in the local rest-frame, hence absence of external torques. To that end we study the change of a purely spatial vector n^μ in the local rest-frame, where $n^\mu = (0, \boldsymbol{n})$ and $u^\mu = (1,0,0,0)$, so that $n^\mu u_\mu = g_{\mu\nu} n^\mu u^\nu = \eta_{\mu\nu} n^\mu u^\nu = 0$, as before. The Christoffel symbols are also zero, and Fermi-Walker transport $\delta n^\mu / \delta s = 0$ implies

$$\frac{dn^\mu}{ds} = (a^\mu u_\nu - u^\mu a_\nu) n^\nu = -u^\mu a_\nu n^\nu . \tag{8.13}$$

It follows that $dn^i/ds = 0$: the instantaneous rate of change of the spatial part of n^μ is zero, so that there is no instantaneous rotation (but there would be one if $H^{\mu\nu} \neq 0$).

This completes the derivation of the *Fermi-Walker transport law* (8.4), with $K^{\mu\nu}$ given by (8.7), (8.11) and $u^\mu = dx^\mu/ds$. It is a differential equation specifying the change of an accelerated vector A^μ on which no torques are exerted in the local rest-frame. We note the following:

(1). The middle term on the right hand side of (8.4) is of special-relativistic origin. In SR the Γ's are zero (in rectangular co-ordinates) but $K^{\mu\nu} \neq 0$. This term is responsible for the Thomas precession.

(2). The last term in (8.4) is a general-relativistic effect. If the only force is gravity, then x^μ is a geodesic $\rightarrow Du^\mu/Ds = 0 \rightarrow a^\mu = 0 \rightarrow K^{\mu\nu} = 0$. And in that case eq. (8.4) is identical to parallel transport. One of the consequences is the geodesic precession. Any additional (non-inertial) force causes an extra Thomas-like precession.

Exercise 8.1: We are using a spacelike unit vector n^μ with $n^\mu n_\mu = -1$. Negative lengths, how is that again?

Hint: Very simple. For example, in the local rest-frame $n^\mu = (0, n^1, n^2, n^3)$ and $n_\mu = \eta_{\mu\nu} n^\nu = (0, -n^1, -n^2, -n^3)$. The value of the scalar $n^\mu n_\mu = -|\boldsymbol{n}|^2 = -1$ is invariant.

Exercise 8.2: Prove the statement between (8.1) and (8.2).

Hint: § 2.6: $g_{\nu\mu} DA^\mu/Ds = g_{\nu\mu} A^\mu{}_{;\sigma} u^\sigma = (g_{\nu\mu} A^\mu)_{;\sigma} u^\sigma = A_{\nu;\sigma} u^\sigma = DA_\nu/Ds$. Furthermore, $g_{\nu\mu} K^\mu{}_\alpha A^\alpha = K_{\nu\alpha} A^\alpha = K_\nu{}^\alpha A_\alpha$.

Exercise 8.3: Show that $a^\mu u_\mu$ is indeed zero.

Hint: $a^\mu u_\mu = \tfrac{1}{2} u_\mu Du^\mu/Ds + \tfrac{1}{2} u_\mu Du^\mu/Ds = \tfrac{1}{2} D(u_\mu u^\mu)/Ds = 0$. This last step requires that $u_\mu Du^\mu/Ds = u^\mu Du_\mu/Ds$. See previous exercise for inspiration.

8.2 Thomas precession

This is a problem from SR, and the qualitive explanation has already been given in § 1.1. An electron moves in a circular orbit in the x^1, x^2 plane. Spacetime is flat and we use Cartesian co-ordinates so that all Γ's are zero. According to (8.4), Fermi-Walker transport of the spin vector s^μ is described by

$$\frac{ds^\mu}{d\tau} = c K^\mu{}_\nu s^\nu , \qquad (8.14)$$

because $d/ds = (1/c)d/d\tau$. To determine $K^{\mu\nu}$ we analyse the circular motion of the electron and take

$$x^1 = r \cos \omega\tau ; \qquad x^2 = r \sin \omega\tau ; \qquad x^3 = 0 , \qquad (8.15)$$

from which

$$\left.\begin{aligned} u^1 &= c^{-1} dx^1/d\tau = -(\omega r/c) \sin \omega\tau ; \\ u^2 &= (\omega r/c) \cos \omega\tau ; \\ u^3 &= 0 . \end{aligned}\right\} \qquad (8.16)$$

Here ω is the orbital frequency measured in the proper time of the electron; u^0 can be obtained from $1 = u^\mu u_\mu = \eta_{\mu\nu} u^\mu u^\nu = (u^0)^2 - (u^1)^2 - (u^2)^2$:

$$u^0 = \sqrt{1 + (\omega r/c)^2} = \text{constant} , \qquad (8.17)$$

and this serves to find the relation between proper time τ and laboratory time t, because $u^0 = \gamma = 1/\sqrt{1-\beta^2}$, see (3.23). Therefore $\omega\tau = \omega t/\gamma \equiv \Omega t$, where Ω = orbital frequency in laboratory time:

$$\left.\begin{aligned} \gamma &= \sqrt{1 + (\omega r/c)^2} ; \qquad \Omega = \omega/\gamma ; \\ \frac{1}{\omega} \frac{d}{d\tau} &= \frac{1}{\Omega} \frac{d}{dt} . \end{aligned}\right\} \qquad (8.18)$$

Since the Γ's are zero, we infer from (8.11) and (2.26) that $a^\mu = Du^\mu/Ds = c^{-1} du^\mu/d\tau$. We may now write (8.14) as:

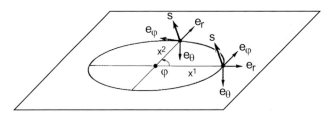

Fig. 8.2. Geodesic precession of the vector s analysed in the equatorial plane $\theta = \pi/2$ of the rotating reference frame e_r, e_θ, e_φ.

$$\frac{ds^\mu}{d\tau} = c\left(a^\mu u_\nu - u^\mu a_\nu\right) s^\nu = -u^\mu \frac{du_\nu}{d\tau} s^\nu \; ; \qquad (8.19)$$

$$u_\nu s^\nu = 0 \; , \qquad (8.20)$$

because we know that $u_\nu s^\nu$ is constant (Fermi-Walker transport), and that $u^\mu = (1, 0, 0, 0)$ and $s^\mu = (0, \boldsymbol{s})$ in the local rest-frame, so that $u_\nu s^\nu = \eta_{\nu\alpha} u^\alpha s^\nu = 0$. Because $u_\nu s^\nu$ is invariant (8.20) holds in any frame. Since $u^3 = 0$ we conclude from (8.19) that $ds^3/d\tau = 0$, or

$$\frac{ds^3}{dt} = 0 \; . \qquad (8.21)$$

Apparently, the z-component of the spin is constant. The behaviour of s^0 follows from (8.20): $0 = \eta_{\nu\sigma} u^\sigma s^\nu = u^0 s^0 - u^1 s^1 - u^2 s^2 \to s^0 = (u^1 s^1 + u^2 s^2)/u^0$. However, s^0 has no physical meaning – its 'function' is to ensure that $u_\nu s^\nu$ and $s^\nu s_\nu$ are constant. The physics is in the behaviour of s^1 and s^2. With (8.16) and $u_i = \eta_{i\nu} u^\nu = -u^i$ we obtain:

$$\frac{d}{d\tau}\begin{pmatrix} s^1 \\ s^2 \end{pmatrix} = \frac{\omega^3 r^2}{c^2} \begin{pmatrix} \sin\omega\tau \cos\omega\tau & \sin^2 \omega\tau \\ -\cos^2 \omega\tau & -\sin\omega\tau \cos\omega\tau \end{pmatrix} \begin{pmatrix} s^1 \\ s^2 \end{pmatrix} . \qquad (8.22)$$

Express this in laboratory time with (8.18):

$$\frac{d}{dt}\begin{pmatrix} s^1 \\ s^2 \end{pmatrix} = (\gamma^2 - 1)\Omega \begin{pmatrix} \sin\Omega t \cos\Omega t & \sin^2 \Omega t \\ -\cos^2 \Omega t & -\sin\Omega t \cos\Omega t \end{pmatrix} \begin{pmatrix} s^1 \\ s^2 \end{pmatrix} . \qquad (8.23)$$

Exercise 8.4: Verify that the solution of (8.23) with initial values $s^1(0) = s$ and $s^2(0) = 0$ is given by

$$s^1 = \tfrac{1}{2}s\left[(1+\gamma)\cos(1-\gamma)\Omega t + (1-\gamma)\cos(1+\gamma)\Omega t\right];$$
$$s^2 = \tfrac{1}{2}s\left[(1+\gamma)\sin(1-\gamma)\Omega t + (1-\gamma)\sin(1+\gamma)\Omega t\right]. \tag{8.24}$$

Expand for $\beta \ll 1$:

$$s^1 \simeq s\left[\cos\tfrac{1}{2}\beta^2\Omega t - \tfrac{1}{4}\beta^2\cos 2\Omega t\right];$$
$$s^2 \simeq -s\left[\sin\tfrac{1}{2}\beta^2\Omega t + \tfrac{1}{4}\beta^2\sin 2\Omega t\right]. \tag{8.25}$$

Verify that the first terms in (8.24) and (8.25) correspond to a rotation of the spin vector with a frequency

$$\Omega_{\text{Thomas}} = (\gamma - 1)\Omega_{\text{orbit}} \simeq \tfrac{1}{2}\beta^2\,\Omega_{\text{orbit}}, \tag{8.26}$$

with $\beta \simeq wr/c \simeq \Omega r/c \ll 1$. The sense of the rotation is opposite to the orbital rotation. Both second terms in (8.25) describe a small, fast modulation that averages to zero.

8.3 Geodesic precession

In § 4.4 we analysed the motion of a test mass moving in the Schwarzschild metric, and found, among other things, that the orbit precesses. This precession of the perihelium is not the only GR effect. If the test mass behaves as a vector, as for example a gyroscope, the (spin) vector will also perform a precession, even when no torque is exerted. We shall now derive this so-called *geodesic precession*. Because the body moves along a geodesic we have that $K^{\mu\nu} = 0$, in which case (8.4) reduces to the equation for parallel transport:

$$\frac{ds^\mu}{d\tau} + c\Gamma^\mu_{\nu\sigma}u^\sigma s^\nu = 0. \tag{8.27}$$

Here s^ν is the unit vector along the spin axis. The following analysis is a sequel of § 4.3, and we shall employ the notation we used there. The 4-velocity $u^\mu = dx^\mu/ds$ is given by:[1]

---

[1] For the geodesic precession the rotation of the Earth is irrelevant. So although Fig. 8.3 suggests otherwise, the satellite may be taken move on the equator $r, \theta =$ constant of the Schwarzschild metric.

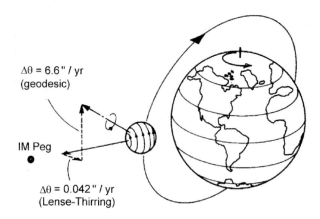

Fig. 8.3. A gyroscope orbiting a rotating mass like the Earth and moving only under the influence of gravity should exhibit a geodesic precession and a Lense-Thirring precession. The experiment is now in progress in the Gravity Probe B satellite, launched in April 2004 into a polar orbit of 640 km altitude. The star IM Pegasi (HR 8703) serves as the pointing reference. See text for details. Adapted from: *Near Zero*, J.D. Fairbank et al. (eds.) Freeman & Co (1988).

$$u^\mu = (c\dot{t}, \dot{r}, \dot{\theta}, \dot{\varphi}) = (c\dot{t}, 0, 0, h/r^2)$$

$$= \left(\left\{1 - \frac{3r_s}{2r}\right\}^{-1/2}, 0, 0, \frac{1}{r^2}\left\{\frac{rr_s/2}{1 - 3r_s/2r}\right\}^{1/2}\right). \qquad (8.28)$$

At the second = sign we choose a circular orbit: r = constant and $\theta = \pi/2$, and we have used (4.34) as well. The last expression in (8.28) follows immediately from (4.32) and (4.45). Next we write out (8.27) explicitly, and obtain the following equations (see exercises):

$$s^0 = \frac{\sqrt{rr_s/2}}{1 - r_s/r} s^3 ; \qquad (8.29)$$

$$\frac{ds^1}{d\tau} = \frac{c}{r}\sqrt{rr_s/2}\sqrt{1 - 3r_s/2r}\, s^3 ; \qquad (8.30)$$

$$\frac{ds^2}{d\tau} = 0 ; \qquad (8.31)$$

$$\frac{ds^3}{d\tau} = -\frac{c}{r^3}\left(\frac{rr_s/2}{1 - 3r_s/2r}\right)^{1/2} s^1 . \qquad (8.32)$$

8.3 Geodesic precession

Take $d/d\tau$ of (8.32) and eliminate $ds^1/d\tau$ with (8.30):

$$\frac{d^2 s^3}{d\tau^2} + \frac{c^2 r_s}{2r^3} s^3 = 0, \tag{8.33}$$

and it is easy to verify that the same equation holds for s^1. The solution with initial value $s^3(0) = 0$ is:

$$\left.\begin{array}{l} s^3 = s^\varphi = -s\sin\omega\tau \; ; \\ s^1 = s^r = sr\sqrt{1 - 3r_s/2r} \, \cos\omega\tau \; ; \\ s^2 = s^\theta = \text{constant} \; , \end{array}\right\} \tag{8.34}$$

where

$$\omega = c\left(\frac{r_s}{2r^3}\right)^{1/2} = \left(\frac{GM}{r^3}\right)^{1/2}. \tag{8.35}$$

The geodesic precession is a consequence of the fact that the precession frequency ω is a little smaller than the orbital frequency, which is equal to

$$\frac{2\pi}{\Delta\tau} = c\left(\frac{r_s}{2r^3}\right)^{1/2}\left(1 - \frac{3r_s}{2r}\right)^{-1/2}. \tag{8.36}$$

Here we have used expression (4.46) for the orbital period $\Delta\tau$. After each orbit the spin vector has rotated over an angle of

$$\omega\Delta\tau = 2\pi\sqrt{1 - 3r_s/2r} \; . \tag{8.37}$$

The spin vector precesses about an axis orthogonal to the orbital plane, but the major part of the precession is caused by the fact that the reference frame itself rotates over an angle of 2π, see Fig. 8.2. When viewed from a non-rotating frame the precession angle per orbit equals

$$\delta\psi = 2\pi\left(1 - \sqrt{1 - 3r_s/2r}\right) \simeq \frac{3\pi r_s}{2r} \; . \tag{8.38}$$

Actually, we must still transform to co-ordinate time, but that gives rise to a correction of higher order. The precession has the same sense of rotation as the orbit. The physical origin of the precession is that a vector that is parallel transported constantly changes its direction, due to the curvature of space-time, see § 2.4. This is visible as a small secular angular rotation. The effect of geodesic precession has been observed in the binary pulsar PSR 1913+16.[2] What if the central object rotates? In that case its exterior metric is replaced by the Kerr metric (in good approximation), and frame-dragging (§ 6.5) induces an additional precession, called the Lense-Thirring effect. The LAGEOS satellites have confirmed the Lense-Thirring effect due to the rotation of the Earth with a precision of 10%.[3]

---
[2] Weisberg, J.M. and Taylor, J.H., *Ap. J.* **576** (2002) 942.
[3] Ciufolini, I. and Pavlis, E.C., *Nature* **431** (2004) 958.

Fig. 8.4. Inside view of a gyroscope of Gravity Probe B and its housing. The rotor has a diameter of 3.8 cm, and is made of fused quartz coated with niobium. Image credit: Don Harley.

8.4 Gravity Probe B

The technology for high-precision measurements of the geodesic precession and the Lense-Thirring effect has been developed in the USA from the beginning of the 1960s. The outcome of this long development programme, the longest in NASA's history to date[4], is Gravity Probe B, launched on April 20, 2004, see Fig. 8.3. The satellite carries 4 precision gyroscopes. The geodesic precession is only 6.6'' per year, and the Lense-Thirring precession is much smaller: 0.04'' per year. The gyros consist of quartz rotors coated with superconducting niobium, suspended in an electrostatic field, see Figs. 8.4 and 8.5. The rotation (about 70 Hz) induces a London magnetic moment that generates a magnetic dipole field aligned with the spin axis. Its direction, and hence the orientation of the spin axis can be measured with high precision.[5] There are many experimental complications. For example, any parasitic torque will cause the gyroscope to precess, and any non-inertial acceleration induces an extra Thomas precession. By using a drag-free satellite that literally follows the inertial motion of one of the the gyroscopes, the residual acceleration will

---

[4] For the programmatic and scientific issues involved see Reichhardt, T., *Nature* *426* (2003) 380.
[5] For more details see *Near Zero*, J.D. Fairbank et al. (eds.), Ch. 6.1–6.3 (Freeman & Co 1988); for theoretical aspects see Will (1993) p. 208; Gravity Probe B website: http://einstein.stanford.edu/

Fig. 8.5. Gravity Probe B carries four gyroscopes, mounted in a single quartz bloc, a prototype of which is shown here. The pointing telescope (not shown) is attached to the flange at the lower end. The whole unit is placed in a much larger helium dewar. Image credit: Gravity Probe B, Stanford University.

be at the 10^{-11}g level. The gyroscopes have a pointing stability of better than 5×10^{-4} arcseconds over a period of a year!

In closing, we draw attention to two issues. The first is the fact that the precession angle (8.38) is independent of the spin rate of the gyroscope, and the same is true for the Lense-Thirring precession.[6] This is a reminder of the physics involved: both effects are a consequence of parallel transport of a vector in the Schwarzschild or Kerr metric. The nature of the vector is immaterial, and so is the existence of mass currents in the gyroscope. A gyroscope is for many reasons by far the best technical solution, but a non-rotating pencil would, as a matter of principle, also do very well – if one could eliminate all parasitic forces and moments.

---

[6] See Will (1993) p. 210.

166   8 Fermi-Walker Transport

The second issue is the pointing reference. Stellar parallaxes and proper motions are generally larger than the accuracy required for Gravity Probe B. Therefore the only suitable pointing references are quasars. Quasars are distant powerful radio sources that are believed to constitute the best available inertial reference frame. But quasars are too dim in visible light for the small pointing telescope (aperture 14 cm). Therefore a relatively bright star had to be found, that is also a strong radio point source, and located sufficiently close to a few reference quasars to permit measuring the relative positions with the method of Very Long Baseline Interferometry (VLBI). The outcome is IM Peg (HR 8703). The proper motion and parallax of IM Peg with respect to the quasars have been accurately measured in a VLBI programme extending over many years. In this way the orientation of the gyroscopes can ultimately be related to the quasar reference frame.

Exercise 8.5: Write down the explicit expression for the Christoffel symbols necessary to elaborate (8.27).

Hint: From (4.29): $2\nu = -2\lambda = \log(1 - r_s/r)$; furthermore $\theta = \pi/2$. Result:

$$(4.10): \quad \Gamma^1_{00} = \frac{r_s}{2r^2}(1 - r_s/r); \qquad \Gamma^1_{33} = -r(1 - r_s/r).$$

$$(4.11): \quad \Gamma^2_{12} = \frac{1}{r}; \qquad \Gamma^2_{33} = 0.$$

$$(4.12): \quad \Gamma^3_{13} = \frac{1}{r}; \qquad \Gamma^3_{23} = 0.$$

Exercise 8.6: Show that $u^\mu s_\mu = 0$ holds here as well, just as in the case of Thomas precession. Use that to derive (8.29).

Hint: $0 = g_{\mu\nu}u^\mu s^\nu = g_{00}u^0 s^0 + g_{33}u^3 s^3$; use (4.29) and $\theta = \pi/2$.

Exercise 8.7: Prove now eqs. (8.30) to (8.32).

Hint: Insert the Γ's, and u^0 and u^3 from (8.28), and use (8.29).

Exercise 8.8: Show that a gyroscope in orbit around the Earth at an altitude of 650 km has a geodesic precession of $6.6''$ per year.

Hint: (8.38) + Keplerian orbit $\to 3(GM_a)^{3/2}/(2c^2 r^{5/2})$ rad s^{-1}, etc.

Exercise 8.9: We wish to compare the precession amplitudes along e_r and along e_φ, see Fig. 8.2. But that is not possible as s^1 and s^3 in (8.34) have different dimensions. How is that?

Hint: Physical lengths follow from (3.7)! Amplitudes along the r-direction: $dl_r^2 = -g_{rr}(s^r)^2 \simeq r^2(1 - r_s/2r)s^2$; $dl_\varphi^2 = -g_{\varphi\varphi}(s^\varphi)^2 = r^2 s^2$.

Exercise 8.10: Does a linearly accelerated electron experience any Thomas-like effect?

Hint: Take the 1-axis in the direction of the acceleration, then $u^2 = u^3 = 0$. According to (8.19) only s^0 and s^1 will change. To see what actually happens, assume that the electron experiences a constant acceleration a, and use that $x^1 = (c^2/a)\cosh(a\tau/c) + \text{const}$, $x^0 = ct = (c^2/a)\sinh(a\tau/c)$, see Rindler (2001), so that $u^0 = \cosh(a\tau/c)$ and $u^1 = \sinh(a\tau/c)$. Now solve (8.19).

9

The Robertson-Walker Metric

Cosmology is the science that addresses the large-scale structure and evolution of the universe. Why would that require the framework of GR? Because the universe as a whole may be regarded as a compact object – in the sense that its 'radius' R is comparable to its Schwarzschild radius! From (4.28) we see that $R \sim r_\mathrm{s}$ if $R \sim 2GM/c^2$. Now take $M = (4\pi R^3/3)\,\rho$ and use for R the Hubble radius c/H_0. This is the distance where the expansion speed becomes formally c according to the primitive Hubble law (9.4). Result:

$$\rho \sim \frac{3H_0^2}{8\pi G} \equiv \rho_\mathrm{c} \;. \tag{9.1}$$

The density of the universe should be comparable to the *critical density* ρ_c, a concept that will be explained later. And Table 9.2 shows that this is indeed the case. This argument, simple as it may be, does indicate that only description in terms of GR may be expected to produce meaningful results. In this chapter we shall review the most important observations, the form of the metric, the spatial structure of the universe, and the equation of motion for the scale factor S.

9.1 Observations

On a cosmological scale the smallest relevant unit is a galaxy. Galaxies occur in aggregates called groups. Our own galaxy and the large spiral galaxy M31 in Andromeda (distance 770 kpc) are the two biggest members of the Local Group, which has approximately 40 members. Groups in turn form clusters. Table 9.1 gives some characteristic sizes and distances. From redshift surveys, Fig. 9.1, it is apparent that matter is distributed in a filamentary fashion, in concentrations of widely varying size, with 90% of the matter located in walls, strings and sheets that occupy a relative volume of the order of 10%, while 90% of space is virtually empty ('voids'). During the past century there has been an intense debate on the relative densities of various forms of matter

Table 9.1. Characteristic length scales

galaxy	1 – 50	kpc
group	1	Mpc
cluster	10	Mpc
distances between clusters	100	Mpc
most distant clusters	3	Gpc
distance to quasars	4.5	Gpc
distance to horizon	10	Gpc

in the universe. This debate has recently culminated in the publication of the results of several surveys among which those of the WMAP mission,[1] see Table 9.2. The densities are expressed in terms of the critical density ρ_c:

$$\rho_c = \frac{3H_0^2}{8\pi G} = 1.88 \times 10^{-29} h^2 \quad \mathrm{g\,cm^{-3}}$$

$$\simeq 10^{-29} \mathrm{\,g\,cm^{-3}}, \tag{9.2}$$

and h is the Hubble constant in units of 100 km s^{-1} Mpc^{-1}, see (9.5). The present matter energy density is

$$\epsilon_m = \Omega_m \rho_c c^2 = 2.4 \times 10^{-9} \mathrm{\,erg\,cm^{-3}}. \tag{9.3}$$

Only ~ 2% of all matter in the universe can actually be seen because it is luminous. The remaining 98% is *dark*, where dark traditionally means *optically dark*. It is only indirectly visible through the gravity it exerts, for example in the rotation curves of galaxies, and in the velocity distribution of galaxies in clusters. To prevent the latter from flying apart they should contain a lot more matter than we see. Dark matter consists partly of baryons (mainly hot gas in and between clusters, but also brown dwarfs, old white dwarfs, etc.). About 80-90% of all baryons is dark. Some of this baryonic dark matter is now beginning to be seen in UV and X-rays.[2] But baryons comprise only a small fraction of all dark matter. Non-baryonic dark matter consists of weakly interacting massive particles (WIMPs) of unknown identity.[3] The largest constituent in Table 9.2 is dark *energy* (not be confused with dark *matter*), associated with the cosmological constant, whose nature is not understood. It seems unlikely that the debate on the values in Table 9.2 has

---

[1] Wilkinson Microwave Anisotropy Probe, see Bennett, C.L. et al., *Ap. J. S. 148* (2003) 1, and following papers.
[2] Nicastro, F. et al., *Nature 433* (2005) 495 and *421* (2003) 719; Kaastra, J.S. et al., *A&A 397* (2003) 445.
[3] WMAP excludes the possibility that they are massive neutrinos, since it finds $\Omega_\nu < 0.015$. Current contenders are the neutralino (the lightest supersymmetric particle) and axions. For detection attempts see Sumner, T.J., *Living Rev. Relativity 5* (2002) 4.

Table 9.2. Relative densities of matter and energy in the universe [a]

Type	$\Omega = \rho/\rho_c$ [b]	Comment
Matter (Ω_m)	0.27 ± 0.04	consists of 3 components
- luminous baryons	0.006 ± 0.003	} total baryons:
- dark baryons	0.038 ± 0.003	} $\Omega_b = 0.044 \pm 0.004$
- non-baryonic dark matter	0.23 ± 0.04	unknown WIMP [c]
Dark energy (Ω_Λ)	0.73 ± 0.04	unknown origin, § 9.5
Total ($\Omega_m + \Omega_\Lambda$)	1.02 ± 0.02	geometry of universe is flat

[a] See [1] and Fukugita, M. and Peebles, P.J.E., *Ap. J. 616* (2004) 643;
[b] $\rho_c = 3H_0^2/8\pi G = 1.88 \times 10^{-29} h^2 \simeq 10^{-29}$ g cm^{-3};
[c] WIMP = Weakly Interacting Massive Particle.

really ended. The WMAP results confirmed the prevailing theoretical prejudice of the day and were quickly canonized. We shall follow suit, but note that the future may hold surprises.

An important observation is that the universe is *isotropic*. The distribution of matter in space is statistically the same in all directions, also as a function of distance, i.e. within redshift subclasses. There are obvious *evolution effects*. The morphology of the systems changes gradually with distance, and at large distances we see only quasars, objects $10^2 - 10^3$ times brighter than the average nearby galaxy. The Hubble Deep Field observations illustrate clearly that the universe did look quite different in the past.[4] Hubble demonstated in 1929 that the universe expands. All galaxies move away from us on average with a velocity proportional to the distance, but independent of direction. This universal expansion is referred to as the Hubble flow:

$$v = H_0 d, \qquad (9.4)$$

with

$$H_0 = 100\,h \text{ km s}^{-1} \text{ Mpc}^{-1} \quad \text{and} \quad h = 0.71 \pm 0.04, \qquad (9.5)$$

as measured by WMAP.[1] In fact one measures a redshift z rather than a velocity. The precise meaning of v and d in (9.4) will be explained in § 11.3. In physical units:

$$H_0 = (2.3 \pm 0.1) \times 10^{-18} \text{ s}^{-1}. \qquad (9.6)$$

The peculiar velocities of the systems, i.e. the deviations from the Hubble flow, are generally small, $\lesssim 500$ km s^{-1}. The Hubble flow is 'cold' and this is because the universe cools adiabatically as it expands.

---

[4] Driver, S.P. et al., *Ap. J. 496* (1998) L93; Ferguson, H.C. et al., *A.R.A.A. 38* (2000) 667.

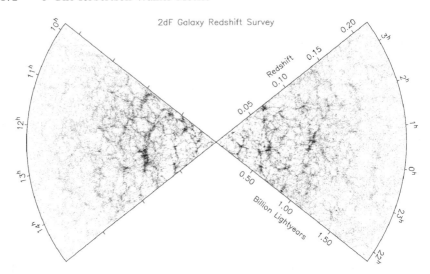

Fig. 9.1. The 2dF galaxy redshift survey comprises about 220,000 galaxies and shows that the distribution of matter in the universe is homogeneous at large, but clumpy on smaller scales. The left slice measures $75° \times 10°$ and is located in the Northern galactic hemisphere, the right slice is $80° \times 15°$ near the galactic South pole. Picture taken from the 2dFGRS image gallery. See Colless, M. et al., *M.N.R.A.S.* **328** (2001) 1039; Peacock, J.A. et al., *Nature* **410** (2001) 169.

Fig. 9.2. 'Baby picture' of the universe: WMAP image of the Cosmic Microwave Background at $\lambda = 3.2$ mm. Monopole and dipole have been subtracted but the galactic foreground has not. Color coding: black = $-200\,\mu K$, red = $+200\,\mu K$. The minute temperature variations indicate clustering of matter in the early universe. This is analysed in §§ 10.4 and 11.4. From Bennett, C.L. et al., *Ap. J. S.* **148** (2003) 1.

In addition to matter, the universe contains all kinds of radiation, of which the *cosmic microwave background* (CMB) has by far the largest energy density. This radiation had been predicted by Gamov and coworkers in 1948 ($T \sim 5\,\text{K}$), as a remnant of a hot early stage of the universe, and was discovered by Penzias and Wilson in 1965. Observations of the COBE satellite have shown that the spectrum is to high accuracy a thermal Planck spectrum in the wavelength range from 10 cm to 0.1 mm with a maximum at $\lambda \sim 2$ mm. Temperature and energy density are:

$$T = 2.725 \pm 0.002 \text{ K} ; \tag{9.7}$$

$$\epsilon_\text{r} = \frac{4\sigma}{c} T^4 \simeq 4.19 \times 10^{-13} \text{ erg cm}^{-3} . \tag{9.8}$$

The CMB has a dipole anisotropy of $|\Delta T| \simeq 3.35 \pm 0.02$ mK, and this is interpreted as a Doppler shift due to the velocity of the solar system of 369 km s^{-1} towards galactic co-ordinates $(\ell, b) = (264°, 48°)$ with respect to the frame defined by radiation.[5] After subtraction of the dipole component the CMB is highly isotropic, $\Delta T / T \simeq 10^{-5}$ on angular scales $\gtrsim 7°$ (COBE), and WMAP has improved that to angular scales $\gtrsim 0.2°$.

On theoretical grounds there should also exist a neutrino background with a temperature and energy density comparable to those of the CMB (§ 12.2). If we add that to (9.8), the total radiation energy density is:

$$\epsilon_\text{r} \simeq 7 \times 10^{-13} \text{ erg cm}^{-3} . \tag{9.9}$$

The conclusion seems obvious: the universe is a space of vast expanse, extremely cold (2.7 K), and to our standards almost empty. It is isotropic and evolves with time. An important aspect of the evolution is the expansion, which should have begun approximately a *Hubble time* $H_0^{-1} \simeq 14$ Gyr ago. The microwave background is a remnant of a hot early stage of the universe, called the *Big Bang*. For an extensive discussion of the observations see for example Peebles (1993) and Peacock (1999). We return to observational issues in Ch. 11.

---

[5] This velocity in turn induces an aligned dipole asymmetry in the observed matter distribution, see Blake, C. and Wall, J., *Nature 416* (2002) 150.

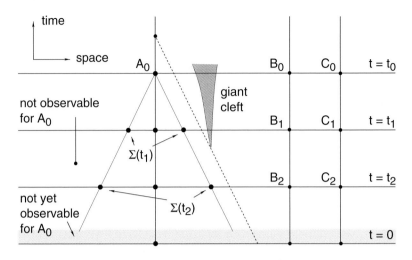

Fig. 9.3. Co-ordinate picture of the spacetime of the universe. Our present position is A_0, and shown are our past light-cone, the worldlines of a few galaxies (vertical lines), and a hypothetical inhomogeneity ('giant cleft') that we might get to see in the future.

9.2 Definition of co-ordinates

Fig. 9.3 shows a spacetime diagram of the universe. We (A_0) are only able to see events located on our past light-cone. We experience our light-cone as a series of nested, ever larger concentric spherical shells around us, showing an increasingly younger section of the universe. Because of the observed isotropy, each shell $\Sigma(t_i)$ must be on average homogenous. Due to our limited technological capabilities we have not yet been able to detect signals from the early universe, i.e. from the most distant shells. We now make an assumption about the part of spacetime that is outside our past light-cone and therefore unobservable. To that end we use the *cosmological principle*, which states that we (A_0) occupy no special position in the universe, and that other observers B_0, C_0 in Fig. 9.3 see on average the same universe as we do. Hence if we translate our light-cone sideways, the aspect of the shells $\Sigma(t_i)$ would not change, apart from statistical fluctuations (the so-called cosmic variance). The implication is that every subspace $t =$ constant is isotropic and homogenous on average. Cosmological principle and the isotropy of the universe imply that it is homogeneous.

We now come to the definition of *rest* ($x^i =$ constant). We are free to adopt any definition we like, but there is one that stands out as very natural: a test mass is at rest if it does not move with respect to the Hubble flow. That means that the spatial co-ordinates of galaxies are constant (we shall

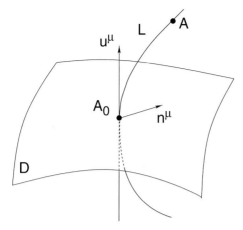

Fig. 9.4. Introducing Gaussian co-ordinates in the spacetime of the universe. The starting point is a 3-dimensional subspace of spacetime, D, which is spacelike but otherwise arbitrary (see text). The tangent space of $A_0 \in D$ is T.

ignore their peculiar velocities). Their worldlines are straight vertical lines in Fig. 9.3. This figure is a co-ordinate picture, see Fig. 2.1, and contains no information about the geometry (the geometrical picture appears in Fig. 11.2). Due to the expansion the *geometrical* distance between B_0 and C_0 is larger than between B_1 and C_1. It remains possible that the spacetime that we shall see in the future contains huge inhomogeneities, and that the cosmological principle will eventually prove to be incorrect,[6] see Fig. 9.3. Presently, however, the assumption that every subspace $t =$ constant is homogeneous and isotropic is adequate. But it should be clear that very little can be said about the future of the universe without extra assumptions such as the cosmological principle.

We assume that spacetime already possesses co-ordinates and a metric, and we now construct new co-ordinates to simplify the metric. Let the subspace D in Fig. 9.4 be spacelike (but not necessarily of the type $t' =$ constant), i.e. for every vector $n^{\mu'}$ in the tangent space we have $n^{\mu'} n_{\mu'} < 0$. The primes denote the old co-ordinates. Consider an event $A_0 \in D$ with tangent space T. We define a vector $u^{\mu'}$ by requiring $u^{\mu'} n_{\mu'} = 0$ for every $n^{\mu'} \in T$. These $u^{\mu'}$ are unique, apart from an overall factor, and timelike (see exercise). We normalize them as $u^{\mu'} u_{\mu'} = 1$. Next we construct a geodesic L tangent to $u^{\mu'}$ in A_0, and we define new spatial co-ordinates $(\tilde{x}^1, \tilde{x}^2, \tilde{x}^3)$ in D (how we do

---

[6] It is a peculiar fact that our universe appears to be homogeneous on the scale of the Hubble radius c/H_0, but inhomogeneous both on much larger scales (prediction of inflation theory) and on much smaller scales (Fig. 9.1).

that is immaterial). Finally, we assign the following co-ordinates to an event A on L:

$$x^i = \tilde{x}^i \, ;$$
$$x^0 = \text{arc length } s \text{ of } A_0 A \text{ along } L \, . \qquad (9.10)$$

This construction is possible because L is timelike. In this way we have defined well-behaved new co-ordinates $\{x^\mu\}$ as long as the different geodesics L do not intersect. For $\mathrm{d}x^i = 0$ (i.e. along L) we have that $\mathrm{d}s^2 = (\mathrm{d}x^0)^2$, and comparing that to $\mathrm{d}s^2 = g_{\alpha\beta}\,\mathrm{d}x^\alpha \mathrm{d}x^\beta = g_{00}(\mathrm{d}x^0)^2$ we conclude that $g_{00} = 1$ on L. In the new co-ordinates the 4-velocity of a point on L equals

$$u^\mu = \dot{x}^\mu \equiv \frac{\mathrm{d}}{\mathrm{d}s}(s,\tilde{x}^1,\tilde{x}^1,\tilde{x}^3) = (1,0,0,0) \, , \qquad (9.11)$$

and $n^\mu \in T$ is of the form $n^\mu = (0, \boldsymbol{n})$, so that $0 = u^{\mu'} n_{\mu'} = u^\mu n_\mu = g_{\mu\alpha} u^\mu n^\alpha = g_{0i} n^i$. It follows that $g_{0i} = 0$, because n^i is arbitrary. On the subspace D the metric now has the form

$$\mathrm{d}s^2 = (\mathrm{d}x^0)^2 + g_{ik}\,\mathrm{d}x^i \mathrm{d}x^k \, . \qquad (9.12)$$

Exercise 9.2 shows that (9.12) holds everywhere. These co-ordinates are called *Gaussian co-ordinates*, after Gauss who invented them.

The essence of Gaussian co-ordinates is that the worldlines L of a selected set of freely falling test masses are taken as the co-ordinate lines of the new co-ordinate system, and these lines L remain always orthogonal to the subspaces $t = $ constant. Because the derivation is completely general, we may use Gaussian co-ordinates in any physical situation, also for example in the Schwarzschild metric. They are not very convenient in that case, but that is another matter. In cosmology, however, they are very useful. The sections $t = $ constant are snapshots of the homogeneous and isotropic universe, and the selected test masses are the galaxies. Because these are at rest ($\mathrm{d}x^i = 0$) it follows from (9.12) that $\mathrm{d}\tau = \mathrm{d}t$: at any time t all clocks of galaxies tick at the same rate. This must be so because otherwise a subspace $t = $ constant would not be homogeneous. In Gaussian co-ordinates the proper time of any galaxy in this subspace serves as the co-ordinate time t. Since we deal mostly with objects at rest (galaxies), the notion of proper time plays a minor role in cosmology. Proper time is only important when we consider motion with respect to the Hubble flow, as in exercise 9.9.

Exercise 9.1: Prove that $u^{\mu'}$ introduced above (9.10) is unique and timelike.

Hint: Timelike is invariant, so employ the local rest-frame $^-$ of A_0. With 3 independent $\bar{n}^\mu$ one may construct 3 orthonormal spacelike unit vectors; $\bar{u}^\mu$

Exercise 9.2: Prove that (9.12) is valid everywhere.

Hint: Work out (2.34) along the geodesic L with (9.11) $\to \Gamma^\mu_{00} = 0$. Now use (2.24) and $g_{00} = 1$ on $L \to g^{\mu\lambda} g_{\lambda 0,0} = 0$; $\det\{g^{\mu\lambda}\} \neq 0 \to g_{\lambda 0,0} = 0$ on $L \to g_{j0}$ constant on $L \to g_{j0} = 0$ on L (i.e. everywhere, q.e.d.).

9.3 Metric and spatial structure

Due to the expansion the metric will depend on x^0, and that dependence must be the same for every g_{ik}, otherwise anisotropies would develop. Therefore (9.12) can be written as

$$ds^2 = (dx^0)^2 + S(t)^2 a_{ik}\, dx^i dx^k, \tag{9.13}$$

with a_{ik} constant. We may simplify $a_{ik} dx^i dx^k$ further by noting that the space is certainly spherically symmetric around an (arbitrarily chosen) origin. The implications of that have been elaborated as we discussed the Schwarzschild metric, § 4.1. The spatial metrics associated with (4.2) and (9.13) at time t_1 are

$$dl^2 = \begin{cases} e^{2\lambda} dr^2 + r^2 d\Omega^2 \\ -S_1^2\, a_{ik}\, dx^i dx^k, \end{cases} \tag{9.14}$$

where $d\Omega^2 = d\theta^2 + \sin^2\theta\, d\varphi^2$ and $S_1 = S(t_1)$. These two metrics describe the same space, as both are spherically symmetric around the origin. We conclude that $-a_{ik}\, dx^i dx^k$ may also be written as $(e^{2\lambda} dr^2 + r^2 d\Omega^2)/S_1^2$. After a rescaling $S(t)/S_1 \to S(t)$ we find that (9.13) reads

$$ds^2 = (dx^0)^2 - S(t)^2 \left(e^{2\lambda} dr^2 + r^2 d\Omega^2 \right). \tag{9.15}$$

To find $\lambda(r)$ we compute the total curvature $R^i{}_i$ of the subspace t = constant of (9.15) when $S(t) = 1$. This $R^i{}_i$ turns out to be equal to $R^\mu{}_\mu$ from (4.19) with ν = constant (see exercise), or

$$R = 2 \left(\frac{2\lambda'}{r} - \frac{1}{r^2} \right) e^{-2\lambda} + \frac{2}{r^2} = \frac{2}{r^2} \left(1 - \frac{d}{dr} r e^{-2\lambda} \right), \tag{9.16}$$

from which it follows that

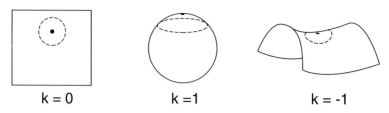

Fig. 9.5. Two-dimensional analogons of a flat, a spherical and a hyperbolic universe. After Berry (1978).

$$\frac{d}{dr} r e^{-2\lambda} = 1 - \tfrac{1}{2} R r^2 . \qquad (9.17)$$

We now argue that R is constant because the space $t = $ constant is homogeneous, and we may integrate:

$$e^{2\lambda} = (1 - \tfrac{1}{6} R r^2 + A/r)^{-1} . \qquad (9.18)$$

The integration constant A should be zero, otherwise the co-ordinates would not be locally flat in $r = 0$. Denoting $R = 6k$ we get:

$$ds^2 = (dx^0)^2 - S(t)^2 \left\{ \frac{dr^2}{1 - kr^2} + r^2(d\theta^2 + \sin^2\theta \, d\varphi^2) \right\} . \qquad (9.19)$$

By a co-ordinate transformation $r \to \tilde{r}$ we may always make k equal to 0, 1 or -1. Henceforth we restrict ourselves to $k = 0, \pm 1$. Robertson and Walker have shown in 1936 that (9.19) is the most general metric of a space-time whose subspaces $t = $ constant are homogeneous and isotropic. Therefore (9.19) is called the *Robertson-Walker metric*.

By means of symmetry arguments we have succeeded to find the metric up to an unknown *scale factor* $S(t)$. The scale factor is determined by the field equations. Before we enter into that we discuss the structure of the spaces defined by (9.19). According to (3.7) the spatial metric is given by:

$$dl^2 = S^2 \left\{ \frac{dr^2}{1 - kr^2} + r^2(d\theta^2 + \sin^2\theta \, d\varphi^2) \right\} . \qquad (9.20)$$

It is important to realise that because of the homogeneity all points in the spaces defined by (9.20) are equivalent, and that the origin $r = 0$ may chosen wherever we like. For $k = 0$ the geometry is Euclidean and the space is flat – a homogeneous, isotropic and flat universe. For $k = \pm 1$ space is no longer flat and it is useful to make a transformation:

$$\frac{dr^2}{1 - kr^2} \equiv d\chi^2 , \qquad (9.21)$$

which integrates to

$$r = \begin{cases} \sin\chi & (k = +1) ; \\ \sinh\chi & (k = -1) . \end{cases} \quad (9.22)$$

As long as one moves *on* a surface $r = $ constant one does not notice anything out of the ordinary, because if we take $dr = 0$ in (9.20) we obtain the usual geometry of the surface of a sphere. The surface (2-volume) O of such a sphere is $4\pi r^2 S^2$. Exercise 9.6 illustrates how we may use that to measure χ and r. We may also construct a θ, φ-grid on the sphere as usual.

Spherical universe with positive curvature

Things are different when the radial direction r comes into play. For $k = +1$ we have:

$$dl^2 = S^2 \left\{ d\chi^2 + \sin^2\chi \, (d\theta^2 + \sin^2\theta \, d\varphi^2) \right\} . \quad (9.23)$$

We may visualise this universe as the boundary of a 4-dimensional sphere of radius S embedded in a 4-dimensional Euclidean space. The boundary of such a 4-sphere may be parametrized as follows (see exercise):

$$\left.\begin{aligned} x &= S \sin\chi \sin\theta \cos\varphi \\ y &= S \sin\chi \sin\theta \sin\varphi \\ z &= S \sin\chi \cos\theta \\ w &= S \cos\chi \end{aligned}\right\} \quad \begin{aligned} 0 &\leq \chi \leq \pi ; \\ 0 &\leq \theta \leq \pi ; \\ 0 &\leq \varphi \leq 2\pi . \end{aligned} \quad (9.24)$$

The advantage of the χ-co-ordinate is that it is monotonous, contrary to r: if $k = 1$, r runs from 0 to 1 and then back again to 0 (see exercise 13.2). The space (9.24) has no boundary and in the exercises it is shown that its 3-volume is finite. This is called a *closed universe*. Note that the embedding space has no physical reality. The fourth (radial) dimension from the origin of the embedding space towards the boundary of the sphere and beyond does not exist: we can only move within the boundary, in the space defined by (9.24), and have no notion of what is lurking outside. There *is* no outside.

Hyperbolic universe with negative curvature

For $k = -1$ we have:

$$dl^2 = S^2 \left\{ d\chi^2 + \sinh^2\chi \, (d\theta^2 + \sin^2\theta \, d\varphi^2) \right\} . \quad (9.25)$$

There is now no natural limit to r; both r and χ run from 0 to ∞. This space is much harder to visualize. The closest analogon is a saddle surface in R_3. However, in a flat R_3 there exists no 2-surface without boundary and with a constant negative curvature.[7] And in a flat R_4 there exists no 3-surface with a constant negative curvature. The space (9.25) also has no boundary, and an exercise shows that its 3-volume is infinite, as in the case of $k = 0$. This is called an *open universe* ($k = 0, -1$).

Exercise 9.3: Prove (9.16).

Hint: according to (3.7) the metric of the subspace $t = $ constant and $S = 1$ is $dl^2 = e^{2\lambda}dr^2 + r^2 d\Omega^2$. But that is also the metric of the subspace $t = $ constant of the Schwarzschild metric (4.2). Therefore we calculate $R = R^i{}_i$ of that 3D space. It is not sufficient to restrict the indices α, β in (4.19) to 1, 2, 3: relation (2.57) shows that we also should set to zero all $\Gamma^\mu{}_{\alpha\beta}$ with one or more zero indices. According to § 4.1 these are $\Gamma^0{}_{10}, \Gamma^0{}_{01}$ and $\Gamma^1{}_{00}$, and these can put to zero by taking $\nu = $ constant. Then $R^0{}_0 = g^{0\mu}R_{\mu 0} = g^{00}R_{00} = 0$ as well according to (4.15).

Exercise 9.4: Prove that (9.24) is a space with metric (9.23) and that (9.24) is the boundary of a 4-sphere with radius S in an Euclidean R_4.

Hint: (9.24) should be a sphere ($x^2+y^2+z^2+w^2 = S^2$) embedded in Euclidean space, that is $dl^2 \equiv dx^2 + dy^2 + dz^2 + dw^2 =$ (9.23).

Exercise 9.5: Prove that for $k = +1$ the 3-volume of space is finite and equal to $2\pi^2 S^3$.

Hint: From (9.23): $\sqrt{g}\, d\chi d\theta d\varphi = S^3 \sin^2\chi \sin\theta \cdot d\chi d\theta d\varphi$ (exercise 4.4).

Exercise 9.6: Consider a sphere with radius $r = \sin(h)\chi$ around the origin. Calculate the 2-volume of the boundary (the surface) and the length of the radius. Prove that

$$\frac{\text{surface sphere}}{4\pi\, (\text{length radius})^2} = \begin{cases} (\sin\chi/\chi)^2 < 1 & (k = +1) \; ; \\ (\sinh\chi/\chi)^2 > 1 & (k = -1) \; . \end{cases} \quad (9.26)$$

Hint: Take for example $k = +1$; $d\chi = 0$ in (9.23) $\to$ metric of the spherical boundary: $dl^2 = S^2 \sin^2\chi\, (d\theta^2 + \sin^2\theta\, d\varphi^2) \to \sqrt{g}\, d\theta d\varphi = S^2 \sin^2\chi \sin\theta \cdot d\theta d\varphi$, then integrate. Length along χ: $dl = S d\chi$ ($d\theta = d\varphi = 0$ in (9.23)).

---

[7] Stillwell, J.: 1992, *Geometry of Surfaces*, Springer-Verlag.

9.4 Equations of motion

The derivation of the equations of motion is quite a bit of work: we have to repeat the entire derivation of §§ 4.1 and 4.2 for the Robertson-Walker metric. However, we shall pass over many details. The starting point is the calculation of the Christoffel symbols. To this end we write down the equation for an arbitrary geodesic with the help of variational calculus, see (2.36) and (9.19): $\delta \int L \, dp = 0$, where L is given by:

$$L = g_{\alpha\beta} \dot{x}^\alpha \dot{x}^\beta$$
$$= (\dot{x}^0)^2 - \frac{S^2 \dot{r}^2}{1 - kr^2} - S^2 r^2 \dot{\theta}^2 - S^2 r^2 \sin^2\theta \, \dot{\varphi}^2 \,, \tag{9.27}$$

with $\dot{} = d/dp$.[8] Note that the co-ordinates $x^0 = ct$, $x^1 = r$, $x^2 = \theta$ and $x^3 = \varphi$ are functions of the parameter p. The scale factor S depends on t, i.e. on x^0. All x^0-dependence of L is in S, and $S' \equiv dS/dx^0$. We elaborate the Euler-Lagrange equations (2.37) for x^0: $\partial L/\partial x^0 = (\partial L/\partial \dot{x}^0)\dot{}$:

$$-2SS' \left(\frac{\dot{r}^2}{1 - kr^2} + r^2 \dot{\theta}^2 + r^2 \sin^2\theta \, \dot{\varphi}^2 \right) = (2\dot{x}^0)\dot{} \,. \tag{9.28}$$

After some rearranging:

$$\ddot{x}^0 + \frac{SS'}{1 - kr^2} \dot{r}^2 + SS' r^2 \dot{\theta}^2 + SS' r^2 \sin^2\theta \, \dot{\varphi}^2 = 0 \,. \tag{9.29}$$

We compare this to (2.34) so that we may read $\Gamma^0_{\alpha\beta}$ from the equation (numbering: $1 = r$, $2 = \theta$, $3 = \varphi$):

$$\Gamma^0_{11} = \frac{SS'}{1 - kr^2} \,; \qquad \Gamma^0_{22} = SS' r^2 \,; \qquad \Gamma^0_{33} = SS' r^2 \sin^2\theta \,, \tag{9.30}$$

and all other $\Gamma^0_{\alpha\beta}$ are zero. An exercise invites the reader to prove that

$$\Gamma^\nu_{0\nu} = \frac{3S'}{S} \,; \qquad \Gamma^\alpha_{00} = 0 \,; \qquad \Gamma^0_{ik} = -\frac{S'}{S} g_{ik} \,. \tag{9.31}$$

According to (9.19) the metric tensor $g_{\alpha\beta}$ is

$$\left. \begin{array}{ll} g_{00} = 1 \,; & g_{11} = -\dfrac{S^2}{1 - kr^2} \,; \\[2ex] g_{22} = -S^2 r^2 \,; & g_{33} = -S^2 r^2 \sin^2\theta \,. \end{array} \right\} \tag{9.32}$$

---

[8] Our notation is not very consistent. Sometimes $\dot{}$ stands for d/dp and sometimes for $\partial/\partial t$. Here we are forced to distinguish d/dp (denoted by $\dot{}$) and $\partial/\partial t$ (denoted as $' = d/dx^0$). Later we switch again to $\dot{S} = dS/dt$.

The (long) technicalities of the computation of $R_{\mu\nu}$ are left aside, and we mention only the final result

$$R_{00} = \frac{3S''}{S} \; ; \qquad R_{0i} = 0 \; ; \qquad\qquad R_{ik} = \frac{SS'' + 2(S')^2 + 2k}{S^2} \, g_{ik} \, , \qquad (9.33)$$

with $' = \mathrm{d}/\mathrm{d}x^0$. Furthermore (see exercise)

$$G_{00} = -\frac{3\{(S')^2 + k\}}{S^2} \, . \qquad (9.34)$$

We have expressed R_{ik} and Γ^0_{ik} in terms of g_{ik} where possible because that will be useful later.

Next we consider the stress-energy tensor $T_{\mu\nu}$. The universe is filled homogeneously with a mixture of matter (galaxies) and radiation (the CMB). The bulk velocity of that mixture with respect to the Hubble flow is zero: $u^\mu = (1,0,0,0) \to u_\mu = g_{\mu\nu}u^\nu = g_{\mu 0} = 0$ for $\mu = 1,2,3$ and 1 for $\mu = 0 \to u_\mu = u^\mu$. With (3.57) we obtain:

$$T_{00} = \rho \; ; \qquad T_{0i} = 0 \; ; \qquad T_{ik} = -\frac{p}{c^2} \, g_{ik} \, . \qquad (9.35)$$

We conclude from (3.58) that $G_{00} + \Lambda = -(8\pi G/c^2) \cdot T_{00}$, or:

$$\left(\frac{S'}{S}\right)^2 = \frac{8\pi G \rho}{3c^2} + \frac{\Lambda}{3} - \frac{k}{S^2} \, . \qquad (9.36)$$

This is the equation of motion for S, first derived by Friedmann in 1922 for the special case that the cosmological constant Λ is zero.

Adiabatic expansion

There is still information in $T^{\mu\nu}{}_{:\nu} = 0$, but only in $T^{0\nu}{}_{:\nu} = 0$. From (2.51):

$$T^{0\nu}{}_{:\nu} = T^{0\nu}{}_{,\nu} + \Gamma^0_{\nu\sigma} T^{\nu\sigma} + \Gamma^\nu_{\sigma\nu} T^{0\sigma} = 0 \, . \qquad (9.37)$$

Now $T^{ik} = g^{i\lambda}g^{k\mu} T_{\lambda\mu} = g^{il}g^{km} T_{lm} = -(p/c^2)g^{ik}$, and is straightforward to see that $T^{00} = T_{00} = \rho$, and $T^{0i} = 0$. This simplifies (9.37) to:

$$\begin{aligned}
0 &= T^{00}{}_{,0} + \Gamma^\nu_{0\nu} T^{00} + \Gamma^0_{ik} T^{ik} \\
&= \rho' + \frac{3S'}{S}\rho + \frac{S'}{S}\frac{p}{c^2} g_{ik} g^{ik} \\
&= \rho' + \frac{3S'}{S}\rho + \frac{3S'}{S}\frac{p}{c^2} \, .
\end{aligned} \qquad (9.38)$$

Here we have made use first of $\Gamma^0{}_{0\alpha} = 0$, then of $\rho_{,0} = \rho'$ and finally of (9.31). On multiplying (9.38) with $c^2 S^3$ we get

$$(\rho c^2 S^3)' + p(S^3)' = 0 . \qquad (9.39)$$

This equation says that the gas in a volume $V \propto S^3$ expands *adiabatically*: $dQ/dt \equiv dU/dt + p dV/dt = 0$ with $U \equiv \rho c^2 V$.

The role of the pressure

Equations (9.36) and (9.39) determine the evolution of the universe once we know the equation of state $p(\rho)$. This is the subject of the next chapter. Note that (9.36) and (9.39) may be combined into the following relation (see exercise):

$$\frac{S''}{S} = -\frac{4\pi G}{3c^2}\left(\rho + \frac{3p}{c^2}\right) + \frac{\Lambda}{3} . \qquad (9.40)$$

This equation carries a few important messages. We take $\Lambda = 0$ first, and deal with $\Lambda \neq 0$ in the next section. In this case $S'' < 0$, i.e. S' decreases. In other words, the expansion of the universe is slowing down. The classical explanation is that this is gravity at work, which is constantly trying to pull the matter together. Another implication is that the expansion must have been faster in the past. Perhaps more astounding is that pressure also acts to reduce the expansion. The intuitive idea that pressure should accelerate expansion is apparently not correct. The explanation is that a pressure *gradient* gives rise to a force, like between the inside and the outside of a balloon. But the universe is homogeneous and there are no pressure gradients. To continue in the spirit of the metaphor, we don't live inside the balloon but on the homogeneous surface (the interior of the 'balloon' does not exist). What remains is that pressure is a form of potential energy and acts as a source of gravity if it is sufficiently large, $p \sim \rho c^2$. A similar thing happened in the case of the TOV equation, § 5.3.

9.5 The cosmological constant

Historically, Einstein introduced the cosmological constant Λ because it was a term that logically should appear in the field equations, and it allowed the existence of a *static*, zero-pressure spherical universe:

$$S = \Lambda^{-1/2} = c(4\pi G\rho)^{-1/2} ; \qquad k = +1 . \qquad (9.41)$$

Remember – this was before it was discovered that the universe expands. The effect of a positive cosmological constant in eq. (9.40) is to increase the expansion rate. Nowadays we believe we know that $\Omega_\Lambda \equiv \Lambda c^2/3H_0^2 \simeq 0.7$ and that the expansion of the universe is actually accelerating. The cosmological constant is a property of the vacuum since Λ remains in eqs. (9.36) and (9.40) after ρ and p have been set to zero. It is possible to explain the term $\Lambda g^{\mu\nu}$ in (3.58) in terms of a stress-energy tensor associated with the vacuum. Following the literature we endow it with a constant energy density $\rho_{\rm v} c^2$ of unknown, probably quantummechanical origin.[9] In special relativity, the stress-energy tensor in the local rest-frame of a fluid is[10]

$$T^{\mu\nu} = \frac{1}{c^2}\begin{pmatrix} \rho_{\rm v} c^2 & & & \emptyset \\ & p_{\rm v} & & \\ & & p_{\rm v} & \\ \emptyset & & & p_{\rm v} \end{pmatrix}. \qquad (9.42)$$

We have replaced $\rho \to \rho_{\rm v}$ and $p \to p_{\rm v}$, in anticipation of (9.42) being the $T^{\mu\nu}$ of the vacuum. Now comes the key observation: the vacuum is physically identical in all inertial frames, so that (9.42) must be the same in all inertial frames, and it must be Lorentz-invariant. This is only possible if $T^{\mu\nu} = {\rm const}\cdot\eta^{\mu\nu}$, which implies that

$$T_{\rm v}^{\mu\nu} = \rho_{\rm v}\eta^{\mu\nu} \ ; \qquad p_{\rm v} = -\rho_{\rm v} c^2 \ . \qquad (9.43)$$

A negative pressure is formally in agreement with energy conservation (9.39): $dU/dt + pdV/dt \equiv d\rho_{\rm v}c^2V/dt + (-\rho_{\rm v}c^2)dV/dt = 0$ as $\rho_{\rm v}$ is constant. The principle of general covariance suggests that in GR we should take

$$T_{\rm v}^{\mu\nu} = \rho_{\rm v} g^{\mu\nu} \ . \qquad (9.44)$$

Next, following eq. (3.59), we write the field equation (3.42) as

$$G^{\mu\nu} = -\frac{8\pi G}{c^2}\left(T_{\rm v}^{\mu\nu} + T_{\rm m}^{\mu\nu}\right), \qquad (9.45)$$

where the index m stands for matter. After insertion of (9.44) we recover the field equation (3.58) with the Λ-term, and

$$\Lambda = \frac{8\pi G \rho_{\rm v}}{c^2}, \quad {\rm or} \quad \Omega_\Lambda \equiv \frac{\Lambda c^2}{3H_0^2} = \frac{\rho_{\rm v}}{\rho_{\rm c}} \ . \qquad (9.46)$$

The parameters Ω_Λ and $\Omega_{\rm m} \equiv \rho/\rho_{\rm c}$ will play an important role in the next chapters.

We handle eq. (9.40) in the same spirit: omit the term $\Lambda/3$ and split the $\rho + 3p/c^3$ term in a vacuum part and a matter part. The former equals again $\Lambda/3$:

---

[9] Carroll, S.M. et al., *A.R.A.A.* **30** (1992) 499.
[10] Set $u^\mu = (1,0,0,0)$ and $g^{\mu\nu} = \eta^{\mu\nu}$ in (3.57).

$$-\frac{4\pi G}{3c^2}\left(\rho_v - 3\rho_v\right) = +\frac{8\pi G \rho_v}{3c^2} \equiv \frac{\Lambda}{3}. \tag{9.47}$$

This also demonstrates that the anti-gravity generated by the negative pressure outweighs the gravity associated with the vacuum density ρ_v.

In summary, we assign to the vacuum a constant energy density $\rho_v c^2$ of unknown origin, referred to as *dark energy*. Formal arguments such as Lorentz invariance force us to assign to it a negative pressure $-\rho_v c^2$ as well. This is then equivalent with the Λ-term in eq. (9.36). The anti-gravity generated by the negative pressure makes the expansion of the universe accelerate. Observations suggest that ρ_v is a little less than ρ_c. The ultimate explanation of Λ and ρ_v must come from a theory of quantum gravity.

9.6 Geodesics

The geodesics of the Robertson-Walker metric are simple in the sense that they are all effectively *radial* geodesics. Given a geodesic, spatial homogeneity permits us to move the origin to a point on the geodesic. Seen from this new origin the geodesic must be a radial geodesic ($d\theta = d\varphi = 0$), on account of symmetry. The situation is therefore simpler than in the case of the Schwarzschild metric. All optical observations, for example, may be analysed with radial null geodesics, and these are simple: $ds^2 = 0$ in (9.19) $\rightarrow$ $dx_0 = \pm S(1-kr^2)^{-1/2} dr$ (§ 11.1). The only non-trivial *material* geodesics are those having a nonzero initial velocity, for example a test mass fired into space, see Fig. 9.6 and exercise 9.9. The outcome may be understood right away: the test mass does not reach spatial infinity, but rather a constant co-ordinate distance r_0. What happens is that the speed of the test mass decreases with respect to the local Hubble flow, and after a (formally infinite) time it finds itself at rest in the Hubble flow. One of the consequences is that the peculiar motion of a galaxy superposed on the Hubble flow is generally damped. This is just a manifestation of adiabatic cooling, which we already encountered in connection with eq. (9.39). It explains why the Hubble flow is cold.

Exercise 9.7: Prove (9.31) and (9.34).

Hint: (9.32) $\rightarrow$ $-g = S^6 r^4 \sin^4\theta/(1-kr^2)$, then (2.33); Γ^i_{00} requires the other geodesic equations from $\partial L/\partial x^i = (\partial L/\partial \dot{x}^i)\dot{}$. According to (2.34) it comes down to showing that there are no terms $\propto (\dot{x}^0)^2$; Γ^0_{ik} from (9.30) and (9.32); G_{00} from (2.60), and $R = g^{\mu\nu}R_{\nu\mu} = R_{00} + g^{ik}R_{ki}$. Then (9.33).

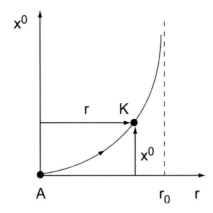

Fig. 9.6. Star wars. Under the pressure of mounting political tension the Upper Master of galaxy A decides to fire a bullet K to an unfriendly neighbour with initial velocity $\beta = v_0/c$. The bullet (think of a jet) moves along a radial geodesic $r(\tau)$, $x^0(\tau)$. Due to the expansion, the bullet reaches a finite co-ordinate distance r_0. The computation of exercise 9.9 is only indicative as it does not allow for the gravitational attraction of A.

Exercise 9.8: Prove (9.40).

Hint: First multiply (9.36) by S^2. Write $\rho S^2 = \rho S^3/S$, in anticipation of the substitution of (9.39).

Exercise 9.9: Test mass K in Fig. 9.6 moves on a material geodesic in the Robertson-Walker metric. Show that

$$\frac{d\chi}{dt} \equiv \frac{1}{\sqrt{1-kr^2}} \frac{dr}{dt} = \frac{\lambda c}{S\sqrt{S^2+\lambda^2}}, \qquad (9.48)$$

with $\lambda = \gamma\beta S_0$ and $\beta = v_0/c$, $v_0 =$ initial velocity of K and $\gamma = (1-\beta^2)^{-1/2}$.

Hint: Nasty problem. Since the Robertson-Walker metric depends on time, u_0 is not a constant of the motion, as it was in the Schwarzschild metric. The constants of the motion are θ and φ. The equation for x^0 is (9.29) with $d\theta = d\varphi = 0$. To obtain the second equation it is easiest to 'divide' (9.19) by ds^2 as we did in (4.35):

$$\ddot{x}^0 + \frac{SS'\dot{r}^2}{1-kr^2} = 0 ; \qquad (9.49)$$

$$(\dot{x}^0)^2 - \frac{S^2\dot{r}^2}{1-kr^2} = 1 , \qquad (9.50)$$

with $\dot{} = d/ds$ and $' = d/dx^0$. In this problem the proper time plays its usual role again. Eliminate $\dot{r}^2/(1 - kr^2)$:

$$\ddot{x}^0 + (S'/S)\{(\dot{x}^0)^2 - 1\} = 0 . \tag{9.51}$$

Multiply with $S^2\dot{x}^0$ and use that $S'\dot{x}^0 = \dot{S}$. The result may be integrated to $S^2(\dot{x}^0)^2 - S^2 = \lambda^2 = $ integration constant:

$$\dot{x}^0 = \sqrt{S^2 + \lambda^2}/S . \tag{9.52}$$

Initial condition: at $t = t_0$ we have $S = S_0$ and according to (1.6) $\dot{x}^0 \equiv dt/d\tau = (1 - \beta^2)^{-1/2}$ with $\beta = v_0/c \to \lambda = \gamma\beta S_0$. Substitute (9.52) in (9.50):

$$(1 - kr^2)^{-1/2}\dot{r} = \lambda/S^2 . \tag{9.53}$$

Finally, $dr/dt = c\dot{r}/\dot{x}^0$. Eq. (9.48) can be integrated once $S(t)$ is known (exercise 11.4). It is easy to see that for $S \propto t^\alpha$ and $k = 0$ the test mass will travel a finite co-ordinate distance if $\alpha > \frac{1}{2}$.

Worlds in collision. A spectacular merger in progress in NGC 4676, at a distance of 92 Mpc. Analysis shows that we are seeing two spirals some 160 Myr after closest encounter. Tidal interaction created long tails that contain many associations of young and hot (blue) stars. The pair will eventually merge into a single elliptical galaxy. The horizontal image size is about $2'$. A similar merger might happen when our galaxy hits its neighbour M31, a few billion years from now. Image taken by the Advanced Camera for Surveys on the HST in April 2002. Credit: NASA, H. Ford et al., and the ACS Science team.

10
The Evolution of the Universe

In the previous chapter we learned that GR opens completely new possibilities for the spatial structure of the universe, even if we restrict ourselves to homogeneous isotropic spaces. Space becomes a dynamic entity whose topology and geometry depend on the matter it contains. This is a major conceptual advance over the Newtonian idea of an absolute, flat and infinite space. This chapter tells the story of the Friedmann-Robertson-Walker (FRW) model. That is, the homogeneous isotropic universe with a Robertson-Walker metric whose evolution is determined by the Friedmann equation (9.36). In 1927 Lemaître proved that Friedmann's solution implies a linear relation between distance and redshift. The discovery of the expansion of the universe in the 1920s by Slipher and Hubble did not come out of the blue, but had been anticipated by the theoretical developments of the time. During the second half of the 20th century it was realised that a FRW universe must have had a hot start, of which the matter and the cosmic microwave background (CMB) are ancient relics. More than anything else, the discovery of this CMB by Penzias and Wilson in 1965 has changed the face of cosmology from a speculative backyard in the 1950s into the quantitative science it is today.

10.1 Equation of state

Equations (9.36) and (9.39) determine the evolution of the universe as soon as we know the equation of state $p(\rho)$, or ρ as function of S. In cosmology it is customary to group all relativistic particles under the name *radiation*, regardless of their mass, and to reserve the term *matter* for all non-relativistic particles. The reason is that these two groups contribute in rather different ways to the dynamics of the universe.[1] The density ρ in (9.36) and (9.39)

---

[1] There are now strong indications that neutrinos have a small mass. The WMAP data indicate that $\Omega_\nu < 0.015$, implying that while some of the neutrinos may actually be 'matter' now, they are likely to be all relativistic at the beginning of the matter era. For that reason the energy density of the neutrino background has been added to the radiation density in (9.9).

Table 10.1. Pressure and density in the universe

	Matter $(\epsilon_m \gg \epsilon_r)$	Radiation $(\epsilon_m \ll \epsilon_r)$
$p = p_m + p_r$	0	$\frac{1}{3}\epsilon_r$
$\rho = (\epsilon_m + \epsilon_r)/c^2$	ϵ_m/c^2	ϵ_r/c^2
S-dependence	$\epsilon_m S^3 =$ const	$\epsilon_r S^4 =$ const

should be interpreted as ϵ/c^2, where ϵ = total energy density, including the rest mass contribution, and p represents the total pressure. As long as the temperature is sufficiently low, $m_0 c^2 \gg \kappa T$, the total energy of a particle with mass hardly exceeds $m_0 c^2$. Such a non-relativistic particle has a constant contribution to ϵ. In the early universe, however, the temperature is very high and $m_0 c^2$ may be much smaller than κT. In that case the rest mass of the particle is effectively zero and it behaves like a photon, whose wavelength scales $\propto S$ (the proof is given in § 11.1). The contribution of such a particle to ϵ is $\propto S^{-1}$. Since the number of particles in a comoving volume $V \propto S^3$ remains constant, we find that the energy density is $\propto S^{-3}$ for matter and $\propto S^{-4}$ for radiation.

A consequence of this matter/radiation definition is that particles with $m_0 \neq 0$ switch gender during the evolution of the universe, from 'radiation' to 'matter', first the heavier particles, subsequently followed by the lighter ones, since the temperature decreases so drastically. It turns out that the evolution of the universe can be described by two limiting cases: (1) the recent history of the universe, during which $\epsilon_m \gg \epsilon_r$ so that the evolution is entirely determined by the matter, and (2) the hot early universe where $\epsilon_m \ll \epsilon_r$ and the radiation determines the evolution. In the former case the pressure is zero, because $p = p_m + p_r \sim n\kappa T + \epsilon_r \ll \epsilon_m + \epsilon_r \simeq \epsilon_m \simeq \rho c^2 \rightarrow p \ll \rho c^2 \rightarrow p \simeq 0$, since we know that p is only relevant when $p \sim \rho c^2$. For particles of zero mass $p = \epsilon/3$ holds generally, see Appendix D. In this way we arrive at the relations in Table 10.1.

Exercise 10.1: Show that $\epsilon_m S^3 =$ constant and $\epsilon_r S^4 =$ constant from (9.39).

Hint: Matter: trivial. Radiation: $(\epsilon_r S^3)' + \frac{1}{3}\epsilon_r (S^3)' = 0 \rightarrow (\epsilon_r S^4)' = 0$.

10.2 The matter era

On comparing (9.3) and (9.9) we see that the matter energy density in the universe is about a factor 3000 larger than the energy density in radiation. This imbalance will remain in the future as S increases, because $\epsilon_m \propto S^{-3}$, while $\epsilon_r \propto S^{-4}$. It is only in the early universe that $\epsilon_r > \epsilon_m$. During most of its life the universe evolves according to the limiting case 'matter'. The equations for this so-called *matter era* follow from (9.36), (9.40) and Table 10.1. We also revert to the notation $\dot{} = d/dt$:

$$\left(\frac{\dot{S}}{S}\right)^2 = \frac{8\pi G \rho}{3} + \frac{\Lambda c^2}{3} - \frac{kc^2}{S^2} \; ; \tag{10.1}$$

$$\frac{\ddot{S}}{S} = -\frac{4\pi G \rho}{3} + \frac{\Lambda c^2}{3} \; ; \tag{10.2}$$

$$\rho S^3 = \rho_0 S_0^3 \; . \tag{10.3}$$

Here and everywhere else the index 0 indicates the value of a quantity at the present epoch $t = t_0$; ρ is the density of matter (the index m has been dropped). The first step is to rewrite (10.1) in a seemingly complicated way for $t = t_0$:

$$1 = \Omega_m + \Omega_\Lambda + \Omega_k \; . \tag{10.4}$$

The constants Ω_m, Ω_Λ and Ω_k are defined as:

$$\left. \begin{array}{l} \Omega_m = \dfrac{8\pi G \rho_0}{3H_0^2} = \dfrac{\rho_0}{\rho_c} \; ; \\[1em] \Omega_\Lambda = \dfrac{\Lambda c^2}{3H_0^2} = \dfrac{\rho_v}{\rho_c} \; ; \\[1em] \Omega_k = -k \left(\dfrac{c}{H_0 S_0}\right)^2 . \end{array} \right\} \tag{10.5}$$

and

$$H_0 \equiv \left(\frac{\dot{S}}{S}\right)_0 = \frac{\dot{S}_0}{S_0} \; ; \qquad \rho_c = \frac{3H_0^2}{8\pi G} \; . \tag{10.6}$$

That (10.4) is the same as (10.1) at $t = t_0$ is just a matter of substitution. The parameters Ω_m, Ω_Λ and Ω_k indicate the relative importance of the density, the cosmological constant, and the curvature of space in the evolution of the universe at the present epoch. We have already met the parameters Ω_m and Ω_Λ and the critical density ρ_c in the previous chapter. The proof that H_0 defined in (10.6) is really the Hubble constant is given in § 11.3. Since $\text{sign}(\Omega_m + \Omega_\Lambda - 1) = \text{sign}(k)$, we arrive at an important conclusion:

$$
\left.\begin{array}{lll}
\text{closed universe} & (k=+1) & \leftrightarrow \quad \Omega_{\text{m}} + \Omega_\Lambda > 1 \,; \\
\text{flat universe} & (k=0) & \leftrightarrow \quad \Omega_{\text{m}} + \Omega_\Lambda = 1 \,; \\
\text{open universe} & (k=-1) & \leftrightarrow \quad \Omega_{\text{m}} + \Omega_\Lambda < 1 \,.
\end{array}\right\} \quad (10.7)
$$

The spatial structure of an FRW universe is fixed by the matter density, the cosmological constant and the Hubble constant. And since $\Omega_{\text{m}} + \Omega_\Lambda = 1.02 \pm 0.02$, see Table 9.2, our universe is very likely to have a flat geometry.

After substitution of (10.3), we may cast the evolution equation for the scale factor (10.1) in a suitable dimensionless form:

$$\dot{u}^2 = H_0^2 \left(\Omega_{\text{m}} u^{-1} + \Omega_\Lambda u^2 + \Omega_{\text{k}} \right), \quad \text{with} \quad u = S/S_0 \,. \quad (10.8)$$

We now discuss the evolution of FRW universes as given by (10.8). At the present epoch we have $\dot{S} > 0$, hence $\dot{u} > 0$ in $u = 1$. Note that $\dot{u}$ can only change sign if the right hand side of (10.8) becomes zero. If we move into the past, i.e. smaller u, the right hand side of eq. (10.8) becomes larger and $\dot{u}$ increases, provided Ω_Λ is not too large.[2] It follows that u will reach zero in a finite time. We arrive at a second important conclusion: the expansion of FRW universes started a finite time ago from a singularity. The density and pressure must have been extremely high at that time. This is called the *Big Bang*. It turns out that isotropy and $\epsilon_{\text{m}} \gg \epsilon_{\text{r}}$ are no essential ingredients. The expansion must have started from a singularity.

FRW models with zero Λ

We now consider the future evolution of FRW models, and take $\Omega_\Lambda = 0$ first. For $\Omega_\Lambda = 0$ we have

$$|\Omega_{\text{m}} - 1|^{1/2} = |\Omega_{\text{k}}|^{1/2} = \frac{c}{H_0 S_0} \,, \quad (10.9)$$

provided $\Omega_{\text{m}} \neq 1$, and

$$\dot{u}^2 = H_0^2 \left(\Omega_{\text{m}} u^{-1} + 1 - \Omega_{\text{m}} \right), \quad u = S/S_0 \,. \quad (10.10)$$

At the present epoch $\dot{u} > 0$ in $u = 1$. For $\Omega_{\text{m}} \leq 1$ the right hand side of (10.10) is positive, so that $\dot{S}$ is always positive. The expansion will continue

---

[2] A singularity may not occur if $\Omega_\Lambda > 1$. For a classification of universe models as a function of Ω_{m} and Ω_Λ we refer to Peacock (1999) § 3.2. The word singularity should not be taken too literally. Quantum gravity will probably prevent it, and even during the earliest phases of the Big Bang the universe never was a 'point', see Ch. 13.

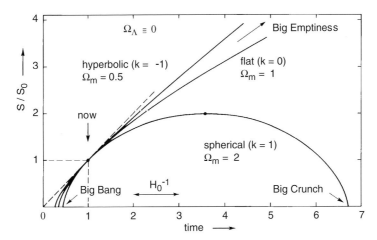

Fig. 10.1. Three solutions of eq. (10.10): an open, a flat and a closed FRW universe with $\Lambda = 0$, tuned to the same size and expansion rate at the present epoch t_0, arbitrarily located at $t = 1$. Time is in units of H_0^{-1}.

forever. The density is too low and the associated gravity not strong enough to stop it. For large S the expansion rate approaches

$$\dot{u} \to \begin{cases} 0 & \Omega_{\rm m} = 1 \,; \\ H_0\sqrt{1 - \Omega_{\rm m}} & \Omega_{\rm m} < 1 \,, \end{cases} \quad (10.11)$$

or, in terms of S: $\dot{S} \to 0$, c for $k = 0, -1$. When $\Omega_{\rm m} > 1$ the right hand side of (10.10) may become zero. A horizontal asymptote $u \to$ const. is not possible because (10.2) requires that $\ddot{S} < 0$. In other words, $\dot{S}(t)$ must decrease all the time. Hence, for $\Omega_{\rm m} > 1$, u will reach a maximum $u = \Omega_{\rm m}/(\Omega_{\rm m} - 1) > 1$ after which a contraction follows; $u(t)$ is symmetric with respect to the maximum (why?). The gravity generated by the matter is sufficient to stop the expansion, after which the universe begins to contract again, and 'the movie is shown in reverse order'. The contraction steadily accelerates and continues until space degenerates formally into a point. This is called the *Big Crunch*. Fig. 10.1 shows the evolution of three FRW universes with $\Omega_\Lambda = 0$.

Eq. (10.10) has a simple solution for $\Omega_{\rm m} = 1$. We have $u^{1/2}\dot{u} = H_0 \to u = {\rm const} \cdot t^{2/3}$:

$$\left. \begin{array}{l} \dfrac{S}{S_0} = \left(\tfrac{3}{2}H_0 t\right)^{2/3} \Rightarrow \\[1em] t_0 = \tfrac{2}{3}H_0^{-1} \simeq 6.5\,h^{-1}\,{\rm Gyr} \end{array} \right\} \quad \text{for} \quad \Omega_{\rm m} = 1 \text{ and } \Omega_\Lambda = 0 \,. \quad (10.12)$$

Table 10.2. The age of a FRW universe as a function of its size.

S/S_0			$H_0 t$		
	Ω_m	0.5	1	2	0.3
	Ω_Λ	0	0	0	0.7
0.01		$9.4 \cdot 10^{-4}$	$6.7 \cdot 10^{-4}$	$4.7 \cdot 10^{-4}$	$1.2 \cdot 10^{-3}$
0.02		$2.7 \cdot 10^{-4}$	$1.9 \cdot 10^{-3}$	$1.3 \cdot 10^{-3}$	$3.4 \cdot 10^{-3}$
0.05		0.010	$7.5 \cdot 10^{-3}$	$5.3 \cdot 10^{-3}$	0.014
0.1		0.029	0.021	0.015	0.038
0.2		0.080	0.060	0.043	0.11
0.5		0.29	0.24	0.18	0.41
1		0.75	0.67	0.57	0.96
2		1.8	1.9	3.1	1.7
5		5.6	7.5	$-^a$	2.8

^{a} The maximum size S/S_0 of this closed universe is 2.

This model serves as a kind of reference model in cosmology. The $t^{2/3}$-dependence can be understood with the help of a classical argument: matter homogeneously filling a flat space under its own gravity moves exactly in the same manner, see exercise. Given that $h = 0.71$, the age t_0 of this universe is about 9 Gyr, which is too young. The radiation era prior to the matter era lasted only $\lesssim 10^5$ yr, and cannot significantly affect the value of t_0. It is possible to increase t_0 by taking $\Omega_m < 1$, but in order to attain a reasonable age, say $t_0 \sim 12.5$ Gyr, Ω_m must be ≤ 0.1, which is excluded by the observations.

FRW models with non-zero Λ

The solution of this age problem came after the WMAP mission had established the parameters of our universe: $(\Omega_m, \Omega_\Lambda) \simeq (0.3, 0.7)$. The cosmological constant produces an extra acceleration, and a universe with $\Lambda > 0$ expands forever (unless Ω_m is large), and is usually older. For large u we infer from (10.8) that $\dot{u} \simeq H_0 \sqrt{\Omega_\Lambda}\, u \to u \propto \exp(H_0 \sqrt{\Omega_\Lambda}\, t)$. The expansion is exponential. The turning point where the expansion rate changes from decreasing to increasing can be obtained from (10.2) in dimensionless form:

$$\ddot{u} = H_0^2 \left(-\tfrac{1}{2}\Omega_m u^{-2} + \Omega_\Lambda u \right), \tag{10.13}$$

and $\ddot{u} = 0$ for $u = (\Omega_m/2\Omega_\Lambda)^{1/3} \sim 0.6$. Such an $(\Omega_m, \Omega_\Lambda) = (0.3, 0.7)$ universe is now forever in a state of accelerating expansion. The time evolution follows from (10.8): $du/dt = H_0(\cdots)^{1/2}$ or $H_0 dt = (\cdots)^{-1/2} du$:

$$H_0 t = \int_0^{S/S_0} \left(\Omega_m u^{-1} + \Omega_\Lambda u^2 + \Omega_k \right)^{-1/2} du . \tag{10.14}$$

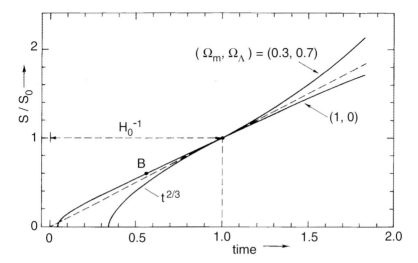

Fig. 10.2. The evolution of a $(\Omega_m, \Omega_\Lambda) \simeq (0.3, 0.7)$ universe, from numerical integration of (10.14). B is the turning point where $\ddot{S} = 0$. The model is compared with the $(\Omega_m, \Omega_\Lambda) = (1, 0)$ reference model, and both are scaled to the same size and expansion rate at the present epoch, arbitrarily located at $t = 1$. Time is in units of H_0^{-1}.

Some values are given in Table 10.2, and Fig. 10.2 shows the $(0.3, 0.7)$ solution together with the $(1, 0)$ reference model. The $(0.3, 0.7)$ universe is older than the $(1, 0)$ model because u begins convex but turns concave later on. The age t_0 of this universe is

$$H_0 t_0 = \int_0^1 \left(\Omega_m u^{-1} + \Omega_\Lambda u^2 + \Omega_k\right)^{-1/2} du$$

$$\simeq \frac{2}{3}\left(0.7\Omega_m - 0.3\Omega_\Lambda + 0.3\right)^{-0.3}. \qquad (10.15)$$

The approximate expression holds for $0.1 \lesssim \Omega_m \lesssim 1$ and $|\Omega_\Lambda| \lesssim 1$ (Peacock, 1999). Taking the WMAP parameters, the age of our $(0.27, 0.73)$ universe would be $0.99 H_0^{-1} \simeq 13.6$ Gyr, almost exactly a Hubble time, in fair agreement with other age indicators such as globular clusters (12 ± 1 Gyr), and nuclear dating (15.6 ± 4.6 Gyr).[3] It seems therefore that we live in a flat universe that is forever flying apart, heading faster and faster towards Big Emptiness. The driving force behind this *cosmic inflation*, the second in the life of the universe,[4] is the anti-gravity associated with an ill-understood vac-

---

[3] Reid, I.N., *A. J.* **114** (1997) 116; Cowan, J.J. et al., *Ap. J.* **521** (1999) 194.
[4] The first inflation phase occurred right after $t = 0$, see Ch. 13.

uum energy (the cosmological constant Λ). We refer to Adams and Laughlin (1999) for an eloquent account of what future of our universe may look like.

Exercise 10.2: Prove that the age of an $\Omega_\Lambda = 0$ universe cannot be larger than H_0^{-1}.

Hint: (10.10) $\to$ $\dot{u} \geq H_0$ for $u \leq 1$ $\to$ $dt \leq du/H_0$ $\to$ $\int_0^{t_0} dt \leq H_0^{-1} \int_0^1 du$, etc. For $\Omega_\Lambda \neq 0$ the argument no longer applies.

Exercise 10.3: Show that an observer in an FRW universe with $\Omega_\Lambda \neq 0$ now, will measure $(\Omega_m, \Omega_\Lambda, \Omega_k) \simeq (1, 0, 0)$ at early times, and $\simeq (0, 1, 0)$ at late times.

Hint: The values of the Ω's in (10.4) depend on time. Write out (10.1) for an arbitrary time:

$$1 = \frac{8\pi G \rho}{3H^2} + \frac{\Lambda c^2}{3H^2} - k\left(\frac{c}{HS}\right)^2. \quad (10.16)$$

For S small (10.1) says $H^2 \equiv (\dot{S}/S)^2 \propto S^{-3}$, i.e. second and third term in (10.16) approach zero. For large S we have $H = \dot{S}/S = $ constant, i.e. first and third term approach zero.

Exercise 10.4: Show that eq. (10.10) describes the dynamics of self-gravitating matter homogeneously filling an infinite flat space.

Hint: Choose an origin O and a point M at distance S. Acceleration of M with respect to O is $\ddot{S} = -G(4\pi\rho S^3/3)S^{-2}$ (Newton's law). Take $\rho S^3 = \rho_0 S_0^3$ and $S/S_0 = u$ $\to$ $2\ddot{u} = -H_0^2 \Omega_m u^{-2}$; multiply with $\dot{u}$ and integrate: $\dot{u}^2 = H_0^2(\Omega_m u^{-1} + \text{const.})$. Integration constant from $\dot{u} = H_0$ in $u = 1$. Weak point: all mass outside the sphere with radius S is ignored.

Exercise 10.5: Conclude from (10.4) and (10.5) that $S_0 = (c/H_0) \cdot |\Omega_m + \Omega_\Lambda - 1|^{-1/2}$. Is S_0 a measurable quantity? Why is that no longer the case when $\Omega_m + \Omega_\Lambda = 1$, in a flat universe?

Hint: For the second question, see (9.19): for $k = 0$ there is a redundancy: S and r appear only in the combination Sr, and there is no room for two independent parameters S and r.

Exercise 10.6: Show that $\rho - \rho_c$ cannot change sign, so that an FRW universe cannot change type. Restrict yourself to $\Lambda = 0$.

Hint: Follows directly from the equation

$$(\rho - \rho_c)\dot{} = -2H(\rho - \rho_c) \,. \tag{10.17}$$

Proof: $(\rho - \rho_c)\dot{} = \dot\rho - (3H^2/8\pi G)\dot{} = \dot\rho - (6H/8\pi G)\dot H$; write (9.38) as $\dot\rho = -3H(\rho + p/c^2)$, and $\dot H = (\dot S/S)\dot{} = (\ddot S S - \dot S^2)/S^2 = -H^2 + \ddot S/S$, then (9.40). What if $\Lambda \neq 0$?

10.3 The radiation era

In the matter era the energy density ϵ_m of matter is much larger than the radiation energy density. But since $\epsilon_m \propto S^{-3}$ and $\epsilon_r \propto S^{-4}$, things must have the other way around in the early universe. During this so-called *radiation era* the universe was an almost perfectly homogeneous, rapidly expanding and cooling fireball. We shall now study this early hot phase of the universe which lasted only some 10^5 yr. We write

$$\epsilon_m = \epsilon_{m0}\left(\frac{S_0}{S}\right)^3 ; \qquad \epsilon_r = \epsilon_{r0}\left(\frac{S_0}{S}\right)^4 . \tag{10.18}$$

And $\epsilon_r = \epsilon_m$ for $a \equiv S/S_0 = \epsilon_{r0}/\epsilon_{m0}$. From (9.3) and (9.9):

$$a \equiv \left(\frac{S}{S_0}\right)_{\text{rad}\to\text{mat}} = \frac{\epsilon_{r0}}{\epsilon_{m0}} = \frac{\epsilon_{r0}}{\Omega_m \rho_c c^2}$$

$$= 4.14 \times 10^{-5}\,(\Omega_m h^2)^{-1} . \tag{10.19}$$

This parameter a determines the evolution of the universe in the radiation era. For $\Omega_m \simeq 0.27$ and $h \simeq 0.71$ we have $a \simeq 3.04 \times 10^{-4}$. The transition from a radiation-dominated to a matter-dominated universe took place when the universe was a factor $S_0/S = (1/a) \sim 3000$ smaller than today. Because ϵ_r is both $\propto S^{-4}$ and $\propto T^4$, we have $T \propto S^{-1}$, or:

$$T_r = T_{r0}\frac{S_0}{S} . \tag{10.20}$$

This presupposes that photons with a Planck distribution will keep a Planck distribution as the scale factor S changes (exercise 10.7). If that were not the case the interpretation of the CMB would be rather problematic. For

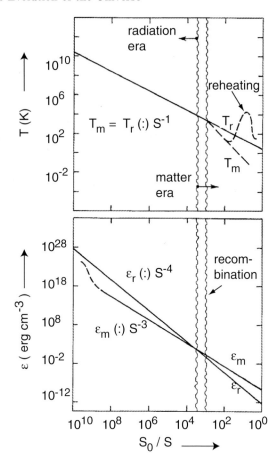

Fig. 10.3. The thermal history of the universe as a function of the scale factor ratio S_0/S. Top panel: the temperature of matter and radiation; bottom panel: the energy densities.

$S_0/S \sim 10^3$ the temperature of the CMB is $T_r \sim 2.73 \times 10^3 \sim 3000$ K, which is about the temperature at which hydrogen gets ionized.[5] The ionization of helium requires a higher temperature and a larger value of S_0/S. Ionized matter and radiation are in thermal equilibrium through frequent Thomson scattering of photons on free electrons. We follow the development in forward

---

[5] Collisional ionization requires $T \sim (1-2) \times 10^4$ K, but since there are 10^9 photons to every hydrogen atom photo-ionization is important. The ionization temperature is now lower because at the same temperature there are more photons in the tail of the Planck distribution than there are particles in the tail of the Maxwell-Boltzmann distribution, see Peebles (1993) p. 165 ff.

direction. When S_0/S decreases below ~ 1200 the plasma begins to recombine (a strange term, because neutral atoms had never existed before), and this process is completed around $S_0/S \sim 10^3$.[6] The photons experience a last Thomson scattering, and the universe becomes transparent. The mean value of S_0/S at recombination is 1100.

The high degree of isotropy of the CMB allows us to draw an important conclusion: the density fluctuations during the recombination must have been equally small. Therefore we know for certain that the universe at that time was practically homogeneously filled with hydrogen and helium. It was a hot mixture of radiation and matter that expanded and cooled down. In the next section we shall see that the existing tiny density fluctuations gradually evolved, in the course of the matter era, into the present structure of the universe, dominated by galaxies.

Fig. 10.3 shows T_r, T_m, ϵ_r and ϵ_m as a function of S. Both ϵ_r and ϵ_m continue to scale as $\propto S^{-4}$ and $\propto S^{-4}$, until energy exchange between radiation and matter begins to play a role in the very early universe.[7] For $S_0/S \gtrsim 10^3$ we have $T_m = T_r$. One might think that after the recombination the matter temperature scales as $T_m \propto \rho_m^{\gamma-1} \propto (S^{-3})^{\gamma-1} \propto S^{-2}$ (adiabatic expansion, $\gamma = 5/3$). But in reality density fluctuations develop into mass concentrations, each with its own, independent thermal evolution. Eventually, the first generation of stars is born, marking the end of what is sometimes referred to as the *Dark Ages*. These stars enrich, reheat and eventually re-ionize the gas, probably in several stages. Around $S_0/S \sim 7$, when the universe was about 1 Gyr old, this re-ionization process had been completed.[8]

Time evolution

To investigate how S depends on time we note that at the beginning of the matter era $u = S/S_0 \ll 1$, so that eq. (10.8) reduces to $\dot{u} = H_0(\Omega_m/u)^{1/2}$. There are no longer three types of universe $k = 0, \pm 1$, but space is virtually flat (even if it were not flat today) and the cosmological constant is effectively zero, see also exercise 10.3. To describe the transition from radiation to matter era we may omit the last two terms in (10.8):

$$\left(\frac{\dot{S}}{S}\right)^2 = \frac{8\pi G \rho}{3}, \qquad (10.21)$$

and replace (10.3) with

---

[6] The fractional ionisation freezes out at a value of $\sim 10^{-4}$.
[7] This is a consequence of the definition of matter and radiation, § 10.1.
[8] Loeb, A. and Barkana, R., *A.R.A.A.* **39** (2001) 19; Fukugita, M. and Kawasaki, M., *M.N.R.A.S.* **343** (2003) L25; Wyithe, J.S.B. and Loeb, A., *Nature* **432** (2004) 194.

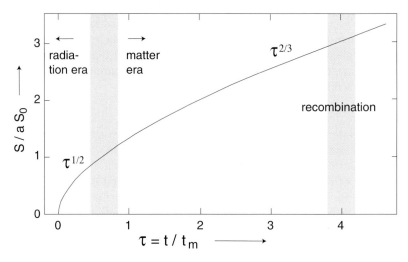

Fig. 10.4. The evolution of the scale factor in the early universe, eq. (10.28). As the radiation era gives way to the matter era, the gravity associated with the pressure disappears and the expansion changes from $\propto \tau^{1/2}$ to $\propto \tau^{2/3}$ and slows down less rapidly. For $\Omega_m = 0.27$ and $h = 0.71$ we have $t_m \simeq 9.4 \times 10^4$ yr, and the matter era begins at $t_{mat} \simeq 0.59 \, t_m$, while recombination is at $t_{rec} \simeq 4 \, t_m$. Note that $\tau = 1$ corresponds to $t \sim 10^{-5} H_0^{-1}$, so that the figure is a huge magnification of the origin of Figs. 10.1 and 10.2.

$$\rho = \frac{\epsilon_{m0}}{c^2}\left(\frac{S_0}{S}\right)^3 + \frac{\epsilon_{r0}}{c^2}\left(\frac{S_0}{S}\right)^4. \tag{10.22}$$

With $u = S/S_0$ and definition (10.19) of a we get

$$\dot{u}^2 = \frac{8\pi G \epsilon_{m0}}{3c^2}\left(\frac{1}{u} + \frac{a}{u^2}\right). \tag{10.23}$$

We introduce the parameter t_m:

$$t_m = \frac{2}{3H_0}\left(\frac{a^3}{\Omega_m}\right)^{1/2} = \frac{1.75 \times 10^3}{(\Omega_m h^2)^2} \text{ yr}, \tag{10.24}$$

which will turn out to be a measure of the age of the universe at the start of the matter era. For the parameters of our universe we find $t_m \simeq 9.4 \times 10^4$ yr. Eq. (10.23) may now be written in terms of dimensionless variables:

$$\left(\frac{dx}{d\tau}\right)^2 = \frac{4}{9}\left(\frac{1}{x} + \frac{1}{x^2}\right) \tag{10.25}$$

where

$$x = \frac{u}{a} = \frac{1}{a}\frac{S}{S_0}, \qquad \tau = \frac{t}{t_m}. \tag{10.26}$$

We may rearrange equation (10.25) as $x\,dx/(1+x)^{1/2} = \frac{2}{3}d\tau$, to obtain

$$\int_0^{u/a} \frac{x\,dx}{\sqrt{1+x}} = \frac{2\tau}{3}. \tag{10.27}$$

The integral is simple after substitution of $x = (1+x) - 1$ in the numerator:

$$\tau = 2 + (1+x)^{3/2} - 3(1+x)^{1/2}. \tag{10.28}$$

The solution is shown in Fig. 10.4. For large x we have $\tau \simeq x^{3/2}$, while $\tau \simeq 3x^2/4$ for small x:

$$x \simeq \begin{cases} (2/\sqrt{3})\,\tau^{1/2} & \tau \ll 1; \\ \tau^{2/3} & \tau \gg 1. \end{cases} \tag{10.29}$$

The expansion begins as $S \propto t^{1/2}$ and changes to $S \propto t^{2/3}$ in the matter era. It is often said that this means that the expansion *accelerates*, but of course what really happens is that the expansion decelerates less rapidly because the pressure becomes effectively zero in the matter era, and the associated gravity disappears.

The age t_{mat} of the universe at the start of the matter era follows from the fact that $u = a$, i.e. $x = 1$, and $\tau = 2 - \sqrt{2} \simeq 0.59$. And the age t_{rec} at recombination when $S_0/S \simeq 1100$ corresponds to $x = (S/S_0)/a \simeq (1100a)^{-1} \simeq 3$ or $\tau \simeq 4$:

$$\left. \begin{array}{l} t_{\text{mat}} \simeq 0.59 t_{\text{m}} \simeq 5.5 \times 10^4 \text{ yr}; \\ t_{\text{rec}} \simeq 4 t_{\text{m}} \simeq 3.8 \times 10^5 \text{ yr}. \end{array} \right\} \tag{10.30}$$

Equation (10.28) becomes invalid in the very early universe, for $t \lesssim 10$ s. The reason is that the extremely high temperature renders some particles relativistic, which then qualify as radiation. This effectively increases the value of ϵ_{r0} and hence of a from (10.19). The electrons are the first to make the switch, when $T_r \gtrsim 6 \times 10^9$ K, for $t \lesssim 10$ s.

Exercise 10.7: Prove the following statements on the Planck distribution:

1. $\epsilon_r = \dfrac{4\sigma}{c} T^4$ erg cm^{-3}; \hfill (10.31)

2. $n_r \simeq 20\,T^3$ cm^{-3}; \hfill (10.32)

3. A Planck distribution remains a Planck distribution as the scale factor S changes.

($\sigma = \pi^2 \kappa^4/(60\hbar^3 c^2)$ and n_r = photon density).

Hint: The photon density equals $n(\nu) = (8\pi\nu^2/c^3)\{\exp(h\nu/\kappa T) - 1\}^{-1}$ cm^{-3} Hz^{-1}, from which

$$n_r = \int_0^\infty n(\nu)\,d\nu = \frac{1}{\pi^2}\left(\frac{\kappa T}{\hbar c}\right)^3 \int_0^\infty \frac{x^2\,dx}{e^x - 1},$$

and $\epsilon_r = \int_0^\infty h\nu\, n(\nu)\,d\nu$. The integrals are tabulated; number of photons dn in a comoving volume V (i.e. $V \propto S^3$) and in a frequency interval $d\nu$ equals $dn = Vn(\nu)d\nu$. Write this as $dn = \text{const} \cdot \nu^2 f(\nu/T)V d\nu$. S changes: $S \to S'$, so $\nu, V, T \to \nu', V', T'$. Use that $\lambda \propto S$ and $\nu \propto S^{-1}$ (proof in § 11.1). Write $\nu = \alpha \nu'$ with $\alpha = S'/S$. Then $V = V'/\alpha^3$. Substitute: $dn = \text{const} \cdot (\nu')^2 f(\alpha\nu'/T)V' d\nu'$. The number does not change: $dn = dn'$, and that implies $T = \alpha T' \to$ a Planck spectrum at temperature T', consistent with (10.20).

Exercise 10.8: Define $\rho_r \equiv \epsilon_r/c^2$ and prove that early in the radiation era

$$\frac{32\pi G}{3}\rho_r t^2 = 1. \qquad (10.33)$$

Hint: Write (10.21) as $(\dot S/S)^2 = 8\pi G \rho_r/3$ (early radiation era $\to$ ignore ϵ_m); $\rho_r \propto S^{-4} \to S = \text{const} \cdot t^{1/2}$, from which $(\dot S/S)^2 = 1/(4t^2)$.

Exercise 10.9: Early in the radiation era the age t and the density ρ_r of the universe are given by

$$t \simeq 1.8 \times 10^{20}\, T_r^{-2}\ \text{s}\,; \qquad (10.34)$$

$$\rho_r \simeq 1.4 \times 10^{-35}\, T_r^4\ \text{g cm}^{-3}. \qquad (10.35)$$

T_r is the photon temperature in K. These relations are valid as long as the radiation consists of photons and neutrinos.

Hint: (10.29): $(S/S_0)^2 = (4a^2/3t_m) \cdot t$, then (10.20); (10.35): combine (10.33) and (10.34).

Exercise 10.10: For $S_0/S \gtrsim 10^3$ we have $T_m = T_r$ due to thermal equilibrium. But why should T_m follow $T_r \propto S^{-1}$? One could also imagine that T_r follows T_m, i.e. $T_r = T_m \propto S^{-2}$, or something like that.

Hint: In cosmology 'matter' is non-relativistic, so that the rest mass is the largest contributor to ϵ_m. That part is not available for energy exchange with photons: $E \simeq m_0 c^2 + \frac{1}{2} m_0 v^2$, and $\frac{1}{2} m_0 v^2 \sim \kappa T \ll m_0 c^2$. For the evolution of $S(t)$ the total ϵ_m matters, but for energy exchange with photons only an energy reservoir $(\kappa T/m_0 c^2)\epsilon_m$ is available. This is much less than the reservoir ϵ_r that the radiation has in stock: $(\kappa T/m_0 c^2)\epsilon_m/\epsilon_r$ is independent of S and therefore equal to $\kappa T/m_0 c^2$ at $S_0/S \sim 3000$ where $\epsilon_m = \epsilon_r$, and is $\ll 1$ for all particles.

Exercise 10.11: Show that the baryon (protons + neutrons) to photon ratio in the universe is constant, and very small. There are many more photons than baryons in the universe:

$$n_b/n_r = \text{constant} \simeq 6.1 \times 10^{-10}. \qquad (10.36)$$

Hint: (10.32) $\to n_r \propto T^3 \propto S^{-3}$ and $n_b \propto \rho_m \propto S^{-3} \to n_r/n_b = \text{const}$; $n_r = 20(2.725)^3 = 405 \text{ cm}^{-3}$; $n_b = \Omega_b \rho_c/m_{\text{proton}} = 0.044 \cdot 1.88 \times 10^{-29} \cdot 0.71^2/1.67 \times 10^{-24} \text{ cm}^{-3}$ (Table 9.2).

10.4 The formation of structure

The formation of structures takes place in matter era, with roots going back as far as the inflation period. This is a very active field of research involving many complex physical processes. We restrict ourselves here to outlining a few basic ideas. A density concentration will collapse under its own gravity if the time for gravitational contraction $(G\rho)^{-1/2}$ is shorter than the time L/v required for a pressure correction (L = size of the region, v = sound speed). This is the Jeans instability (1902). Equating the two gives

$$L_J = v \left(\frac{\pi}{G\rho}\right)^{1/2} ; \quad T_J = (G\rho)^{-1/2} . \qquad (10.37)$$

Mass concentrations larger than the *Jeans length* L_J will collapse on a timescale T_J, smaller ones will oscillate with a period T_J. The factor $\sqrt{\pi}$ emerges from a detailed calculation.

The Jeans instability is slowed down drastically by the expansion. Consider a homogeneous spherical mass concentration with a density ρ_i and scale factor S_i that differs from the rest of the universe. The idea is that the outside world keeps evolving unperturbed as a $k = 0$ universe – the outside

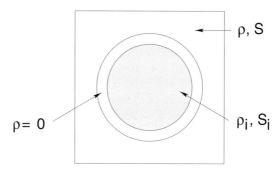

Fig. 10.5. A spherical region (density ρ_i) contracts in an expanding flat FRW universe (density ρ).

world will not notice the inner mass concentration as long as it is spherically symmetric – while the inner region evolves as a $k = +1$ universe. It does not expand as fast as the external Hubble flow, reaches maximal expansion and collapses, leading eventually to the formation of (a cluster of) galaxies. We may describe the evolution of the disturbance by two equations of the form (10.2):

$$\ddot{S}/S = -a\rho \ ; \qquad \ddot{S}_i/S_i = -a\rho_i \ . \tag{10.38}$$

For brevity we write $a \equiv 4\pi G/3$. The cosmological constant can be ignored in the early universe. Conservation of mass demands that

$$\rho_i S_i^3 = \rho S^3 \ . \tag{10.39}$$

Put $\rho_i = \rho + \delta\rho$ and $S_i = S + \delta S$ and linearise (10.39) for small $\delta\rho$ and δS:

$$0 = (\rho + \delta\rho)(S + \delta S)^3 - \rho S^3 \simeq S^3 \delta\rho + 3\rho S^2 \delta S \ ,$$

or

$$\delta\rho/\rho = -3\,\delta S/S \stackrel{\mathrm{D}}{=} x \ . \tag{10.40}$$

From (10.38):

$$\ddot{S} + \delta\ddot{S} = -a(\rho + \delta\rho)(S + \delta S)$$

$$\simeq -a\rho S - aS\delta\rho - a\rho\delta S \ . \tag{10.41}$$

Because $\ddot{S} = -a\rho S$, we are left with $\delta\ddot{S} = -aS\delta\rho - a\rho\delta S$. Now insert $\delta\rho = \rho x$ and $\delta S = -Sx/3$:

$$(Sx)^{\cdot\cdot} = 2aS\rho x = -2\ddot{S}x \ . \tag{10.42}$$

After some rearranging we find

$$\ddot{x} + \frac{2\dot{S}}{S}\dot{x} + \frac{3\ddot{S}}{S}x = 0 \ . \tag{10.43}$$

For a $k = 0$ universe in the matter era $S \propto t^{2/3}$, so that $\dot{S}/S = 2/3t$ and $\ddot{S}/S = -2/9t^2$:

$$\ddot{x} + \frac{4}{3t}\dot{x} - \frac{2}{3t^2}x = 0 \ , \tag{10.44}$$

from which we see that x must be $\propto t^\alpha$. Substitution: $\alpha^2 + \frac{1}{3}\alpha - \frac{2}{3} = 0 \rightarrow \alpha = \frac{2}{3}$ or -1. It follows that $x = \delta\rho/\rho \propto t^{-1}$ or $\propto t^{2/3}$, i.e. $\propto S$. The former solution is a more rapidly expanding perturbation connecting to the Hubble flow at $t = \infty$. The second solution is the one we are looking for, and the conclusion is that in the matter era $\delta\rho/\rho$ grows $\propto t^{2/3} \propto S$ as long as $\delta\rho/\rho \ll 1$.

The twist in the story is that the CMB gives information on the value of $\delta\rho$ at the time of recombination. Prior to decoupling, adiabatic compression generates a temperature perturbation in response to a baryon density variation $\delta\rho$:

$$\frac{\delta T}{T} = \frac{1}{3}\left(\frac{\delta\rho}{\rho}\right)_b \ . \tag{10.45}$$

If $\delta\rho$ is located on the last scattering surface we observe this δT in the CMB today, provided the density perturbation is smaller than the horizon size at recombination, i.e. for angles $\lesssim 1°$, see exercise. The observed CMB temperature difference between two directions separated by $1°$ or less is $(\delta T/T) \sim 3 \times 10^{-5}$ (§ 11.4). Consequently $(\delta\rho/\rho)_{b,\mathrm{rec}} \sim 10^{-4}$, which shows that the universe was very homogeneous during the recombination. We may now compute the present value of $(\delta\rho/\rho)_b$:

$$\left(\frac{\delta\rho}{\rho}\right)_{b,0} \simeq 10^3 \left(\frac{\delta\rho}{\rho}\right)_{b,\mathrm{rec}} \sim 0.1 \ . \tag{10.46}$$

It follows that the density contrast is only ~ 0.1, so that we would have no galaxies today, in obvious conflict with the facts. The conclusion is that structure formation in a universe filled with baryons, electrons and photons does not proceed as observed. The missing link is the non-baryonic dark matter, which turns out to be able to enhance the initial value of $(\delta\rho/\rho)_b$.

Structure and dark matter

The evolution of density perturbations in the dark matter and the baryon-electron-photon gas in the early universe is a complicated affair where only numerical simulations can provide reliable answers. We present here a much simplified description of the main issues. In this and the next section 'dark matter' is understood to be non-baryonic dark matter, assumed to be *cold*,[9]

---

[9] Non-baryonic matter is said to be hot / cold when the WIMP in question is relativistic / non-relativistic at the moment of its own decoupling.

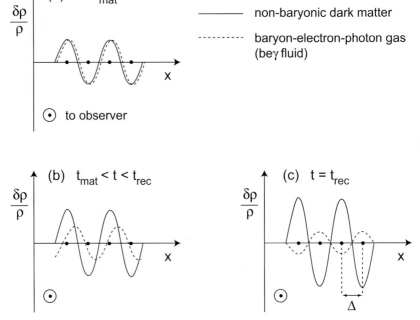

Fig. 10.6. The evolution of one Fourier component of dark matter and $be\gamma$ fluid density perturbations of the same wavelength. The x-axis is along an arbitrary direction in the surface of last scattering, and comoving with the Hubble flow along the line of sight. The expansion along the x-axis is suppressed. (a) At the beginning of the radiation era the modes have the same phase and amplitude, and fast expansion prevents growth. (b) During the matter era the dark matter mode amplitude grows while the $be\gamma$ mode is damped. (c) At recombination the photons propagate freely in all directions and carry a characteristic angular temperature modulation pattern that we observe today. The path length difference Δ between dark matter and $be\gamma$ mode, observed from a distance of the last scattering surface, determines the position of the maxima in the angular power spectrum of the CMB. See text and § 11.4.

so that the signal speed v is small. The baryon-electron-photon gas is referred to as the $be\gamma$ fluid, and includes of course all baryons, also those that develop into dark baryons later. Frequent electron-photon Thomson scattering and charge neutrality render the $be\gamma$ fluid a tightly coupled system with a very high signal speed $v = c/\sqrt{3}$, see exercise. The photons provide the pressure and the baryons the inertia. The only communication between dark matter and $be\gamma$ fluid is through perturbations $\delta\phi$ in the gravity potential. The latter are mainly generated by the density perturbations in the dark matter as it has a much higher density than the baryons.

10.4 The formation of structure

At the end of the inflation period the energy of the scalar field ψ that governs the evolution of the universe is converted into (dark) matter, § 13.3, and fluctuations $\delta\psi$ appear as density fluctuations. Dark matter and $be\gamma$ fluid have the same initial relative density distribution, so that the Fourier modes of $\delta\rho/\rho$ of both have initially the same phase and amplitude (at a given wave number). But the expansion in the radiation era is so fast that the density perturbations cannot grow (without proof), Fig. 10.6a.

In the matter era, dark matter modes with wavelengths smaller than the horizon size grow $\propto S$ as derived earlier. But $be\gamma$ modes of similar wavelengths cannot grow, because their Jeans length is much larger since $v = c/\sqrt{3}$. Computations show that $be\gamma$ mode amplitudes *decrease* and that they are outrunning their virtually stationary dark counterparts, Fig. 10.6b. In configuration space the dark matter perturbations are seen to grow, and the associated gravity perturbation tries to pull the baryons into the dark matter concentrations. But photon pressure is able to prevent that, and the $be\gamma$ fluid perturbations are actually damped.

At recombination the $be\gamma$ fluid desintegrates. Free photons depart in all directions and the baryons now do fall into the gravity wells of the dark matter, after which $(\delta\rho/\rho)_b$ grows $\propto S$. But since $(\delta\rho/\rho)_{DM}$ has grown relative to $(\delta\rho/\rho)_b$, the initial value of $(\delta\rho/\rho)_b$ is larger than the value 10^{-4} derived previously. The observed $\delta T/T$ of the CMB is further reduced by the required summing over all waves and directions, the Doppler effect due to modes that cross the last scattering surface (which we neglected sofar), perturbations along the path to the observer, etc. The upshot is that in a universe with non-baryonic dark matter the observed CMB temperature fluctuations correspond to a larger value of $(\delta\rho/\rho)_{b,rec}$ than what one would naively infer from (10.45).

Simulations[10] show that the matter distribution evolves into a cosmic web of filaments and voids, more or less as observed, see Fig. 10.7. Models in which the dark matter is hot (HDM models) predict a preponderance of large mass concentrations because the small-scale density fluctuations formed during the inflation period are largely wiped out by the fast WIMPs. Models with cold dark matter (CDM models) with slow WIMPs produce more small-mass concentrations, and agree better with the observed mass distribution.[11] The identity of the dark matter WIMP(s) is unknown.

---

[10] Bertschinger, E., *A.R.A.A.* 36 (1998) 599.
[11] See further Börner (1988) Ch. 10 ff; Kolb and Turner (1990) Ch. 9; Padmanabhan (1993); Peacock (1999) Ch. 15 ff.

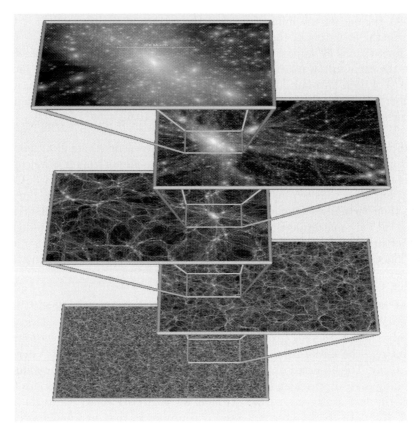

Fig. 10.7. Large-scale structure simulations have become a sophisticated industry. The computational volume of this CDM (Cold Dark Matter) simulation is a cube with sides $500/h$ Mpc at $z = 0$, about 45 times the distance to the Virgo cluster. The figure shows the projected dark matter distribution in a slice of $15/h$ Mpc (at $z = 0$) cut from the periodic simulation volume at an angle, to avoid replicating structures in the lower two images. The zoom sequence displays consecutive enlargements by factors of four, centered on one of the many galaxy clusters present in the simulation. The top frame shows several hundred gravitationally bound dark matter structures orbiting the cluster. The bottom frame shows a virtually homogeneous isotropic *cosmic web* of cold dark matter clusters, filaments and voids of characteristic size $100/h$ Mpc. The challenge for observational cosmology in the coming decades is to detect and chart the baryonic component of this cosmic web.

The bottom frame measures $3/h$ Gpc horizontally, the top frame $11/h$ Mpc. Colour coding: brightness indicates the relative density with respect to ρ_c, and colour the velocity dispersion. The simulation comprises 10^{10} particles of $8.6 \times 10^8/h\, M_\odot$ and began at $z = 127$ ($t \simeq 14\,\mathrm{Myr}$). Parameters: $\Omega_m = 0.25$, $\Omega_\Lambda = 0.75$, $h = 0.73$. Credit: the Virgo consortium. From Springel, V., et al., *Nature* **435** (2005) 629.

Imprints on the CMB

Because the last scattering takes place in a relatively short time interval, the CMB provides a snapshot of the acoustic waves in the $b e \gamma$ fluid catching modes of different wavelengths in different phases of their oscillation. At recombination the $b e \gamma$ modes have travelled a distance Δ with respect to the dark matter, and Fig. 10.6c shows the mode for which this distance corresponds to $\lambda/2$, together with its dark matter counterpart. CMB photons coming from direction A will have a higher temperature because they emerge from a region with underdense dark matter, i.e. $\delta\phi > 0$, which changes their temperature by $\delta T/T = \delta\phi/c^2$, while their initial $\delta T/T = \frac{1}{3}(\delta\rho/\rho)_b$ is also positive. Likewise, photons from direction B exhibit a lower temperature because $(\delta\rho/\rho)_b < 0$ and $\delta\phi < 0$. The result is a spatial modulation of the CMB temperature along the x-direction. Modes of different wavelength will produce a less pronounced spatial modulation. The weak side of the story is that photons react to the net gravity perturbation of all modes, and summing over all modes and directions x smoothes $\delta T/T$. Nevertheless we expect a maximum temperature difference in the CMB between directions that subtend a distance $\lambda/2$ at recombination, where λ is constrained by $(n + \frac{1}{2})\lambda = \Delta$. If d is the distance to the last scattering surface at the time of recombination, the corresponding angles are $\theta_n \simeq \{(\lambda/2)/d\}(\Omega_m + \Omega_\Lambda)^{1/2}$, or

$$\theta_n \simeq \frac{\Delta}{(2n+1)d}(\Omega_m + \Omega_\Lambda)^{1/2}. \qquad (10.47)$$

For completeness we have included the factor $(\Omega_m + \Omega_\Lambda)^{1/2}$ to allow for the fact that curvature affects the apparent angles θ_n. Since Δ/d is of the order of 0.7°, relation (10.47) predicts a grainy structure in the CMB temperature at sub-degree scales, which is clearly visible in Fig. 9.2. The corresponding peaks and their positions θ_n have now been observed in the angular power spectrum of the CMB. Relation (10.47) is not as simple as it looks because Δ depends on λ, i.e. effectively on n, and we refer to Appendix E for details. But the bottom line is that Δ can be accurately computed since it depends on linear mode physics, and so we have a yardstick of known length that we observe from a distance d, and we may use (10.47) to determine the cosmological parameters (§ 11.4).

As explained in the next chapter, points whose mutual distance is larger than the horizon size[12] cannot have exchanged any signal yet. The above applies therefore to density perturbations smaller than the horizon size $\sim 2ct_{\rm rec}$ at recombination, i.e. for angles smaller than $\sim 2ct_{\rm rec}/d \sim 1°$. Temperature differences between directions subtending larger angles are solely due to pre-existing gravity perturbations $\delta\phi$. As adiabatic compression no longer operates for these long-wavelength perturbations, relation (10.45) becomes

---

[12] i.e. about $2ct$ in the radiation era – factor 2 due to expansion.

210   10 The Evolution of the Universe

invalid and needs to be replaced. A $\delta\phi$ in the region of emergence induces a $\delta T/T = \delta\phi/c^2$. But the distance the photons have to cover also changes (this is comparable to the Shapiro delay of radio signals, § 4.4). Consequently they start their trip to us at some other instant, whence we see a different temperature. It turns out that $\delta T/T = -\frac{2}{3}\delta\phi/c^2$. The net result is called the *Sachs-Wolfe* effect:

$$\frac{\delta T}{T} = \frac{1}{3}\frac{\delta\phi}{c^2} = -\frac{1}{3}\left(\frac{H}{kc}\right)^2\left(\frac{\delta\rho}{\rho}\right)_{\text{DM}}. \qquad (10.48)$$

It is the dominant effect for density perturbations with wavelengths $2\pi/k$ larger than the horizon size at recombination. The potential perturbations are linked to dark matter density perturbations by Poisson's equation $-k^2\delta\phi = \nabla^2\delta\phi = 4\pi G\delta\rho$, and G is eliminated with (10.21) in the form $H^2 = 8\pi G\rho/3$. So $\delta\rho/\rho$ in (10.45) refers to baryons, but in (10.48) to dark matter. Recall that in this section 'dark matter' stands for non-baryonic dark matter. The final step is again a summation over all waves. For a so-called scale-free spectrum of dark matter perturbations, $\langle(\delta\rho/\rho)_k^2\rangle \propto k$, the result is that the r.m.s. CMB temperature difference between two directions subtending an angle $\theta \gg 1°$ is approximately independent of θ. [13]

Exercise 10.12: Show that a region with a diameter equal to the horizon size at recombination is now seen under an angle of $\sim 1°$.

Hint: This exercise and the next require some knowledge of the next chapters. Let's work in the subspace $t = t_0$. The horizon size at t_{rec} is about $2ct_{\text{rec}}$ (a more precise value is given in (11.20)); at t_0 this has expanded by a factor $1 + z = 1100$. In a flat universe the angle is $2ct_{\text{rec}}(1+z)/d_0 \equiv 2ct_{\text{rec}}/d$ where $d_0 = 3.3ct_0 = 3.3 \cdot 0.96c/H_0$ is the distance to the last scattering surface, Table 11.1. For the influence of expansion on the viewing geometry see Fig. 13.2.

Exercise 10.13: Show that the diameter of a sphere containing $10^{15} M_\odot$ at recombination is now seen under an angle of about $0.25°$.

Hint: Work again in the flat subspace $t = t_0$. The angle is $2R/d_0$ with R fixed by $(4\pi/3)R^3\Omega_b\rho_c = 10^{15} \cdot 2 \times 10^{33}$; see previous hint for d_0.

Exercise 10.14: Show that the signal speed in the $b e \gamma$ fluid is $c/\sqrt{3}$.

---

[13] For more information on the physics of CMB temperature fluctuations see Peacock (1999) Ch. 18; Hu and Dodelson, *A.R.A.A.* **40** (2002) 171.

Hint: Ignore the baryons and electrons as there are very few of them, see (10.36). The speed v of small perturbations in a medium with pressure p and density ρ is $v^2 = \partial p/\partial \rho$ (e.g. a gas with $p \simeq \rho v_{\rm th}^2$, so $v \simeq v_{\rm th} =$ thermal speed). A photon gas has $p = \epsilon/3$ and $\rho = \epsilon/c^2$. In reality the influence of the baryons can only be neglected in the early radiation era. In the matter era prior to recombination the signal speed is noticeably smaller than $c/\sqrt{3}$.

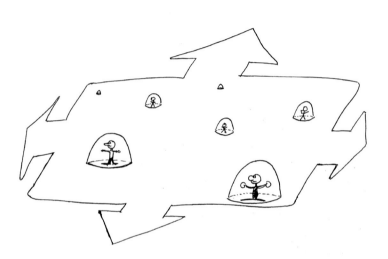

11
Observational Cosmology

The two previous chapters dealt with the properties of universes of the FRW type, which in all likelihood includes our own. The perspective was the behaviour of the homogeneous subspaces $t =$ constant, as a function of t. However, since we are located inside the universe we cannot observe these spaces. We observe events located on our past light-cone, and that gives us a totally different perspective on the universe. Our view is restricted to a small section of the universe, as epitomized in the cartoon on the left. The situation resembles observers on Earth who cannot look beyond the horizon. The question arises how the properties of these spaces $t =$ constant may be determined observationally. To this end it is necessary consider the meaning of distance in an expanding universe and to obtain the theoretical form of the Hubble relation. Attention is paid to the recent breakthrough in the determination of the cosmological parameters H_0, $\Omega_{\rm m}$ and Ω_Λ by the observation of distant type Ia supernovae and the angular correlation spectrum of the CMB. Finally, we consider the computation of observable quantities by integration over the light-cone.

11.1 Redshift and distance

The act of observing is analysed in Fig. 11.1. Our worldline is AA_0, and BB_0 is the worldline of a distant source B, at a constant co-ordinate distance r_0 from us. The *geometrical* distances of $A_0 B_0$ and AB are called d_0 and d, respectively. These are the distances of B to us in the subspace $t =$ constant at time t_0 ('now'), and at an earlier time t. The expansion makes that $d_0 > d$, but that is not visible in a co-ordinate picture. The fact that we observe B means that it emits light propagating to us on a null geodesic, arriving in A_0 at time t_0. We cannot observe B_0 because it is not on our light-cone. The shape of this light-cone is given by $d(t)$. First we determine d_0. Take $d\theta = d\varphi = 0$ in (9.20), to find that $dl = S\,dr/\sqrt{1 - kr^2}$, and integrate:

$$d_0 = S_0 \int_0^{r_0} \frac{dr}{\sqrt{1-kr^2}} = \begin{cases} S_0 f(r_0) & (k = \pm 1)\,; \\ S_0 r_0 & (k = 0)\,, \end{cases} \qquad (11.1)$$

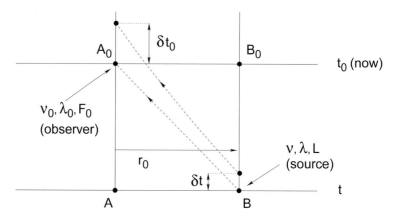

Fig. 11.1. Co-ordinate picture showing the vertical worldlines of two objects participating in the Hubble flow, at a fixed co-ordinate distance r_0. At time t (t_0) the *geometrical* distance of A and B is d (d_0). Photons emitted by B travel along null geodesics (dotted lines) and are detected in A_0. For simplicity B is assumed to be a monochromatic source (wavelength λ, frequency ν, luminosity L), while A_0 sees a wavelength λ_0, frequency ν_0 and a flux density F_0.

with $f(x) = \arcsin(\text{h})\,x$, but we need that only in § 11.3. If we define $v_0 \equiv \dot{d}_0$ then $v_0 = \dot{S}_0 f(r_0) = (\dot{S}_0/S_0)d_0$, or

$$v_0 = H_0 d_0\,, \qquad \text{with} \qquad H_0 \equiv \dot{S}_0/S_0\,. \tag{11.2}$$

Apparently, the 'geometrical speed' v_0 and the geometrical distance d_0 obey the Hubble relation. But this is rather useless as neither v_0 nor d_0 can be measured. We can only measure distances of sources that we see, i.e. are connected to us by a null geodesic, like for example B. But since B and A_0 are not in the same subspace $t = $ constant, their distance is not a well defined concept. Moreover we do not measure a velocity but rather B's redshift z:

$$z = \frac{\lambda_0 - \lambda}{\lambda}\,, \tag{11.3}$$

where λ, λ_0 = wavelength at emission by the source B, and at detection in A_0, respectively. We shall now first express z in terms of the scale factor S, and return to the distance issue later.

Fig. 11.1 shows two neighbouring null geodesics from B to A_0. These are given by[1]

$$\frac{dt}{S} = -\frac{1}{c}\frac{dr}{\sqrt{1-kr^2}}\,, \tag{11.4}$$

---

[1] Put $ds^2 = d\theta = d\varphi = 0$ in (9.19), and $dt > 0$ for $dr < 0$.

from which it is inferred that

$$\int_t^{t_0} \frac{dt}{S} = \frac{1}{c}\int_0^{r_0} \frac{dr}{\sqrt{1-kr^2}} = \frac{1}{c} f(r_0) . \qquad (11.5)$$

This relation determines the time of emission t for given r_0. By comparing (11.5) and (11.1) we see that

$$d_0 = cS_0 \int_t^{t_0} \frac{dt}{S} . \qquad (11.6)$$

We now have two expressions for d_0. In (11.1) we know only the co-ordinate distance r_0, but in (11.6) we have exploited the extra information of 'eye contact' to eliminate r_0. Because the right hand side of (11.5) is constant we have

$$\int_t^{t_0} \frac{dt}{S} = \int_{t+\delta t}^{t_0+\delta t_0} \frac{dt}{S} \quad \rightarrow \quad \frac{\delta t_0}{S_0} = \frac{\delta t}{S} . \qquad (11.7)$$

Furthermore we know that $\nu \delta t = \nu_0 \delta t_0$, so that $\lambda_0/\lambda = \nu/\nu_0 = \delta t_0/\delta t = S_0/S$, and

$$z = \frac{\lambda_0 - \lambda}{\lambda} = \frac{S_0}{S} - 1 . \qquad (11.8)$$

We observe that $z > 0$ and this is now seen to be a consequence of the expansion: the scale factor increases, $S(t_0) > S(t)$. Apparently, the wavelength of the photon is stretched in proportion to the expansion of the universe.

It is illuminating to derive the redshift from a different perspective. According to (9.23) or (9.25) a radial distance $d\ell$ from the origin is equal to $d\ell = S d\chi$. The local velocity v of a particle is therefore $v = d\ell/dt = S d\chi/dt = \lambda c/\sqrt{S^2 + \lambda^2}$, according to (9.48). A little algebra shows that $Sv/\sqrt{1-(v/c)^2} = $ constant, or

$$pS = \text{constant} , \qquad (11.9)$$

where p is the particle's momentum. This says that the De Broglie wavelength h/p of the particle scales $\propto S$, and expresses the fact that particles are subject to adiabatic cooling as the universe expands. Note that relation (11.9) holds also for photons since $p = \hbar k = \hbar \omega/c$.

The redshift z is a key observable in cosmology, and astronomers are habitually given to jargon like 'the universe at redshift z'. This expression indicates the spherical shell around us denoted as $\Sigma(t)$ in Fig. 9.3, containing all sources at that redshift, assuming that they follow the Hubble flow. It is the cross section of our past light-cone and the subspace $t = $ constant, where t is fixed by (11.8) and $S = S(t)$. However, the phrase is also used to indicate the entire homogeneous subspace $t = $ constant – a space that we cannot observe (but of course a very convenient theoretical concept).

Alternative explanations for the redshift have been advanced, such as the tired light concept. The idea is that photons would be subject to a small systematic energy loss as they propagate through space. That would mimic Hubble's law, in the absense of a real expansion. The main problem with this explanation is that any mechanism that changes the energy of a photon will also affect its momentum. That is, to some degree it is a scattering process. Distant objects would be blurred – contrary to what is observed. Furthermore, in the standard interpretation of the redshift, light curves of distant supernovae should broaden with z, as is observed, but tired light would produce no such broadening. Other explanations suffer from similar objections, and the conclusion that the universe expands seems inescapable.

Cosmological models

We are now in a position to construct a cosmological model, that is, a listing of the age t of the universe, of d, d_0 and the luminosity distance d_L (a concept defined in § 11.3) as a function of redshift, see Table 11.1. This table is constructed as follows. We begin by rewriting (11.8):

$$u = \frac{S}{S_0} = \frac{d}{d_0} = \frac{1}{1+z}. \tag{11.10}$$

Here we have used that $d = Sf(r_0)$ so that $d/d_0 = S/S_0$. Relation (11.10) fixes d/d_0, and t since $S = S(t)$. To make this more explicit, start with $dt = (dt/du)du$, so that $t = \int_0^u du/\dot{u}$, and:

$$\frac{t}{t_0} = \frac{\int_0^u du/\dot{u}}{\int_0^1 du/\dot{u}}. \tag{11.11}$$

The upper integration limit u equals $1/(1+z)$. The explicit expression for t and t_0 is given in (10.14) and (10.15). Next, we write (11.6) in dimensionless form with the help of $cS_0 dt/S = c dt/u = cdu/(u\dot{u})$:

$$\frac{d_0}{ct_0} = \frac{\int_u^1 (u\dot{u})^{-1} du}{\int_0^1 du/\dot{u}}. \tag{11.12}$$

And finally, $d = ud_0$. The next step is to substitute $\dot{u}$ from (10.8), after which numerical evaluation of (11.11), (11.12) and (11.27) is straightforward. We have normalised distances to ct_0, that is, the light distance corresponding to the age of the universe. The advantage of using relative quantities in Table 11.1 like t/t_0 and distances/ct_0 is that there is no longer a big difference between the models. This is why the reference model $(\Omega_m, \Omega_\Lambda) = (1, 0)$ remains very useful even if $(\Omega_m, \Omega_\Lambda) \neq (1, 0)$.

11.1 Redshift and distance

Table 11.1. Two FRW universe models [a]

	$\Omega_m = 1$; $\Omega_\Lambda = 0$; $H_0 t_0 = 0.67$				$\Omega_m = 0.3$; $\Omega_\Lambda = 0.7$; $H_0 t_0 = 0.96$			
z	t/t_0	d/ct_0	d_0/ct_0	d_L/ct_0	t/t_0	d/ct_0	d_0/ct_0	d_L/ct_0
0	1	0	0	0	1	0	0	0
0.2	0.76	0.22	0.26	0.31	0.82	0.16	0.20	0.24
0.5	0.54	0.37	0.55	0.83	0.63	0.30	0.46	0.69
1	0.35	0.44	0.88	1.8	0.43	0.40	0.80	1.6
2	0.19	0.42	1.3	3.8	0.24	0.42	1.3	3.8
5	6.8-2	0.30	1.8	1.1+1	8.6-2	0.31	1.9	1.1+1
10	2.7-2	0.19	2.1	2.3+1	3.5-2	0.21	2.3	2.5+1
30	5.8-3	7.9-2	2.4	7.6+1	7.3-3	8.9-2	2.7	8.5+1
100	9.9-4	2.7-2	2.7	2.7+2	1.2-3	3.0-2	3.1	3.1+2
1000	3.2-5	2.9-3	2.9	2.9+3	4.0-5	3.3-3	3.3	3.3+3
∞	0	0	3	∞	0	0	3.4	∞

[a] Notation: $a \pm b \equiv a \times 10^{\pm b}$;
t = age of universe at the time the object emits the light we now see;
d = geometrical distance of object at time t;
d_0 = geometrical distance of object now, at time t_0;
d_L = luminosity distance (11.27) of the object.

Exercise 11.1: Show that the invariant definition of the redshift is:

$$1 + z = \frac{(k_\alpha u^\alpha)_e}{(k_\alpha u^\alpha)_o}, \tag{11.13}$$

where the index e, o indicates the emittor and the detector, respectively; u^α = 4-velocity (of the emitter or the detector), and k^α is the photon wavevector.

Hint: $1 + z = \lambda_o/\lambda_e = \nu_e/\nu_o = E_e/E_o$. Then (3.55) with $p^\alpha = \hbar k^\alpha$.

Exercise 11.2: Show that the age of an FRW universe at $z \gg 1$ is independent of Ω_Λ:

$$H_0 t = \tfrac{2}{3}\Omega_m^{-1/2}(1+z)^{-3/2}. \tag{11.14}$$

Hint: For $u = S/S_0 \ll 1$ only the Ω_m/u-term in (10.14) matters. If we compute $t_{\rm rec}$ of our universe with (11.14) the result is $t_{\rm rec} = 4.8 \times 10^5$ yr. Why is this larger than (10.30)?

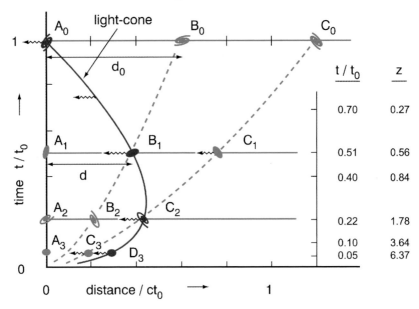

Fig. 11.2. Scale model of an $(\Omega_m, \Omega_\Lambda) = (1, 0)$ FRW universe. On the vertical axis the age t of the universe in units of the present age t_0. On the horizontal axis, in green, a 1D cross section along an arbitrary line of sight with equidistant galaxies (for simplicity). Distance scale: we arbitrarily adopt $A_0C_0 = 1.2\,ct_0$. The galaxies partake in the universal expansion (- - -) and evolve with time as they do so. Also indicated are the geometrical distances d and d_0 from Table 11.1, and A_0's past light-cone $d(t)$ in red. The wiggly lines are photons travelling locally with speed c. Adapted from Hoyng, P., Zenit, July/August 1998, p. 340.

11.2 The visible universe and the horizon

We shall now take a closer look at the properties of FRW universes as given in Table 11.1. Since we discuss issues here that most FRW universes share, we focus on the $(\Omega_m, \Omega_\Lambda) = (1, 0)$ model as a typical example. Age and expansion of this universe are given by (10.12): $S/S_0 = (t/t_0)^{2/3}$ and $t_0 = \frac{2}{3}H_0^{-1}$. For the shape of the past light-cone $d(t)$, it is easiest to use (11.6) because we know $S(t)$: $d = (S/S_0)d_0 = cS \int_t^{t_0} dt/S = ct^{2/3} \int_t^{t_0} dt/t^{2/3}$, or

$$\frac{d}{ct_0} = 3\left(\frac{t}{t_0}\right)^{2/3}\left\{1 - \left(\frac{t}{t_0}\right)^{1/3}\right\}. \tag{11.15}$$

This leads to the scale model shown in Fig. 11.2. The horizontal axis of this figure is a 1D cross section through the universe along an arbitrary line of

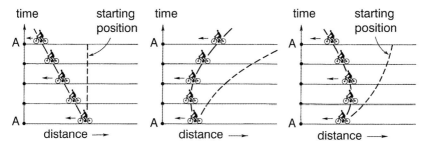

Fig. 11.3. Photon propagation in an expanding universe may be understood with the example of a cyclist moving at constant speed c with respect to the local road, while the road is being stretched like a rubber band. Left: no expansion, the path is a straight line with inclination c. Middle: the expansion is initially slow, but accelerates with time. Right: expansion is initially fast, but slows down with time, as in a real universe (for $\Omega_\Lambda = 0$). Adapted from Hoyng, P., Zenit, July/August 1998, p. 340.

sight. On this axis are located our system A, then B, next C, etc. The broken lines show how the universe expands $\propto t^{2/3}$. Each 1D cross section may be generalised to a 3D image of the universe at that age. This is the green section of Fig. 11.2. It is effectively an *external* point of view: the observer is located outside the universe and surveys the entire universe at a glance, as if one is studying a map.

However, due to the finite speed of light we (A_0) do not see our neighbours at the same time t_0 but at some earlier time. All light that we receive at t_0 must have travelled along the past light-cone, given by (11.15). Some photons come from far and began their journey long ago, while others enjoyed only a brief trip. But all have travelled along the path marked light-cone, indicated in red in Fig. 11.2. Hence, we see the systems B_1, C_2, D_3,..., behind each other, at progressively larger redshift. These systems are juvenile forms of B_0, C_0, D_0,.. located in the universe at time t_0. The upshot is that we experience the universe as a series of nested spherical shells, each showing a different piece of an increasingly younger universe. This is the *internal* point of view, that of an observer inside the universe. Note that Fig. 11.2 is also the geometrical picture corresponding to the co-ordinate picture in Fig. 9.3.

The shape of the past light-cone may be understood with the help of Fig. 11.3. A photon in an expanding universe is like a cyclist on a road that is being stretched like a rubber band. The cyclist moves always at constant speed c *with respect to the road* (locally special relativity holds). The right panel corresponds to the situation in the universe. The expansion is initially fast and slows down gradually. The cyclist is initially 'drawn away' from A

(us), but may eventually reach any position in the direction of cycling. This picture is an exact model of photon propagation, as we shall now show. The distance d between A and the cyclist obeys

$$\dot{d} = Hd - c , \qquad (11.16)$$

where $H = H(t)$. The first term describes the homogenous stretching of the road, the second term the motion with respect to the road. Substitute $H = \dot{S}/S$, and (11.16) may be written as

$$\left(\frac{d}{S}\right)^{\cdot} = -\frac{c}{S} . \qquad (11.17)$$

Integration yields $d/S = -c \int_0^t dt/S + \text{const}$. Initial condition: $d = 0$ at $t = t_0$. Result: $d = cS \int_t^{t_0} dt/S$, which coincides with (11.6) since $d = (S/S_0)d_0$.

We draw attention to two remarkable properties of FRW models. The first is that according to Fig. 11.2 distant sources at large z were relatively near to us at the time they emitted the radiation we now see. Formally $d \to 0$ as $t \to 0$. In spite of this proximity, the light could not reach us any sooner because the universe was expanding so much faster than it does today – otherwise it would have long since recollapsed. One might say that new space is created at a very high rate, which makes that the photon 'moves away from us as it travels in our direction'. Only later, when the expansion has slowed down, the photon is able to reach us. The inward bending of our past light-cone at large z is therefore caused by the extremely rapid expansion of the early universe.

The particle horizon

The second feature is that $d_0 \to 3ct_0$ or thereabout for $z \to \infty$, see Table 11.1. The present distance of the remotest objects that we can see is apparently not larger than $\sim 3ct_0$. Let's check that for the $(\Omega_m, \Omega_\Lambda) = (1, 0)$ universe: $d_0 = (S_0/S)d = (t_0/t)^{2/3}d$, according to (10.12), and with (11.15): $d_0 = 3ct_0\{1 - (t/t_0)^{1/3}\} \to 3ct_0$ for $t \to 0$. This boundary is called the *horizon*, more precisely the particle horizon. Since light travels locally at speed c one may say that a photon has traversed a distance ct_0 from the moment of the Big Bang. The expansion increases the distance between starting and arrival point of the photon apparently by another $2ct_0$. This extra amount depends on the details of the expansion, i.e. on $S(t)$, but not very strongly. For a $(1, 0)$ reference universe $t_0 = \frac{2}{3}H_0^{-1}$, and the horizon distance is $2c/H_0$. And for an $(\Omega_m, \Omega_\Lambda) = (0.3, 0.7)$ universe the horizon is at $3.4ct_0$ (Table 11.1), which is equal to $3.4c \cdot 0.96/H_0 \sim 3.3c/H_0$.

The horizon distance in a FRW universe is apparently a few times the

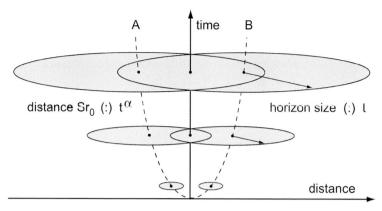

Fig. 11.4. The visible universe is the space inside the horizon of an observer. It contains all matter from which the observer may have received a light signal. The visible universes of any two observers A and B comoving with the Hubble flow overlap progressively, but were disjunct at some point in the past. A can only see B and vice versa after they have entered each other's horizon. This leads to the so-called *horizon problem*: why do A and B begin to participate in the expansion at the same moment?

Hubble radius c/H_0. The space inside the horizon is called the *visible universe*, sometimes just *horizon space*. Note that each observer has its own visible universe, see cartoon on p. 212. The name horizon derives from the analogy with the terrestrial horizon. An object can only have interacted with objects inside its horizon – anything outside can have had no influence.[2]

Consider two point A and B at a fixed co-ordinate distance r_0, Fig. 11.4. In a $k = 0$ universe their geometrical distance d at time t is $S(t)r_0$. It follows that $d \propto t^\alpha$ with $\alpha \simeq 1/2$ in the radiation era, $\alpha \simeq 2/3$ in the matter era. The horizon distance at that time is (put $t \to 0$ and $t_0 \to t$ in (11.6)):

$$d = cS \int_0^t \frac{dt}{S} = \frac{ct}{1-\alpha}, \qquad (11.18)$$

or $3ct$ in the matter era and $2ct$ in the radiation era. The general expression is given in (11.20). It follows that the horizon distance grows eventually faster than the geometrical distance, so that the visible universes of A and B will overlap more and more in the future. Conversely, regardless of the distance of A and B, if we go back in time, there comes a moment that their horizon spaces were disjunct. This leads to the so-called *horizon problem*, a fundamental defect shared by all FRW universes, that has only been remedied by the advent of inflation theory, see Ch. 13.

---

[2] An $\Omega_\Lambda \neq 0$ universe possesses also an event horizon, see exercise 11.6.

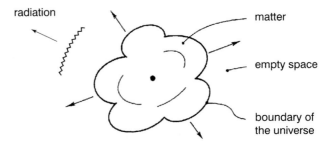

Fig. 11.5. The Big Bang is often misinterpreted as a point explosion, with matter expanding into a pre-existing empty space. Adapted from Hoyng, P., Zenit, July/August 1998, p. 340.

A common mistake

The nature of the Big Bang is often misunderstood. The very name suggests an analogy with a point explosion, Fig. 11.5. This seems a rather natural idea, and that may explain why it appears to be so popular. But it is in conflict with the observations. Briefly, the argument is as follows. From the average density in the universe and the fact that the edge of the explosion is at most ct_0 away, the optical depth to the boundary is inferred to be much smaller than 1, so that we should be able to see it. But the universe is also observed to be highly isotropic. These two statements are incompatible unless we are located at the centre of a spherically symmetric explosion, which is highly unlikely. Accepting that would mean a relapse to some kind of a geocentric world model. It would, incidentally, also be impossible to explain the CMB as a remnant of the Big Bang since any radiation emitted by the explosion is necessarily ahead of the matter.

The correct picture is that the universe has no boundary, that space is homogeneously filled with matter, and that space itself is swelling. This picture emerges clearly from the derivation of the Robertson-Walker metric in §§ 9.2 and 9.3. The galaxies have constant co-ordinates ('do not move') and are rather like currants in a rising bun. This picture of a swelling space should be used with care. The wavelength of a photon (more generally, the De Broglie wavelength of a particle) is stretched proportional to S, indeed, stretched with the swelling of space, but that does not imply that extended material objects expand as well. That would only happen if the various parts of the object move along geodesics of the Robertson-Walker metric. But this is usually not the case due to extra forces, for example internal elastic forces in a measuring rod, or local gravity in galaxies.

11.2 The visible universe and the horizon

Exercise 11.3: Show that in an $(\Omega_m, \Omega_\Lambda) = (1, 0)$ universe the photons we see today (including those of the CMB) have never been farther away from us than $d = (2/3)^2 ct_0 \simeq 0.44 ct_0$. This happened at $t/t_0 = (2/3)^3 \simeq 0.30$. At that point the photon just beats the expansion and its geometrical speed $\dot{d}$ to us is zero. Any conflict with SR?

Hint: See Fig. 11.2, and determine the maximum of (11.15). There is no conflict with SR: the locally measured speed of the photon is always c. The co-ordinate speed of a photon falling into a black hole also becomes zero near the horizon, § 6.3.

Exercise 11.4: Continue exercise 9.9 and prove that in an $(\Omega_m, \Omega_\Lambda) = (1, 0)$ universe the bullet travels a co-ordinate distance

$$\Delta r = \frac{2\beta}{S_0} \frac{c}{H_0}, \qquad (11.19)$$

provided the initial velocity is small, $\beta \ll 1$. This result may be interpreted as follows. Mark the position that the bullet will eventually reach as A. The geometrical distance between the point of firing and A is now $2\beta(c/H_0)$.

Hint: (9.48) becomes $dr/dt \simeq \beta c S_0/S^2$ or $\Delta r \simeq \beta c S_0 \int_{t_0}^\infty dt/S^2$; (10.12): $S/S_0 = (t/t_0)^{2/3}$ and integrate. Then use (11.1).

Exercise 11.5: Show that the horizon distance in the early universe is given by:

$$d = 3ct_m x \left(\sqrt{x+1} - 1 \right), \qquad (11.20)$$

in the notation of § 10.3. Show that $d = 2ct$ for early times, $3ct$ for late times, and $2.25 ct_{rec}$ at recombination. Does this imply that the speed of light is larger than c, or that the horizon actually moves at superluminal speed?

Hint: $d = cS \int_0^t dt/S = cu \int_0^u du/(u\dot{u}) = cx \int_0^x dx/(x\dot{x}) = ct_m x \int_0^x dx/(x\, dx/d\tau)$; insert (10.25): $d = 3ct_m x \int_0^x dx/(2\sqrt{1+x}) \to$ (11.20). For early times ($x \ll 1$): $d \simeq \frac{3}{2} ct_m x^2$; then (10.29): $d \simeq 2ct_m \tau = 2ct$. For large x: $d \simeq 3ct_m x^{3/2} = 3ct_m \tau = 3ct$. At recombination $x = 3 \to d = 9ct_m = \frac{9}{4} ct_{rec}$. Twice no.

Exercise 11.6: Show that an $\Omega_\Lambda \neq 0$ universe has also an event horizon and compute its size.

Hint: Horizons delineate spheres of influence. The particle horizon embraces all points (at time t_0) that have been able to interact with us in the past. Points inside the event horizon will interact with us in the future (however distant). Geometrical distance of starting position of a photon that reaches us at T is $d_0 = cS_0 \int_{t_0}^{T} dt/S$. Let $T \uparrow \infty$: $d_0 = c\int_1^\infty du/u\dot{u} = (c/H_0)\int_1^\infty du(\Omega_m u + \Omega_\Lambda u^4 + \Omega_k u^2)^{-1/2}$ which converges if $\Omega_\Lambda \neq 0$. For $\Omega_k = 0$ (flat universe): $d_0 < \{c/(H_0\sqrt{\Omega_\Lambda})\}\int_1^\infty du/u^2 = c/(H_0\sqrt{\Omega_\Lambda})$. Photons departing to us from beyond d_0 will *never* reach us due to the exponential expansion.

11.3 Luminosity distance and Hubble relation

The geometrical distances d and d_0 in Fig. 11.2 are convenient theoretical concepts but they cannot be measured. We shall not dwell on the issue of distance determination here, as it is a large topic in its own right. We restrict ourselves to illustrating how d and d_0 can be determined through the method of *standard candles*, a time-honoured method to find distances of remote objects. The idea is that there are classes of objects whose members all have about the same absolute luminosity. For example, Cepheid variables with the same oscillation period, the brightest member of a cluster, type Ia supernovae, etc. Once the distances to a subset of objects have been determined independently, we only have to recognise a source as a member of its class, and its absolute luminosity L is known, at least in principle. This leads to the concept of *luminosity distance* d_L, a measurable quantity, defined as $L = 4\pi d_L^2 F_0$ where F_0 is the flux density of the source measured at t_0, and L the luminosity of the source at emission, Fig. 11.1.

For convenience we assume that the source is monochromatic. Number of photons emitted in δt seconds: $\delta N = (L/h\nu)\delta t$. These are spread over a spherical surface of area $O = 4\pi S_0^2 r_0^2$, see below relation (9.22), so that

$$F_0 = \frac{h\nu_0 \, \delta N}{O \, \delta t_0} = \frac{L}{4\pi S_0^2 r_0^2} \frac{\delta t}{\delta t_0} \frac{\nu_0}{\nu}$$

$$= \frac{L}{4\pi S_0^2 r_0^2} \frac{1}{(1+z)^2} \, , \qquad (11.21)$$

from which it follows that

$$d_L = \left(\frac{L}{4\pi F_0}\right)^{1/2} = r_0 S_0 (1+z) \, , \qquad (11.22)$$

11.3 Luminosity distance and Hubble relation

In particular for $k=0$:
$$d_{\rm L} = (1+z)d_0 \ . \tag{11.23}$$

The luminosity distance is a formal quantity in the sense that is not possible to indicate a space 'in which $d_{\rm L}$ lies', as we could in case of d and d_0. The point is, however, that $d_{\rm L}$ can be measured, and then d_0 is also known through (11.23) or (11.26).[3]

We shall now derive the Hubble relation, i.e. the relation between the two observable quantities $d_{\rm L}$ and z. We start from (11.1), and note that the function $f(x)$ equals arcsin x, x, arcsinh x for $k = 1, 0, -1$. Relation (11.1) may now be inverted:
$$r_0 = {\rm sinn}(d_0/S_0) \ , \tag{11.24}$$

with
$$\operatorname{sinn} x = \begin{cases} \sin x & (k = 1) \ ; \\ x & (k = 0) \ ; \\ \sinh x & (k = -1) \ . \end{cases} \tag{11.25}$$

This means that $r_0 = {\rm sinn}\big(|\Omega_{\rm k}|^{1/2} H_0 d_0/c\big)$, because $c/H_0 S_0 = |\Omega_{\rm k}|^{1/2}$ according to (10.5). Insert that in (11.22):
$$\frac{H_0 d_{\rm L}}{c} = |\Omega_{\rm k}|^{-1/2} (1+z) \, {\rm sinn}\!\left(|\Omega_{\rm k}|^{1/2} \frac{H_0 d_0}{c} \right), \tag{11.26}$$

and we have found the generalization of (11.23) for $k \neq 0$. Next we use (11.6) or (11.12) to obtain $d_0 = c\int_u^1 du/u\dot u$ and take $\dot u$ from (10.8):
$$\frac{H_0 d_{\rm L}}{c} = |\Omega_{\rm k}|^{-1/2} (1+z) \, {\rm sinn}\bigg\{ |\Omega_{\rm k}|^{1/2} \int_u^1 dx$$
$$\left(\Omega_{\rm m} x + \Omega_\Lambda x^4 + \Omega_{\rm k} x^2 \right)^{-1/2} \bigg\} \ . \tag{11.27}$$

Since $u = 1/(1+z)$ we have found the theoretical form of the Hubble relation. Simplification is possible if z is small: the integration limits 1 and $1/(1+z)$ are close to each other, so that the argument of the sinn-function becomes small and we may use ${\rm sinn}\, x \simeq x$:
$$\frac{H_0 d_{\rm L}}{c} \simeq (1+z) \int_{(1+z)^{-1}}^{1} dx \left(\Omega_{\rm m} x + \Omega_\Lambda x^4 + \Omega_{\rm k} x^2 \right)^{-1/2} . \tag{11.28}$$

For $k = 0$ this relation is exact for all z. For small z (11.27) may be approximated as (see exercise):

---

[3] There are also other distance measures in use, such as the angular diameter distance. See Lightman et al. (1975), exercise 19.9.

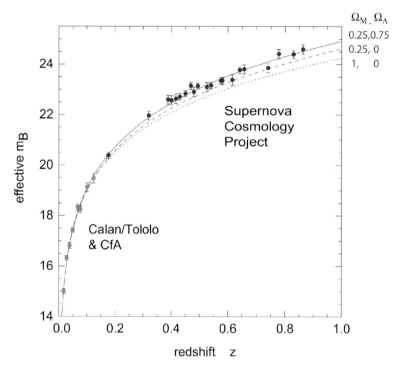

Fig. 11.6. Hubble diagram of type Ia supernovae. Datapoints within $\Delta z < 0.01$ have been grouped together into a single average datapoint. Also shown are the theoretical curves for three FRW universe models. These data provide direct evidence for the existence of dark energy, i.e. a positive cosmological constant Λ. From Knop, R.A. et al., *Ap. J. 598* (2003) 102.

$$\frac{H_0 d_L}{c} \simeq z\left\{1 + \tfrac{1}{2}(1 - \tfrac{1}{2}\Omega_m + \Omega_\Lambda)z + \cdots\right\}, \qquad (11.29)$$

The parameters H_0, Ω_m and Ω_Λ determine the structure and the evolution of the universe, and large efforts have been undertaken to determine their values, in particular during the last decades. The principle is straightforward. A fit of observations of z and d_L to (11.29) yields H_0 and $\tfrac{1}{2}\Omega_m - \Omega_\Lambda$. But the method is plagued by many problems such as selection effects, a limited redshift range ($z \lesssim 0.3$) and the fact that standard candles are not perfect. There is always a spread in intrinsic luminosities. For a long time this caused astronomers to be at loggerheads about the value of H_0,[4] while $\tfrac{1}{2}\Omega_m - \Omega_\Lambda$ could not really be determined. The negative correlation between Ω_m and Ω_Λ, incidentally, is easy to understand: more matter ($\Omega_m \uparrow$) means more

---

[4] Weinberg (1972) p. 441 ff; Börner (1988) § 2.2; Peebles (1993) Ch. 5; Fukugita, M. et al. *Nature 366* (1993) 309.

gravity, and that may be compensated by adding antigravity, i.e. more vacuum energy ($\Omega_\Lambda \uparrow$). It is unfortunate that quasars which have redshifts up to $z \sim 5$ are no good as standard candles: in a given redshift interval their apparent magnitudes vary greatly. Otherwise the values of H_0, Ω_m and Ω_Λ would have long since been known.

These efforts have culminated in an HST Key Project to measure H_0, which has led to the value $H_0 = 72 \pm 8$ km s^{-1}Mpc^{-1}.[5] The subsequent measurement of H_0 by the WMAP mission has confirmed this value with improved accuracy: $H_0 = 71 \pm 4$ km s^{-1}Mpc^{-1}. It is encouraging that this value is now being confirmed by independent techniques, such as the Sunyaev-Zeldovich effect, a method that does not rely on the classic (and slippery) distance ladder.[6] Since the first measurements of H_0 around 1930 its value has come down by almost an order of magnitude.[7]

The Supernova Cosmology Project has used distant Type Ia supernovae, bright objects that may be detected out to $z \lesssim 1$, and have been shown to be rather reliable standard candles. They are believed to be white dwarfs with progenitor masses in the range $4 - 6\, M_\odot$, pushed over the Chandrasekar limit by mass transfer. When they explode they all have (hopefully) the same mass and composition, which explains the standard candle property. The project managed to measure Ω_m and Ω_Λ independently (with low accuracy), Fig. 11.6. The supernova data clearly demonstrate that the cosmological constant of our universe is nonzero.

Exercise 11.7: Show that the redshift is given by the Doppler formula $z = v/c$ for small z, but GR corrections become important at larger z.

Hint: From (9.4) and (11.29); $d \simeq d_L$ for small z.

Exercise 11.8: Provide the details of the derivation of (11.29).

Hint: Write $(\cdots)^{-1/2} \equiv g(x)$ in (11.27); $g(1) = 1$ on account of (10.4) and $a \equiv g'(1) = -1 + \frac{1}{2}\Omega_m - \Omega_\Lambda$. Put $x = 1 - y$ and expand the integral to second order in z: $\int_{(1+z)^{-1}}^{1} g(x)\,dx \simeq \int_0^{z-z^2} g(1-y)\,dy \simeq \int_0^{z-z^2}(1-ay)\,dy \simeq z(1 - z - \frac{1}{2}az)$. And $\sinn x = x$ to second order in x.

---

[5] Freedman, W.L. et al., *Ap. J.* **553** (2001) 47.
[6] Mason, B.S. et al., *Ap. J.* **555** (2001) L11.
[7] Trimble, V., *P.A.S.P.* **108** (1996) 1073.

Exercise 11.9: Prove the Hubble relation for an $(\Omega_m, \Omega_\Lambda) = (1, 0)$ universe:

$$\frac{H_0 d_L}{c} = 2(1+z)\left(1 - \frac{1}{\sqrt{1+z}}\right), \qquad (11.30)$$

and show once more that the distance to the horizon is $3ct_0$.

Hint: $(11.28) \rightarrow H_0 d_L/c = (1+z) \int_{(1+z)^{-1}}^{1} dx/\sqrt{x}$; (11.23): $d_0 = d_L/(1+z) \rightarrow 2c/H_0 = 3ct_0$ for $z \rightarrow \infty$.

Exercise 11.10: Given an object at $z = 0.5$ in an $(\Omega_m, \Omega_\Lambda) = (2, 0)$ FRW universe. What are the values of: (1) the co-ordinate r, (2) the distance d of the object at the time of emission of the light we receive from it today, and (3) the temperature of the CMB at that particular time.

Hint: For large values of z there is no alternative but to make a new Table 11.1 by numerical integration. For small z (11.29) is an option $\rightarrow H_0 d_L/c \simeq 0.5$. Then (11.26): $H_0 d_0/c \simeq \arcsin(1/3) \simeq 0.34$. Expression for r_0 above (11.26): $r_0 \simeq 1/3$. Furthermore $d = (S/S_0)d_0 = d_0/(1+z) = 2d_0/3$. Temperature CMB: $(1+z)2.725\,\text{K} = 4.09\,\text{K}$.

11.4 The microwave background

The COBE satellite has measured the spectrum and the angular distribution of the temperature of the CMB on angular scales of $7°$ and larger, and $\Delta T/T$ was found to be of order 10^{-5}. The CMB is therefore highly isotropic. The angular distribution of the CMB temperature is a very important issue as it carries information on the clustering of matter in the universe at decoupling, $z \sim 1100$. Various groups have measured $\Delta T/T$ down to spatial scales of $\sim 0.1°$ in a section of the sky.[8] The WMAP mission launched in 2001 has mapped the entire sky with a resolution of $\sim 0.2°$. The maps are cleaned from foreground effects, and the resulting temperature distribution is decomposed in spherical harmonics $Y_{\ell m}(\theta, \varphi)$:

$$\Delta T(\theta, \varphi) \equiv T(\theta, \varphi) - T_0 = \sum_{\ell=1}^{\infty} \sum_{m=-\ell}^{\ell} a_{\ell m} Y_{\ell m}, \qquad (11.31)$$

---

[8] E.g. De Bernardis, P. et al., *Nature* **404** (2000) 955; Lee, A.T. et al., *Ap. J.* **561** (2001) L1.

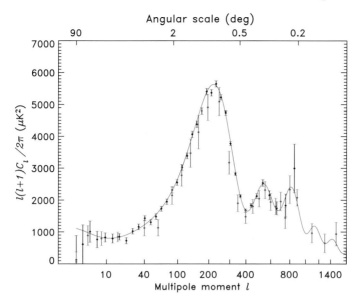

Fig. 11.7. Angular power spectrum of the CMB temperature as measured by WMAP (points in black), obtained by processing the data through (11.32) and (11.35). The angular scale (top) is added afterwards for convenience. The black line is the best fit to a ΛCDM model (= CDM model with $\Lambda \neq 0$). The red points are previously published results. From Hinshaw, G. et al., *Ap. J. S.* **148** (2003) 135.

where $T_0 = \langle T \rangle$ is the average temperature, and $a_{\ell m}$ is given by

$$a_{\ell m} = \int d\Omega \, \Delta T \, Y^*_{\ell m} \, . \tag{11.32}$$

We consider two averages: (1) an average over an ensemble of maps i (all possible realisations of the CMB sky), in terms of which, for example, $\langle T \rangle = \lim_N N^{-1} \sum_i T_i(\boldsymbol{n})$; (2) an angular average over one CMB map, and $\langle \cdot \rangle = (4\pi)^{-1} \int d\Omega$.

The three dipole coefficients a_{1m} are dominated by the Doppler signal due to a net velocity of the solar system of $371 \pm 1 \text{ km s}^{-1}$ with respect to the surface of last scattering. The intrinsic dipole anisotropy of the CMB is presumably much smaller, but cannot be separated from the total signal. The angular correlation function $C(\theta)$ is defined as:

$$C(\theta) = \langle \Delta T(\boldsymbol{n}_1) \Delta T(\boldsymbol{n}_2) \rangle \, | \, \boldsymbol{n}_1 \cdot \boldsymbol{n}_2 = \cos\theta \tag{11.33}$$

$$= \frac{1}{4\pi} \sum_\ell (2\ell + 1) C_\ell \, P_\ell(\cos\theta) \, , \tag{11.34}$$

with
$$C_\ell = \langle |a_{\ell m}|^2 \rangle \simeq \frac{1}{2\ell+1} \sum_m |a_{\ell m}|^2 \ . \tag{11.35}$$

$P_\ell(x)$ is the Legendre polynomial of order ℓ. The first = sign in (11.35) defines C_ℓ as an ensemble average; $\langle |a_{\ell m}|^2 \rangle$ does not depend on m (spherical symmetry). C_ℓ may also be estimated by the second expression, obtained by angular averaging, where $a_{\ell m}$ are the expansion coefficients of the one CMB sky we have. At small ℓ the values of the two expressions differ appreciably due to *cosmic variance*.[9] For completeness we mention that relation (11.34) may be inverted with the help of the orthogonality of the Legendre polynomials:

$$C_\ell = 2\pi \int_0^\pi C(\theta) P_\ell(\cos\theta) \sin\theta \, d\theta \ . \tag{11.36}$$

The proof of relations (11.34), (11.35) and (11.36) is somewhat technical and deferred to Appendix E.

Fig. 11.7 shows the measured values of $c_\ell \equiv \ell(\ell+1)C_\ell/2\pi$, referred to as the *angular power spectrum*. We recognize a flat plateau at low ℓ, followed by a series of peaks at larger ℓ, thus confirming the physical explanation given in § 10.4. We have found earlier that the directions of maximal CMB temperature difference subtend angles θ_n given by (10.47). In Appendix E it is shown that this implies that the first peak in the power spectrum is at

$$\ell_0 \simeq \pi/\theta_0 \simeq 277 \ , \tag{11.37}$$

while the observed value is $\ell = 220 \pm 1$. The origin of the discrepancy is that our treatment ignores two aspects of the physics of $be\gamma$ modes, see Appendix E.

A comparison of the WMAP data in Fig. 11.7 with model simulations of the c_ℓ allows a precise determination of the cosmological parameters, see Table 9.2. In brief outline the story is as follows. The WMAP data, HST Key Project and supernova data together determine $\Omega_{\text{tot}} \equiv \Omega_{\text{m}} + \Omega_\Lambda$, i.e. the geometry of space. The height ratio of the first and second peak fixes $\omega_{\text{b}} \equiv \Omega_{\text{b}} h^2$, while $\omega_{\text{m}} \equiv \Omega_{\text{m}} h^2$ follows from the height ratio of the first peak and the flat plateau at low ℓ. And h follows from (10.47): t_{rec} is known since we know $\Omega_{\text{m}} h^2$ and z_{rec} (by modelling) and $\epsilon_{\text{r}0}$ (standard neutrinos), see § 10.3. Since $\Omega_{\text{m}} = \omega_{\text{m}}/h^2$ and $\Omega_\Lambda = \Omega_{\text{tot}} - \omega_{\text{m}}/h^2$, the value of d at z_{rec} depends effectively only on h, cf. Table 11.1. But the position θ_0 of the first

---

[9] Cosmic variance is cosmologist's jargon indicating the effect that observed and theoretically computed mean values of a cosmological quantity may differ considerably because our visible universe is only one possible realisation out of many. The r.m.s. difference between the two expressions in (11.35) is $(\Delta C_\ell)^2_{\text{r.m.s.}}/C_\ell^2 = 2/(2\ell+1)$.

peak is measured, hence h and then also $\Omega_{\rm b}$, $\Omega_{\rm m}$ and Ω_Λ are known. Readers interested in the (complex) details are referred to the literature.[10]

In 2007 ESA's PLANCK mission will be launched carrying a third generation CMB experiment with a much improved angular resolution and sensitivity. This mission is expected to determine the cosmological parameters H_0, $\Omega_{\rm m}$ and Ω_Λ with a precision of 1%.

Exercise 11.11: Show that $(\Delta T)_{\rm r.m.s.}/T_0 \simeq 3 \times 10^{-5}$.

Hint: (11.34): $C(0) = (\Delta T)^2_{\rm r.m.s.} = \sum_\ell (2\ell+1)C_\ell/4\pi \simeq \sum_\ell c_\ell/\ell$ with $c_\ell \equiv \ell(\ell+1)C_\ell/2\pi$ plotted in Fig. 11.7. The sum is dominated by the low-ℓ plateau. Intelligent handwaving: $C(0) \simeq c_{\rm low\,\ell} \cdot \sum_\ell \ell^{-1} \simeq c_{\rm low\,\ell} \log L$; take the cut-off at $L = 10^3$: $(\Delta T)_{\rm r.m.s.}/T_0 \simeq (1000 \cdot 10^{-12} \log 10^3)^{1/2}/2.725$.

11.5 Light-cone integrals

The computation of observable quantities requires integration over the past light-cone, and we consider here a few simple problems. The first is what is the volume of our past light-cone, i.e. what is the proper volume of the space that we see as we look into the universe? The light-cone may be thought of as a series of nested shells, but the volume of the shells will ultimately decrease with z because the expansion was less advanced.

Draw two subspaces $t = $ constant in Fig. 9.3 intersecting the light-cone at t and $t + dt$. The 2-volume of an intersection is $4\pi S^2 r^2$, see below (9.22). The proper volume of the shell is now $4\pi S^2 r^2 \times$ the light distance $c\,dt$, and the proper volume V of the light-cone follows by integration:

$$V = 4\pi c \int_0^{t_0} S^2 r^2 \, dt = 4\pi c \int_0^{t_0} S^2 \sinn^2\left(\frac{d}{S}\right) dt \,. \tag{11.38}$$

Here we have applied (11.1): $d = Sf(r)$ or $r = f^{-1}(d/S)$ and $f^{-1} = \sinn$. To avoid the complications of non-Euclidean geometry we assume a flat universe, and then (11.38) reduces to the transparent expression $V = 4\pi c \int_0^{t_0} d^2 \, dt$. To keep the calculations simple, we consider the reference model $(\Omega_{\rm m}, \Omega_\Lambda) =$

---

[10] Hu, W. and Dodelson, S., *A.R.A.A.* **40** (2002) 171; in particular Fig. 4; Page, L. et al., *Ap. J. S.* **148** (2003) 233; Spergel, D.N. et al., *Ap. J. S.* **148** (2003) 175.

(1, 0). With the help of $d = (S/S_0) d_0$, $u = S/S_0$, $dt = du/\dot{u}$ and $\dot{u}$ from (10.8) we obtain:

$$d = cu \int_u^1 \frac{du}{u\dot{u}} = \frac{cu}{H_0} \int_u^1 \frac{du}{\sqrt{u}} = \frac{2cu}{H_0} (1 - \sqrt{u}). \qquad (11.39)$$

The calculation may now be completed:

$$\begin{aligned}
V &= 4\pi c \int_0^1 \frac{d^2}{\dot{u}} du \\
&= 2\pi \left(\frac{2c}{H_0}\right)^3 \int_0^1 u^{5/2} (1 - \sqrt{u})^2 du \\
&= 4\pi \left(\frac{2c}{H_0}\right)^3 \int_0^1 x^6 (1 - x)^2 dx \\
&\simeq \frac{4\pi}{3} \left(\frac{2c}{H_0}\right)^3 \cdot 1.19 \times 10^{-2}.
\end{aligned} \qquad (11.40)$$

It follows that in a (1, 0) universe the proper volume of the past light-cone is about 1% of the volume inside the horizon $(4\pi/3)(2c/H_0)^3$ – a number one would not easily have guessed otherwise.

Next, we compute the number of objects N that are located on the light-cone (i.e. how many objects do we see regardless of their brightness), assuming that the universe is homogeneously filled with objects and that their density is now n_0. Obviously, the past density is $n = n_0 (S_0/S)^3 = n_0/u^3$, and the answer is found by inserting n in the integrand of (11.40):

$$N = 2\pi n_0 \left(\frac{2c}{H_0}\right)^3 \int_0^1 \frac{(1 - \sqrt{u})^2}{\sqrt{u}} du = \frac{4\pi n_0}{3} \left(\frac{2c}{H_0}\right)^3, \qquad (11.41)$$

which is the present density × the volume inside the horizon. This should come as no surprise because there is, by definition, a one-to-one correspondence between objects on the past light-cone and ojects inside the horizon, see Fig. 11.2.

Olbers's paradox

How bright is the sky if the objects of the previous example all have a constant luminosity L_0? In a flat universe the number of objects in a shell $c\,dt$ and in a solid angle $\delta\Omega$ is $n \cdot d^2 \cdot \delta\Omega \cdot c\,dt$, and the flux density at the observer of one object is $L_0/4\pi d_L^2$, by definition. The shell contributes therefore an amount $\delta I_0 \delta\Omega = n d^2 \, \delta\Omega \cdot c\,dt \cdot (L_0/4\pi d_L^2)$ to the total brightness I_0 ($\mathrm{W\,m^{-2}\,sr^{-1}}$ or $\mathrm{erg\,cm^{-2}\,s^{-1}\,sr^{-1}}$), or, ignoring absorption by intervening matter:

$$\delta I_0 = \frac{cL_0}{4\pi}\left(\frac{d}{d_{\rm L}}\right)^2 n\,{\rm d}t\;. \tag{11.42}$$

We integrate (11.42) using that ${\rm d}t = {\rm d}u/\dot u$, $\dot u = H_0 u^{-1/2}$ for $(\Omega_{\rm m},\Omega_\Lambda) = (1,0)$ and $d = ud_0 = u^2 d_{\rm L}$ according to (11.10) and (11.23):

$$I_0 = \frac{n_0 L_0}{4\pi}\frac{c}{H_0}\int_0^1 u^{3/2}\,{\rm d}u = \frac{n_0 L_0}{10\pi}\frac{c}{H_0}\;. \tag{11.43}$$

The extragalactic background intensity in the visible and infrared is estimated to be $I_0 \sim 5\times 10^{-5}\,{\rm erg\,cm^{-2}\,s^{-1}\,sr^{-1}}$, and it would follow that the extragalactic luminosity density is $n_0 L_0 \sim 10^9\,h\,L_\odot\,{\rm Mpc}^{-3}$. More realistic computations including source evolution, extinction, spectral range, etc., confirm that I_0 is finite.[11]

The historical roots of the sky brightness problem date back, one might say, to the days when Newton introduced universal gravity. In correspondence with Bentley[12] he concluded that a stationary universe would have to be infinite (and that it required a supernatural power to subsist). It was gradually understood that a stationary infinite universe suffered from another problem. The sky would be as bright as the Sun, because any line-of-sight must eventually hit a stellar surface, no matter in which direction one looks. There would not be a spot in the sky that is not covered by a stellar surface (non-astronomers are reminded that the brightness of a stellar disc is independent of its distance). This is known as Olbers's paradox.[13] The problem disappears in relativistic cosmology as it allows for an expanding universe with a beginning in time. The night sky is dark because arbitrarily long lines-of-sight no longer exist, and the stars within our horizon (the visible universe) cover only a minute fraction of the sky. Contrary to what is often stated in the older literature,[14] the decisive factor is the finite age of the universe – the redshift merely causes an additional reduction of I_0, see exercise.

---

[11] Wesson, P.S., *Ap. J. 367* (1991) 399.
[12] Bentley, a priest, was after proving the existence of God by the classic argument of design, and he took the precaution to ask Newton to comment on his ideas in the light of the then new theory of universal gravity, see *The correspondence of Isaac Newton,* H.W. Turnbull (ed.), Cambridge U.P. (1961), Vol III.
[13] The physician and amateur astronomer H.W. Olbers published this paradox in 1826, but others had raised the issue before him. He also discovered a number of comets and the asteroids Pallas and Vesta.
[14] E.g. Gamov, G., in *Theories of the Universe*, M.K. Munitz (ed.), The Free Press (1957), p. 390.

Exercise 11.12: Give an alternative computation of I_0 by considering the radiation energy density stored in the subspace $t = t_0$, and show that the redshift just adds an extra reduction factor.

Hint: The energy emitted by one source is $\int_0^{t_0} L_0 \, dt$, but that is not the energy that is stored in the subspace $t_0 = $ constant, as the redshift reduces the energy by an amount S/S_0. The stored radiation energy density is $\epsilon_0 = n_0 \int_0^{t_0} L_0 (S/S_0) \, dt$. Since the radiation is isotropic the intensity is $I_0 = c\epsilon_0/4\pi$:

$$I_0 = \frac{cn_0}{4\pi} \int_0^{t_0} L_0 \frac{S}{S_0} \, dt , \qquad (11.44)$$

which is the same as (11.43) since $u = S/S_0$ and $dt = du/\dot{u} = u^{1/2} du/H_0$. The argument is purely local and shows that (11.43) is also valid for $k \neq 0$. The redshift may be switched off by dropping S/S_0 in (11.44) $\rightarrow I_0 = (n_0 L_0/6\pi)(c/H_0)$, a factor $5/3$ more.

12
The Big Bang

During the radiation era the universe was a perfectly homogeneous, rapidly expanding space filled with dense, hot matter, but other than that it was a rather dull period. The universe just expanded and cooled, and that was it – nothing of importance happened. For more exciting times we have to go back to the first 1000 seconds, when temperature and density were so high that nuclear reactions took place. The universe started its life as a gigantic fusion reactor that produced the matter we observe today. Traditionally, this period is referred to as the Big Bang. The matter and the cosmic microwave background (CMB) are the two main relics of the hot Big Bang. Very soon after the discovery of the CMB by Penzias and Wilson in 1965 it was shown[1] how nuclear reactions could explain the observed chemical composition of the universe (H, D, ^{3}He, ^{4}He, ^{7}Li). The idea of nucleosynthesis in the early universe and the concept of a relic thermal background radiation goes back, however, to Gamov and co-workers.[2] Weinberg's book *The First Three Minutes* remains one of the best accounts of this period of the universe, despite the fact that it was written before inflation theory and astro-particle physics made their impact on cosmology. Although some of the details may be complex and still unknown, the story of the Big Bang remains, in broad outline, one of sublime and almost capricious simplicity.

12.1 Nuclear reactions

We shall only summarise the main points, and not engage in explicit calculations. More information can be found in Weinberg (1977), Börner (1988), Padmanabhan (1993), Kolb and Turner (1990), and Peacock (1999). We begin with a brief review of the three kinds of elementary particles and the composite particles.

---

[1] Wagoner, R.V. et al., *Ap. J. 148* (1967) 3.
[2] Alpher, R.A. et al., *Phys. Rev. 73* (1948) 803.

Table 12.1. Temperature, density and age of the universe as a function of the energy scale

particle	m_0c^2 (MeV)	$T_r = m_0c^2/\kappa$ (K)	ρ (g cm^{-3})	t (s)
$W^\pm, Z^0$	9×10^4	10^{15}	10^{25}	10^{-10}
p, n	940	10^{13}	10^{17}	10^{-7}
π, μ	120	10^{12}	10^{13}	10^{-4}
$e^\pm$	0.5	6×10^9	10^3	10

Quarks. There are 6 types, called *up* (u), *down* (d); *charm* (c), *strange* (s) and *top* (t), *bottom* (b). They carry one of the three positive colour charges ('red, green or blue') responsible for the strong nuclear force. In addition they have a fractional electric charge. The electric charge of u, c, t is $\frac{2}{3}$, that of d, s, b is $-\frac{1}{3}$. Together with their antiparticles (that carry a negative colour charge) they number 36 in total. They are fermions with rest masses ranging from $m_u \simeq 1$ MeV to $m_t \simeq 175$ GeV.

Leptons. There are also 6 types of these: e^- (~ 0.5 MeV), ν_e; μ^- (~ 100 MeV), ν_μ; τ^- (~ 1.8 GeV), ν_τ. Together with their antiparticles (e^+, $\bar{\nu}_e$, ..) 12 in total. They are fermions that do not feel the strong nuclear force. The neutrinos have no electric charge, and zero mass according to the standard model. Experimental upper limits: $\nu_e < 4.7$ eV, $\nu_\mu < 160$ keV and $\nu_\tau < 24$ MeV). Measurements of atmospheric and solar neutrinos indicate that neutrinos switch flavour as they propagate. These so-called neutrino oscillations imply that they should have a nonzero mass.

Gauge bosons take care of the interaction between these particles. There is 1 graviton g (gravity); 3 vector bosons $W^\pm$, Z^0 mediating the weak interaction; the photon γ (electromagnetic force) and 8 gluons for the strong interaction. The vector bosons have a rest mass of about 90 GeV and an electric charge of $\pm 1, 0$. The other gauge bosons are massless. The gluons carry a colour charge.

Hadrons. Free quarks cannot exist – they occur only in combinations of two or three quarks called *hadrons* (= heavy particles). Accordingly, there are two kind of hadrons. The *mesons* are colour-free particles consisting of a quark and an antiquark, for example $\pi^+ = u\bar{d}$, $\pi^- = d\bar{u}$, $\pi^0 = (u\bar{u} - d\bar{d})/\sqrt{2}$ (~ 140 MeV). *Baryons* are colourless combinations of 3 quarks. The lightest are the proton $p = uud$ and the neutron $n = udd$ (~ 940 MeV). The mesons and all heavier baryons ($\Lambda = uds$, $\Sigma^+ = uus$, ..) are unstable.

During the extremely hot and dense initial phase of the universe, the particles it contains are continuously subject to interactions of the type

12.1 Nuclear reactions

$$A + B \leftrightarrow C + D \,;$$
$$D \leftrightarrow P + Q \,,$$
(12.1)

etc. As long as $\kappa T_r > m_0 c^2$ there is enough energy to create particles of rest mass m_0. The time available for these reaction is of the order the time scale on which the universe changes due to expansion, $\tau_S = S/\dot{S}$. Because $S \propto t^{1/2}$ we get

$$\tau_S \sim H^{-1} \equiv (\dot{S}/S)^{-1} = 2t$$
$$= 2 \times \text{age of the universe} \,.$$
(12.2)

The available time is therefore of the order of the age of the universe (the factor 2 should not be taken too seriously). In view of the values of ρ and t in Table 12.1 we may suspect that τ_S is generally longer than the characteristic reaction times between the elementary particles, and detailed calculations confirm this suspicion. This has a very important consequence: *matter and radiation are in thermal equilibrium*. If we wish to know the abundances of the particles at a certain temperature we may just as well ignore the expansion, as the reactions proceed much faster anyway, and compute the thermal equilibrium state.

In a non-equilibrium calculation hundreds of rate equation must be advanced in time. That is not really a big deal, but the problem is that many reaction cross sections are not well known. For equilibrium calculations simpler and reliable techniques are available. It is no longer necessary to know the reaction cross section. A typical example is relation (12.9) which shows how the density ratio of protons and neutrons in thermal equilibrium depends only on their mass difference and the temperature, but not on the details of the weak interactions that maintain the equilibrium. It follows that the material composition of the universe is not strongly dependent on previous states. Even if we make a mistake in the early universe because the particle physics at these high energies is not well known, it would have little effect on the material composition at a later time. That is why it is at all possible to make statements on the material evolution of the early universe with some degree of confidence.

In broad outline, the situation is as follows. All particles with rest mass energy $m_0 c^2$ smaller than κT_r are continuously being created and destroyed, usually by many different types of reaction. They have a thermal Fermi-Dirac or Bose-Einstein energy distribution, and they have number densities of the order of those of the massless particles (for example photons). However, as the universe evolves, κT_r becomes smaller than $m_A c^2$, and then things get a little complicated. It may happen that particle A vanishes completely from the scene because reactions such as (12.1), top, and the annihilation reaction

240   12 The Big Bang

$$A + \overline{A} \rightarrow 2\gamma \qquad (12.3)$$

proceed entirely to the right. Free neutrons for example ultimately disappear because they are unstable, though the majority of them gets locked up in ^{4}He, as we shall see. History plays a role in two ways:

- There are a number of conserved quantities, such as the net electric charge (probably zero) and the *baryon number* (= number of quarks minus number of antiquarks). These quantities are simply passed on from early times to later evolutionary stages.

- Because temperature and density decrease, all reaction times increase, and they do so faster than the universe ages. As a result, in the whole network of reactions creating and/or destroying particle A some connections become sterile. These paths effectively disappear from the network. This has no immediate influence on the number of particles A. That happens only when the last path disappears – assuming A did not vanish earlier due to (12.3), for example. The jargon is that particle A *decouples* or *freezes out*. What remains must be calculated for every species individually by solving rate equations.

12.2 The first 100 seconds

The thermal history of the early universe evolves through several stages that we briefly describe here, with reference to the overview in Table 12.2.

Quark-gluon plasma

The story begins when the universe was not yet 10^{-7} seconds old. The temperature was 10^{13} K or more, and the density was 10^{17} g cm^{-3} or higher. The universe consisted of a *quark-gluon plasma*, an extremely dense and heavy stew of quarks, leptons and gauge bosons, all in comparable amounts. The beginning of the quark era is believed to be at $t \sim 10^{-30}$ s, when the temperature had the impressive value of $10^{24} - 10^{25}$ K, and the universe was a linear factor $S/S_0 \simeq 5 \times 10^{24}$ smaller than it is today. The space within our current horizon (radius ~ 10 Gpc, Table 9.1) would, at that time, fit in a sphere with a radius of 100 meter!

Baryogenesis

The quark-gluon plasma is believed to be subject to a phase transition and to condense into hadrons at a few times 10^{12} K. A heavy-ion collision programme at CERN and later at Brookhaven National Laboratory (the Relativistic Heavy Ion Collider (RHIC)) has given hints about the properties

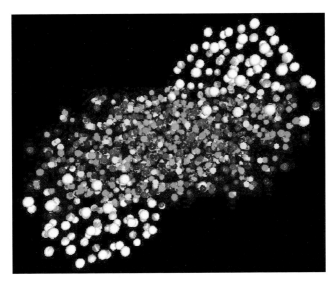

Fig. 12.1. A Little Bang. Snapshot from a simulation of a collision of two lead nuclei 5×10^{-24} s after an off-centre impact at 17.4 GeV per nucleon pair. Unaffected 'spectator' nucleons are white and grey. Colliding hadrons are advanced with a hadron transport model (UrQMD) that handles the first collisions and their hadronic products. At full overlap these hadrons are decomposed into (supposedly deconfined) quarks, which are then advanced with a quark molecular dynamics model (qMD). The colours above indicate the six (anti)colour charges. During the subsequent evolution the quarks quickly team up in colour-neutral clusters that decay into hadrons. The figure is stretched in the beam direction by a factor γ (of order 10) to undo the Lorentz contraction, but time dilation effects are still there. Credit: S. Scherer, University of Frankfurt. See Scherer, S. et al., *New J. Phys. 3* (2001) 8.1.

of the quark-gluon plasma and the phase transition.[3] Various groups have supported these experimental efforts with simulations, one of which is shown in Fig. 12.1.

Very soon after the phase transition only the lightest hadrons remain ($p, \overline{p}$ and $n, \overline{n}$ and some mesons). As the temperature drops further $p, \overline{p}$ and $n, \overline{n}$ annihilate according to (12.3). Calculations show that the baryon density drops to $n_b/n_r = n_{\overline{b}}/n_r \sim 10^{-18}$. Therefore we have a conflict with the observations, which tell us that $n_b/n_r \sim 6 \times 10^{-10}$ and that there is no antimatter, exercise 10.11. Attempts to resolve this conflict include, for example, models with spatial fluctuations which may result in regions having a slight excess of matter, alternated by places with a small antimatter excess. After

---

[3] For a non-technical account see Schwarzschild, B., *Phys. Today, May 2000*, 20; Ludlam T. and McLerran, L., *Phys. Today, October 2003*, 48.

Table 12.2. Overview of the material evolution of the universe

age (s)	temperatur (K)	size (S/S_0)	composition[a] baryons	lepton	gauge bosons
$< 10^{-7}$	$> 10^{13}$	$< 2 \times 10^{-13}$	$\mathbf{q\bar{q}}$	$\boldsymbol{\ell\bar{\ell}}$	$\gamma, g, W^{\pm}, ..$
10^{-6}	5×10^{12}	5×10^{-13}	$p\bar{p}, n\bar{n}, ..$	$\boldsymbol{\ell\bar{\ell}}$	γ, g
10^{-4}	10^{12}	3×10^{-12}	p, n	$\mathbf{e^-e^+}, \nu\bar{\nu}$	γ, g
10^2	10^9	3×10^{-9}	p, n	$e^-, \boldsymbol{\nu\bar{\nu}}$	γ, g
10^3	3×10^8	10^{-8}	$^1H, {}^4He$	$e^-, \boldsymbol{\nu\bar{\nu}}$	γ, g
$> 10^{13}$	< 3000	$> 10^{-3}$	H, He atoms	$\boldsymbol{\nu\bar{\nu}}$	γ, g
4×10^{17}	3	1	⇓ galaxies	⇓ neutrino, microwave and graviton background	⇓

[a] Boldface printed particles have approximately the same density, which is about 10^9 times larger than the other particles on the same line.

annihilation in the hadron era there remain regions with matter and antimatter, and we happen to live in a matter region. The idea has been abandoned because the regions are small and contain much less than a galactic mass. Moreover, the boundaries produce much more annihilation radiation than is actually observed. It is now believed that a small quark-lepton excess of the order of

$$\frac{n_q - n_{\bar{q}}}{n_q + n_{\bar{q}}} \sim 6 \times 10^{-8} \tag{12.4}$$

was created *everywhere* in the universe. Computations show that after the annihilations have taken place, $n_b/n_r \sim 6 \times 10^{-10}$ and $n_{\bar{b}} \ll n_b$, as observed.[4]

---

[4] One might think that $n_b/n_r \sim 6 \times 10^{-8}$. This is correct if there were only one non-relativistic quark gas instead of 36 extremely relativistic ones. The photons and each individual quark type have initially about equal abundance. Moreover, a sizeable fraction of the kinetic energy of the quarks is ultimately converted into photons as well. As a result there are about 100 times more photons after the annihilations than one might think. The proper attack to this type of problem is to require that the total entropy is constant, as in exercise 12.2.

The origin of this excess (12.4), to which we owe our existence, is unknown. A popular speculation is asymmetric decay of leptoquarks X that play a role in Grand Unified Theories (GUTs). These supermassive bosons ($\sim 10^{15}$ GeV) may have existed in the very early universe from $\sim 10^{-43}$ s to $\sim 10^{-34}$ s. Around $t \sim 10^{-34}$ s they decay into two quarks or a quark-lepton pair:

$$X \to \begin{cases} q + q & (r) ; \\ \bar{q} + \bar{l} & (1-r) ; \end{cases} \quad (12.5)$$

$$\overline{X} \to \begin{cases} \bar{q} + \bar{q} & (\bar{r}) ; \\ q + l & (1-\bar{r}) . \end{cases} \quad (12.6)$$

Between parentheses the branching ratios for each decay channel; X and $\overline{X}$ decay at the same net rate, but when r is a little larger than $\bar{r}$ a small matter excess will arise, see exercise.[5] One way to check this scenario would be to measure the induced instability of the proton: the uu in $p = uud$ fuse into a leptoquark by the inverse of the top channel of (12.5), which decays again into $\bar{d} + e^+$ through the lower channel. The remaining d and the new $\bar{d}$ form π^0. Net result: $p \to \pi^0 + e^+$. The predicted decay time is very long, of the order of 10^{32} yr, because the intermediate leptoquarks are so massive. The Japanese Kamiokande facility, well-known for its detection of neutrinos, was originally designed to measure the lifetime of the proton.

Returning to the Big Bang, at the end of the hadron era, around $t = 10^{-4}$ s, the last mesons and the heavier leptons have decayed as well. The universe is now a rapidly expanding fireball consisting of photons, neutrinos, e^+, e^- in approximately equal profusion, with a tiny admixture ($\sim 6 \times 10^{-10}$) of protons and neutrons.

The lepton era

Thermal equilibrium between $e^\pm$, $\nu, \bar{\nu}$ and photons is maintained by scattering of photons and neutrinos off $e^\pm$, and through reactions such as

$$\nu + \bar{\nu} \leftrightarrow e^+ + e^- \leftrightarrow 2\gamma . \quad (12.7)$$

The equilibrium of these particle with p and n (and between p and n) is maintained by the weak reactions

---

[5] More information on the matter-antimatter symmetry problem in Börner (1988) Ch. 8; Kolb and Turner (1990) Ch. 6, and Peacock (1999) § 9.6. For a summary of history and current ideas see Ellis, J., *Nature 424* (2003) 631.

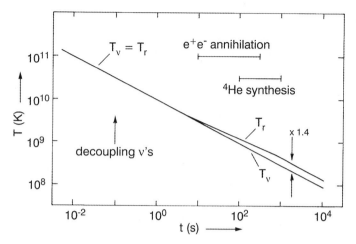

Fig. 12.2. The evolution of the radiation temperature during the decoupling of the neutrinos and the $e^\pm$ annihilation.

$$p + e^- \leftrightarrow n + \nu_e ;$$
$$n + e^+ \leftrightarrow p + \bar{\nu}_e ;$$
$$n \leftrightarrow p + e^- + \bar{\nu}_e .$$
(12.8)

These reactions leave the number of protons plus neutrons invariant. The physical state of the matter and the radiation is entirely determined by the temperature. The previous history of the universe is only relevant in that it determines (a) the ratio $(n_n + n_p)/n_r$ and (b) the time t at which a particular temperature is attained. Three important events take place during the lepton era:

- At $T_r \sim 3 \times 10^{10}$ K the neutrinos decouple because the interaction times between $e^\pm$ and the neutrinos become of the order of τ_S. For the time being T_ν and T_r remain equal as both continue to scale $\propto S^{-1}$.

- The ratio n_n/n_p is determined by thermal equilibrium:

$$\frac{n_n}{n_p} = \exp\left\{-\frac{(m_n - m_p)c^2}{\kappa T_r}\right\} ;$$
(12.9)

$$(m_n - m_p)c^2 \simeq 1.3 \text{ MeV} .$$

At the beginning of the lepton era we have $\kappa T_r \gg (\Delta m)c^2$ so that $n_n \simeq n_p$; the mass difference between p and n plays no role yet. Around $T_r \simeq 3 \times 10^{10}$ K the ratio n_n/n_p begins to decrease, and soon the reaction rates of (12.8) become larger than the age of the universe so that thermal equilibrium (12.9) can no longer be maintained. Calculations show that $n_\mathrm{n}/n_\mathrm{p}$ freezes out at $T_\mathrm{r} \sim 3 \times 10^9$ K at value of (Kolb and Turner (1990) § 4.3; Peacock (1999) Ch. 9):

$$\frac{n_\mathrm{n}}{n_\mathrm{n}+n_\mathrm{p}} \simeq 0.16 \ . \tag{12.10}$$

- Electrons and positrons begin to disappear by annihilation when T_r drops below $\sim 6 \times 10^9$ K. A small fraction of the e^- remains, equal to the fraction of protons. The effect of the $e^\pm$ annihilation is that the photon temperature[6] T_r decreases for some time less rapidly than $\propto S^{-1}$. In the end T_r becomes a factor $(11/4)^{1/3} \simeq 1.4$ larger than the neutrino temperature T_ν, Fig. 12.2. During the subsequent evolution of the universe the energy distribution of the neutrinos remains a thermal (Fermi-Dirac) distribution with $T_\nu \propto S^{-1}$. The present temperature of the neutrino background is therefore predicted to be $2.725\,\mathrm{K}/1.4 = 1.95\,\mathrm{K}$. A measurement of this neutrino temperature would be a powerful check on the hot Big Bang scenario (and would also secure your fame in cosmology).

Exercise 12.1: Show that the decay of an X, $\overline{X}$ pair causes the baryon number B to increase by $r - \overline{r}$, and that a matter excess will arise when $r > \overline{r}$.

Hint: X and $\overline{X}$ are field quanta and have $B = 0$, as do the leptons l; quarks have $B = \frac{1}{3}$, and three quarks compose a baryon with $B = 1$. Antiparticles have opposite B, hence $\Delta B = 2 \cdot \frac{1}{3} r - \frac{1}{3}(1-r) - 2 \cdot \frac{1}{3}\overline{r} + \frac{1}{3}(1-\overline{r})$ for each decaying X, $\overline{X}$ pair.

Exercise 12.2: Explain that $T_\mathrm{r} = (11/4)^{1/3} T_\nu \simeq 1.40\, T_\nu$ at the end of the e^+e^- annihilation.

Hint: During the annihilation the state of the matter is no longer given by a simple limiting case as in Table 10.1, but by relation (9.39) which says that the entropy $\mathcal{S}$ in a volume S^3 is constant. In the calculation below only extremely relativistic gases play a role for which $p = \frac{1}{3}\epsilon = \frac{1}{3}\rho c^2$ and $\mathcal{S} = S^3(p+\rho c^2)/T = (4c^2/3)S^3\rho/T$ (without proof). Let $\rho = aT^4$, for example (10.31) for photons, then for a mixture $\Sigma\, a_i (TS)_i^3$ is constant. The entropy of each neutrino gas remains constant (the neutrinos have no interaction and play no role), while p, n do not contribute significantly to $\mathcal{S}$ due to their relatively small density. What remains is photons, e^+, e^- prior to annihilation, and only photons

---

[6] It is customary to denote the photon temperature as T_r, and to identify it with the temperature of (the radiation in) the universe, even though some components of the radiation, such as the neutrinos, have a different temperature.

thereafter:

$$a_r(T_rS)_b^3 + a_+(T_+S)_b^3 + a_-(T_-S)_b^3 = a_r(T_rS)_a^3 . \tag{12.11}$$

b, a = before, after annihilation; $+, - = e^+, e^-$. Now $a_- = a_+ = \frac{7}{8}a_r$ (see literature), and $T_{-b} = T_{+b} = T_{rb}$ whence $\frac{11}{4}(T_rS)_b^3 = (T_rS)_a^3$. But since $T_\nu \propto S^{-1}$ we have $(T_rS)_b = (T_\nu S)_b = (T_\nu S)_a$, so that after the annihilation $\frac{11}{4}T_\nu^3 = T_r^3$. Details in Peebles (1993) p. 160; Padmanabhan (1993); Peacock (1999) Ch. 9.

Exercise 12.3: Demonstrate that the last scattering surface of the neutrino background is located at $z \sim 10^{10}$. Explain that the sooner a background freezes out, the lower its temperature will be today.

Hint: The neutrino temperature now and at freeze-out are known, Fig. 12.2; furthermore $T \propto S^{-1}$. The earlier a particle A freezes out the more the photon temperature will rise with respect to that of A due to later annihilations.

12.3 The synthesis of light elements

At the end of the lepton era the structure of the universe is very simple. It has a flat geometry (k is effectively zero), and it contains a homogeneous mix of photons and neutrinos, 'doped' one might say with a tiny fraction of e^-, p and n. During the next and longest phase of the Big Bang elements heavier than hydrogen are 'cooked'. Helium could already have existed in the lepton era, because its binding energy is so large (28 MeV $\cong 3 \times 10^{11}$ K). However, the lighter nuclei that are needed to get helium fusion going are not available, because their binding energies are smaller than κT_r at that time. And formation of helium through four-particle collisions is extremely rare. The upshot is that heavier nuclei may only be generated in sequential two-particle collisions. The first step in this process, deuterium (D), determines the rate of the synthesis due to its small binding energy (2.2 MeV) and large cross section for photo-dissociation. Only when $T_r \sim 10^9$ K ($t \sim 100$ s) the equilibrium

$$n + p \leftrightarrow D + \gamma \tag{12.12}$$

begins to shift to the right, Fig. 12.3. Once D is available, other fusion reactions follow immediately:

$$\left.\begin{array}{r} D + D \rightarrow {}^3He + n\ ;\\ \\ D + {}^3He \rightarrow {}^4He + p\ ; \end{array}\right\} \tag{12.13}$$

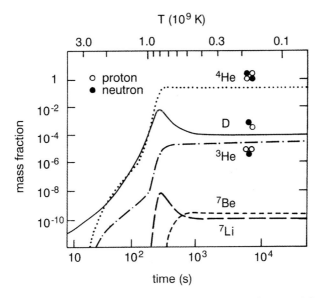

Fig. 12.3. Synthesis of the light elements, after Boesgaard, A.M. and Steigman, G., *A.R.A.A.* 23 (1985) 319.

$$\left.\begin{array}{l} D + D \rightarrow {}^3H + p\,; \\ D + {}^3H \rightarrow {}^4He + n\,, \end{array}\right\} \quad (12.14)$$

and the result is that virtually all neutrons end up in ^{4}He, and only a small fraction in ^{3}He and D. We are now in a position to estimate the abundance of ^{4}He in the universe. The value of $n_n/(n_n + n_p)$ was 0.16 at the freeze-out, and decreased slowly thereafter to about 0.13 at the beginning of the helium synthesis due to β decay of the neutrons. Because almost all neutrons end up in ^{4}He, the mass fraction of ^{4}He equals $Y = 2 \times n_n/(n_n + n_p) \simeq 0.26$. Calculations give a result between 0.20 and 0.28, dependending on the assumed value of n_b/n_r. The mass fraction of the remaining deuterium equals roughly 10^{-4}, and that of ^{3}He is a bit lower. Tritium (^{3}H) reaches a level of $\sim 10^{-7}$, but decays in 18 years and disappears.

The formation of heavier elements is hampered by the absence of stable nuclei with mass number $N + Z = 5$ and 8. Some ^{7}Li and ^{7}Be is formed by the reactions

$$\left.\begin{array}{l} {}^4He + {}^3H \rightarrow {}^7Li + \gamma\,; \\ {}^4He + {}^3He \rightarrow {}^7Be + \gamma\,; \\ {}^7Be + e^- \rightarrow {}^7Li + \nu_e\,, \end{array}\right\} \quad (12.15)$$

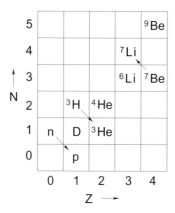

Fig. 12.4. Stable nuclei with $Z \leq 4$. A free neutron and tritium are subject to slow β decay, last line of (12.8), with e-folding times of 900 s (n) and 18 years (^{3}H); ^{7}Be disappears eventually because it is unstable to electron capture, the last reaction in (12.15).

but it is very little because they require the rare nuclei ^{3}H and ^{3}He. As the universe reaches the respectable age of 10 minutes the nucleosynthesis is drawing to a close and the radiation era begins. Since ^{7}Be disappears too, Fig. 12.4, we conclude that the final product of the nucleosynthesis in the early universe is ^{4}He plus a little D, ^{3}He and ^{7}Li. Heavier elements were not formed, broadly speaking, because there was no time. The early universe expanded very fast and the reaction rates soon became vanishingly small due to decreasing densities and Coulomb barriers getting too large. The universe had to wait until the arrival of the stellar era. Stellar interiors have the right density and temperature for the synthesis of carbon and heavier elements.[7] And they have lots of time.

Primordial abundances are difficult to observe because abundances change with time due to evolutionary effects. The best value for ^{4}He is $Y \simeq 0.24 \pm 0.015$, observed in isolated extragalactic H II regions with little contamination from stellar nucleosynthesis. This agrees well with the theoretical prediction. An important point is that stars cannot deliver these large quantities of helium. Stellar nucleosynthesis could have produced $Y \sim 0.04$ at most, and the spatial distribution would be clumpy and cluster around regions of star formation. However, the observed ^{4}He distribution is rather homoge-

---

[7] There is actually a third production process: spallation by cosmic rays. A fraction of the ^{6}Li, ^{7}Li, ^{11}B, and *all* ^{9}Be and ^{10}B in the universe has been formed in this way. See Geiss, J. and Von Steiger, R., in *Fundamental Physics in Space*, ESA SP-420 (1997), p. 99.

neous. It follows that the helium in the universe must be primordial.[8]

The abundance of deuterium in the interstellar medium is D/H $\simeq$ $(1.6 \pm 0.1) \times 10^{-5}$. Deuterium is special in that it is only destroyed during stellar evolution and never created. Hence all measured abundances are lower limits to the primordial abundance. The extragalactic deuterium abundance has recently been measured from absorption lines in the light of a quasar that passes through a gas cloud at $z = 3.6$.[9] The result is $4 \times 10^{-5} <$ D/H $< 2.4 \times 10^{-4}$, nicely consistent with the theoretical prediction. The primordial ^{3}He abundance is very difficult to get hold of. The solar ^{3}He/H value is $(1.5 \pm 0.4) \times 10^{-5}$. A 20-year programme of galactic H II region observations yielded ^{3}He/H $< (1.1 \pm 0.2) \times 10^{-5}$ for the primordial abundance.[10] The measured ^{7}Li abundance in some 100 metal poor Population II halo stars is ^{7}Li/H $= (1.6 \pm 0.07) \times 10^{-10}$, and this number is believed to be indicative of the primordial ^{7}Li abundance.[11]

The correct prediction of the abundances of the light elements is a resounding success for the theory of the hot Big Bang. We saw that the abundance of ^{4}He does not depend strongly on the assumed value of n_b/n_r, but that of D, ^{3}He and ^{7}Li does. This provides a sensitive method to determine the value of n_b/n_r, and because that ratio is constant and n_{r0} is known, we may infer the current baryon density ρ_{b0}. The conclusion is that the outcome of the light element synthesis agrees with the observed abundances if $\rho_{b0} = (3 \pm 1.5) \times 10^{-31}$ g cm^{-3}, or $\Omega_b = 0.03 \pm 0.015$. The light element synthesis scenario is therefore in accordance with the recent WMAP measurements ($\Omega_b = 0.044 \pm 0.004$, Table 9.2).

More details on these topics may be found in the (extensive) literature, e.g. Boesgaard, A.M. and Steigman, G., *A.R.A.A. 23* (1985) 319; Börner (1988) Ch. 3; Kolb and Turner (1990) Ch. 4 (FORTRAN code: p. 96); Padmanabhan (1993) Ch. 3 and 11. There exist also simplified models of the light element synthesis.[12]

---

[8] Quasi-steady-state cosmologists, on the other hand, maintain that *all* ^{4}He has been produced in stars. The energy released by the relevant fusion reactions has a density which is now equal to that of the microwave background. Therefore they interpret the CMB as thermalised starlight (Burbidge, G., et al., *Physics Today*, April 1999, 38).

[9] Songaila, A., et al., *Nature 385* (1997) 137.

[10] Bania, T.M., et al., *Nature 415* (2002) 54.

[11] Molaro, P., et al., *A&A 295* (1995) L47.

[12] Bernstein, J., et al., *Rev. Mod. Phys. 61* (1989) 25; Eskridge, B. and Neuenschwander, D.E., *Am. J. Phys. 64* (1996) 1517.

Exercise 12.4: Prove that at the beginning of the helium synthesis $n_n/(n_n + n_p) \simeq 0.13$.

Hint: $n_n = n_{n0}\exp(-t/\tau)$; $t \simeq 200\,\text{s}$, $\tau \simeq 900\,\text{s}$; $n_n + n_p$ remains constant in β decay $(n \to p + e^- + \bar{\nu}_e)$.

Exercise 12.5: During the helium synthesis the universe was a fusion physicist's dream: a gigantic fusion reactor that converted some 13% of all hydrogen into helium in about 1000 seconds. In comparison, stars need 10^{10} year to fuse a few percent of their hydrogen into helium. Explain why the enormous amount of energy liberated during the helium fusion had no influence on the evolution of T_r – unlike the e^+e^- annihilation during the lepton era.

Exercise 12.6: Neutron star model builders have a hard time in finding a reasonable equation of state $p(\rho)$ at $\rho \sim 10^{15}$ g cm^{-3}. Cosmologists, however, who study the universe at far greater densities couldn't care less. Why is life so much easier on them?

Hint: at comparable densities the matter in the universe is much hotter than neutron star matter. If we increase T at constant density, the interaction energy between nuclei becomes progressively less important, and that simplifies the equation of state. Ultimately, the matter behaves as an ideal gas.

Exercise 12.7: What would be the ^{4}He abundance if deuterium had a higher binding energy?

Hint: It could be as large as $Y \sim 2 \times 0.16 = 0.32$.

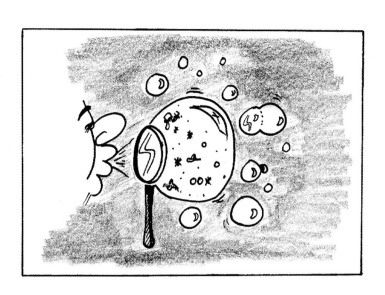

13

Inflation

The standard model of the Friedmann-Robertson-Walker (FRW) universe with a hot beginning is very successful and provides a natural explanation for:

1. the observed expansion velocities of distant galaxies;
2. the microwave background radiation as a relic of the hot Big Bang;
3. the chemical composition of the universe (H, D, ^{3}He, ^{4}He and ^{7}Li) as a relic of nuclear fusion during the Big Bang.

However, a number of problems remain, and the most important of these will be investigated here. For example, an obvious question is why does the universe expand? The only answer we have at this stage is: 'because it expanded faster in the past'. Other issues are the horizon problem, and the question why the geometry of the universe is flat. To illustrate the flatness problem, we know that $\Omega_m + \Omega_\Lambda = 1.02 \pm 0.02$, so that the universe is flat within the observational errors. But the universe must have been much flatter in the past. In exercise 10.3 it was shown that $\Omega_m(t) + \Omega_\Lambda(t) + \Omega_k(t) = 1$ and

$$\lim_{t \to 0} [\Omega_m(t),\ \Omega_\Lambda(t),\ \Omega_k(t)] = [1,\ 0,\ 0] \ . \tag{13.1}$$

To ensure that $\Omega_m \simeq 0.3$ and $\Omega_\Lambda \simeq 0.7$ *now*, the density ρ in the early universe must have been very close to the critical density ρ_c at that time (but not exactly equal). And Ω_Λ must have have been minimally different from zero in the past, by just the right amount to achieve that $\Omega_\Lambda \simeq 0.7$ now. The fact that Ω_k appears to be zero within the error bounds means that Ω_k must have been almost exactly zero in the past. The universe was flat then, and it still appears to be flat today. Why was the universe born with these special initial conditions?

In quest for a solution of these problems, cosmologists and particle physicists have increasingly joined forces. The early universe is an ideal place for particle physicists to test their theories under conditions that can never be attained in a laboratory. They go to the Great Accelerator in the Sky rather

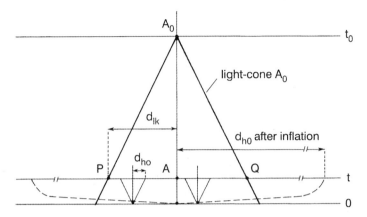

Fig. 13.1. The horizon problem. We observe that the early universe, i.e. a spherical shell around us at large z ($t \ll t_0$), is isotropic. The distance of the shell to us at time t is d_{lk} and the size of causally connected regions is d_{ho}, the horizon distance at that time. Since $d_{ho} \ll d_{lk}$, the early universe consists of many causally unconnected regions that don't know about each other's existence because they have not yet been able to exchange a light signal. If, however, the very early universe has gone through a period of inflation, then $d_{ho} \gg d_{lk}$.

than to CERN. This has led to the discovery of the possibility of *inflation*, a brief period of extremely rapid expansion immediately after the birth of the universe. Designed originally to alleviate the problem that the universe would contain too many magnetic monopoles, inflation soon turned out to be a panacea providing a solution for the horizon and flatness problem as well. In addition, it explained why the universe expands, and it provided the primordial energy density fluctuations from which the large-scale structure in the universe may develop later. In view of these impressive achievements, and in spite of its speculative character, the inflation concept appears to be the most important theoretical development in cosmology of the last decades. Here we shall explain the basic idea of inflation with the help of a simple model due to Linde.[1]

13.1 The horizon problem

The Friedmann-Robertson-Walker (FRW) universe has the nasty property that it consists of many different regions that are outside each other's horizon. And, as explained in § 11.2, the younger the universe is, the worse it gets. And yet, according to observations, our universe is on average homogeneous

---

[1] See Linde, A.D., *Physics Today*, September 1987, 61.

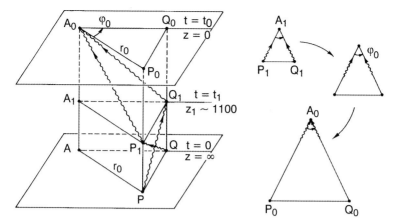

Fig. 13.2. We detect photons of the CMB in the plane $\theta = \pi/2$ from two directions subtending an angle φ_0. Our worldline is AA_0. The photons originate from P_1 and Q_1 on the surface of last scattering at $z \sim 1100$. We arrange things so that P_1 and Q_1 are just inside each other's horizon, so that the physical conditions in P_1 and Q_1 may in principle be the same. Assuming that space is flat, the photons travel along the sides of a *flat* isosceles triangle that has expanded a factor ~ 1100 when they reach the observer. In the text it is shown that $\varphi_0 \sim 1°$. This leaves the observed high degree isotropy of the CMB over the entire sky unaccounted for.

and isotropic. Let P and Q in Fig. 13.1 be two distant objects at large z. We observe that the surroundings of P and Q have the same properties, within the error bounds. The properties in P and Q depend only on the space inside their respective horizons. The figure displays our light-cone, and the geometrical distance d_{lk} between us and an object at time t,[2] as well as the geometrical distance d_{ho} to the horizon at time t. These are given by

$$d_{lk} = cS \int_t^{t_0} \frac{dt}{S} \; ; \quad d_{ho} = cS \int_0^t \frac{dt}{S} \; . \qquad (13.2)$$

In Chap. 11 both distances had been indicated with the same symbol d, for example in (11.15), (11.18) and (11.20), but here a distinction is necessary. In an $(\Omega_m, \Omega_\Lambda) = (1, 0)$ universe the angular size of a causally connected region that we observe at a redshift z is (see exercise):

$$\varphi_0 = \frac{d_{ho}}{d_{lk}} = \frac{\int_0^t dt/S}{\int_t^{t_0} dt/S} = \frac{1}{\sqrt{1+z}-1} \simeq \frac{1}{\sqrt{z}}, \qquad (13.3)$$

for large z. It follows that $d_{ho}/d_{lk} \ll 1$, so that the early visible universe consists of many causally unconnected regions. The problem is innate to all

---

[2] In the notation of § 11.2 d_{lk} equals $d = (S/S_0)d_0$.

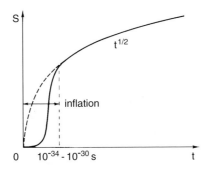

Fig. 13.3. Behaviour of the scale factor in the very early universe. The dotted line $S \propto t^{1/2}$ causes all problems.

FRW models and the outcome of (13.3) is only weakly dependent on Ω_m and Ω_Λ. The horizon spaces of P and Q have had no opportunity to interact. What mechanism is responsible for the neighbourhood of P and Q having similar physical properties (i.e. why is the universe isotropic)? Why indeed would P and Q begin to participate in the expansion at the same moment? The issue is one of causality. An FRW universe seems to behave like someone who walks along the street although the various parts of his body are unable to exchange signals.

Let's take the CMB in our own universe as an example, and t in Fig. 13.1 is the time of recombination $t_{\rm rec}$. We have seen in exercise 11.5 that the horizon distance at recombination is $2.25 c t_{\rm rec}$. The distance $d_{\rm lk}$ of the last scattering surface to us at that time is $3.3 \times 10^{-3} \cdot 0.96 c / H_0$ (Table 11.1). In fact we are repeating exercise 10.12, with a better value for $d_{\rm ho}$, and the result is that the angular size of causally connected regions at recombination is $\varphi_0 = d_{\rm ho}/d_{\rm lk} \simeq 1.1°$ in our universe, while (13.3) predicts $\varphi_0 \simeq 1.7°$ for a $(1, 0)$ universe. It follows that the observed isotropy of the CMB on angular scales $> \varphi_0$ is accidental, as there is no causal connection possible on larger angular scales. The viewing geometry is further explained in Fig. 13.2.

Origin and remedy of the horizon problem

The physical origin of the horizon problem is that the expansion is arbitrarily fast near $t = 0$: $\lim_{t \to 0} \dot{S} = \infty$. Since the signal speed is finite ($\lesssim c$) the universe immediately breaks up into regions that have had no time to communicate, which is unphysical. The fact that $\lim_{t \to 0} \dot{S} = \infty$ is an inevitable consequence of (10.1). If $\rho \neq 0$ the ρ-term in (10.1) is $\propto S^{-3}$ or S^{-4} for radiation, and it follows that $\dot{S} \to \infty$ for $S \to 0$.

13.1 The horizon problem

The problem would disappear if $S(t)$ approaches zero in a different way, as in Fig. 13.3. The numerator $\int_0^t dt/S$ of (13.3), convergent for $S \propto t^{1/2}$, would now become much larger, while the denominator remains unaffected. The value of the denominator $\int_t^{t_0} dt/S$ is difficult to tinker with anyhow, because the shape of $S(t)$ is fixed once the radiation era is underway. But right after $t = 0$ we can't be so sure anymore. For example, let's suppose for the sake of argument that $S \propto t^2$ near $t = 0$. The expansion is then initially slow, $\lim_{t \to 0} \dot{S} = 0$, so that the various regions may interact and have an opportunity to 'homogenize'. The subsequent expansion becomes increasingly rapid. And now $d_{ho} \simeq \infty$ according to (13.2). Thus we would achieve that the horizon distance d_{ho} in the early universe was already much larger than d_{lk} (= size of our visible universe scaled down to an early time t), Fig. 13.1, right.

We have now discovered the essence of the inflation concept. The scale factor $S(t)$ is subject to a very rapid accelerating growth just after $t = 0$, as in Fig. 13.3. An implication is that the very early universe is extremely small, much smaller than one would expect on the basis of $S \propto t^{1/2}$.

Exercise 13.1: Verify the details of (13.3).

Hint: $\varphi_0 = [\int_0^t dt/S] / [\int_t^{t_0} dt/S] = [\int_0^{S/S_0} du/(u\dot{u})] / [\int_{S/S_0}^1 du/(u\dot{u})]$; (10.10): $\dot{u} \propto u^{-1/2} \to \varphi_0 = [\sqrt{u}]_0^{S/S_0} / [\sqrt{u}]_{S/S_0}^1$. Finally $S_0/S = 1 + z$.

Exercise 13.2: The horizon problem in a closed $\Omega_\Lambda = 0$ FRW universe, Fig. 13.4. Assume that photon F starts at the moment of the Big Bang, and show that since that time it has travelled a co-ordinate distance

$$\chi = \sqrt{\Omega_m - 1} \int_0^u \frac{dx}{x(\Omega_m x^{-1} + 1 - \Omega_m)^{1/2}}$$

$$= 2 \arcsin \sqrt{S/S_m}, \qquad (13.4)$$

where S_m is the value of the scale factor at maximum expansion, $S_m/S_0 = \Omega_m/(\Omega_m - 1)$, § 10.2. Prove the following statements (1) complete causal contact of all parts is only attained in the contraction phase, after maximal expansion, and in the expansion phase the universe has causally disconnected parts; (2) after maximal expansion an observer begins to see double images, diametrically opposite on his sky; (3) A will never see his own image.

Hint: Radial null geodesic from (9.19): $(dx_0)^2 = S^2 d\chi^2$ or $d\chi/dt = c/S$, then (10.10); substitute $x = \Omega_m y^2/(\Omega_m - 1)$ to get rid of Ω_m in the integral; (1)

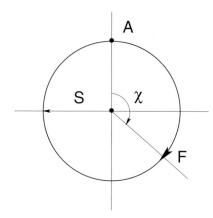

Fig. 13.4. Snapshot of a great circle in a closed $\Omega_\Lambda = 0$ FRW universe, showing the radial co-ordinate $r = \sin\chi$, § 9.3. As time progresses, the circle expands together with the universe and then contracts again. Object A, located at the origin $r = 0$, has emitted photon F at the moment of the Big Bang. The photon returns to A at the moment of the Big Crunch.

complete causal contact requires that photons emitted by A have covered the entire universe $\rightarrow \chi = \pi$; (2) photons travelling in opposite directions may reach the same observer as soon as $\chi > \pi$; (3) A sees his own image at the moment of the Big Crunch. This shows how fast the expansion really is: in a closed $\Omega_\Lambda = 0$ universe photons just manage to make one round trip!

13.2 Evolution of a universe with a scalar field

We have seen that the horizon problem may be solved if the scale factor $S(t)$ behaves differently near $t = 0$. In the 80ies of the last century it was discovered that scalar fields that may have been present in the early universe can do the magical trick. Scalar or Higgs fields had originally been used in particle physics because they could endow mass to the quanta of an otherwise massless vector field, without destroying the possibility of renormalization. Scalar fields have been very popular since that time, and it was only a natural development to investigate their role in cosmology. The bosons in question are believed to be the hypothetical supermassive X-bosons ($\sim 10^{15}$ GeV) that occur in Grand Unified Theories, see § 12.2. Such particles may be abundant just after the Big Bang, and play a role in many inflation models. In practice the scalar field is just postulated and one doesn't worry too much about its place in the grander scheme of things. We shall now derive the equations of motion for the simplest possible model: one scalar field, minimally coupled

13.2 Evolution of a universe with a scalar field

to gravity. We start from the relativistic expression of the energy E of a free particle:

$$E^2 = m^2 c^4 + (\boldsymbol{p}c)^2 , \tag{13.5}$$

where $\boldsymbol{p}$ is the particle's momentum. This is quantized in the usual way by replacing E and $\boldsymbol{p}$ by operators $E \to i\hbar \partial/\partial t$ and $\boldsymbol{p} \to -i\hbar \nabla$:

$$-\hbar^2 \frac{\partial^2 \psi}{\partial t^2} = m^2 c^4 \psi - \hbar^2 c^2 \nabla^2 \psi . \tag{13.6}$$

This is the Klein-Gordon equation for a field ψ of bosons with spin zero and rest mass m. After some cleaning up:

$$(\Box + \mu^2)\psi = 0 ; \qquad \mu = \frac{mc}{\hbar} . \tag{13.7}$$

But (13.7) is is not an acceptable equation because $\Box\psi = \eta^{\mu\nu}\psi_{,\mu\nu}$ is not an invariant scalar. The simplest generalization is: $\eta^{\mu\nu}\psi_{,\mu\nu} \to g^{\mu\nu}\psi_{:\mu:\nu}$. So, we replace (13.7) by

$$g^{\mu\nu}\psi_{:\mu:\nu} + \mu^2 \psi = 0 , \tag{13.8}$$

which is properly invariant. This type of reasoning, incidentally, is another example of how the principle of general covariance is used in practise.

The explicit expression for $g^{\mu\nu}\psi_{:\mu:\nu}$ may be found with the help of (2.47): $\psi_{:\mu} = \psi_{,\mu}$, and then (2.43):

$$g^{\mu\nu}(\psi_{,\mu\nu} - \Gamma^\alpha_{\mu\nu}\psi_{,\alpha}) + \mu^2 \psi = 0 . \tag{13.9}$$

We show later that we may restrict ourselves to $\psi_{,i} = 0$, i.e. to homogeneous ψ:

$$g^{00}\psi_{,00} - g^{\mu\nu}\Gamma^0_{\mu\nu}\psi_{,0} + \mu^2 \psi = 0 . \tag{13.10}$$

But $g^{00} = 1$ and from (9.31): $g^{\mu\nu}\Gamma^0_{\mu\nu} = g^{ik}\Gamma^0_{ik} = -(S'/S)g^{ik}g_{ik} = -3\dot{S}/cS$:

$$\psi_{,00} + \frac{3\dot{S}}{cS}\psi_{,0} + \mu^2 \psi = 0 . \tag{13.11}$$

We have landed on familiar territory: ψ evolves as an harmonic oscillator that is damped by the expansion of the universe.

The field equation is $G_{00} = -(8\pi G/c^2)T_{00}$, or, with (9.34):

$$\left(\frac{\dot{S}}{S}\right)^2 + \frac{kc^2}{S^2} = \frac{8\pi G}{3} T_{00} . \tag{13.12}$$

T_{00} is the total energy of the harmonic oscillator:[3]

---
[3] Landau, L.D. and Lifshitz, E.M.:1971, *Relativistic Quantum Theory*, Pergamon Press, § 12.

$$T_{00} = \tfrac{1}{2}(\psi_{,0}^2 + \mu^2\psi^2)\ . \qquad (13.13)$$

Actually there is another term $\tfrac{1}{2}|\nabla\psi|^2$ on the right hand side of (13.13) which is omitted because of the assumed homogeneity of ψ. The equation for S becomes:

$$\left(\frac{\dot S}{S}\right)^2 + \frac{kc^2}{S^2} = \frac{4\pi G}{3}(\psi_{,0}^2 + \mu^2\psi^2)\ . \qquad (13.14)$$

We now have a closed set of equations (13.11) and (13.14) for S and ψ.

Before we proceed it is useful to write these equations in dimensionless form, with the help of Planck units. The Planck mass M_p is the mass of a black hole whose Schwarzschild radius $2GM/c^2$ and Compton wavelength $\hbar/Mc$ are equal:

$$M_p = \left(\frac{\hbar c}{G}\right)^{1/2} \simeq 2.2 \times 10^{-5}\ \mathrm{g}\ , \qquad (13.15)$$

a macroscopic mass of 22 μg, and $M_p c^2 \simeq 1.2 \times 10^{19}$ GeV. The Compton wavelength $\hbar/M_p c$ of this hole is the Planck length L_p:

$$L_p = \frac{\hbar}{M_p c} = \left(\frac{\hbar G}{c^3}\right)^{1/2} \simeq 1.6 \times 10^{-33}\ \mathrm{cm}\ . \qquad (13.16)$$

The Planck density $\rho_p \equiv M_p/L_p^3$ and the Planck time $t_p \equiv L_p/c$ are

$$\rho_p = \frac{c^5}{\hbar G^2} = M_p^4\left(\frac{c}{\hbar}\right)^3 \simeq 5.2 \times 10^{93}\ \mathrm{g\ cm}^{-3}\ ; \qquad (13.17)$$

$$t_p = \left(\frac{\hbar G}{c^5}\right)^{1/2} = M_p^{-1}\frac{\hbar}{c^2} \simeq 5.4 \times 10^{-44}\ \mathrm{s}\ . \qquad (13.18)$$

We now substitute $G = \hbar c/M_p^2$ in (13.14) and then $\hbar = c = 1$ in (13.11) and (13.14):

$$\ddot\psi + 3H\dot\psi + m^2\psi = 0\ ; \qquad (13.19)$$

$$H^2 + \frac{k}{S^2} = \frac{4\pi}{3M_p^2}(\dot\psi^2 + m^2\psi^2)\ , \qquad (13.20)$$

with $H = \dot S/S$ and $\dot{} = d/dt$. The original units may be restored as follows. From (13.16) – (13.18) we see that [length] = [time] = [mass]$^{-1}$; [density] = [mass]^4. The dimension of ψ follows by requiring that $c^2 T_{00}$ is the energy density of the field → $[(\mu c\psi)^2] = [\rho c^2]$ → $[\psi]$ = [mass]. This means that if we compute a time, we may for example find that $t = m^{-1}$. Since $M_p t_p = 1$ according to (13.18), $t = m^{-1} M_p t_p = M_p/m$ in units of t_p, or $t = (M_p/m)\cdot M_p^{-1}(\hbar/c^2) = \hbar/mc^2$ sec. Analogous results may be derived for other quantities.

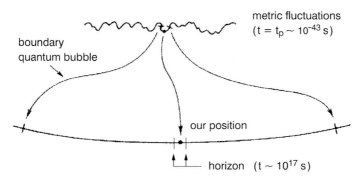

Fig. 13.5. In the chaotic inflation scenario a quantum fluctuation of characteristic size L_p in the metric of an existing spacetime, casually referred to as a 'quantum bubble', will inflate to huge proportions in about 10^{-34} s, if the energy it may contain on the basis of the uncertainty relation resides in one scalar field. During the subsequent reheating phase, the energy of the scalar field is converted into particles, the quark-gluon plasma, marking the beginning of the hot Big Bang around $t \sim 10^{-30}$ s (depending on the inflation scenario). Our visible universe is a minute fraction of the original bubble, and therefore homogeneous and flat.

Equations (13.19) and (13.20) do not allow for interaction with other fields, because we assumed a free particle. The equations become more complicated when these interactions are included, and their mathematical form becomes strongly dependent on the details of the particle physics at the highest energies, about which little is known. This is where the appeal of the *chaotic inflation model* proposed by Linde comes in. Chaotic inflation assumes that the universe is born out of a quantum fluctuation in which the energy in one scalar field ψ dominates over all other fields. The evolution is then presumably well described by (13.19) and (13.20) for a free field.

13.3 Chaotic inflation

We start off from an existing spacetime. On a microscopic scale there are abundant quantum fluctuations in the metric, Fig. 13.5. A causally connected part (a 'quantum bubble') has a characteristic size $L_p = ct_p$ and contains a characteristic energy $M_p c^2$ that is restricted by Heisenberg's uncertainty relation to $M_p c^2 \cdot t_p \sim \hbar$. This energy is more or less equally divided among various fields. These regions are not of interest to us because they do not inflate. We concentrate on one of those rare places and times where one scalar field dominates over all others, and analyse its evolution with (13.19) and (13.20). Because almost all energy resides in the scalar field we get $T_{00} = \frac{1}{2}(\dot{\psi}^2 + m^2\psi^2) \sim \rho_p = M_p^4$ ($\hbar = c = 1$). Since the ignored term $\frac{1}{2}|\nabla\psi|^2$ in (13.13) may also not be larger than $\sim M_p^4$, we infer a restriction on the typical variation $\delta\psi$ of ψ:

$$\delta\psi \sim |\nabla\psi|L_p \leq M_p^2 M_p^{-1} = M_p . \tag{13.21}$$

But $T_{00} \simeq m^2\psi^2 \sim M_p^4$ (we shall show later that $\dot{\psi} \ll m\psi$), so that

$$\psi \sim \frac{M_p^2}{m} ; \quad \frac{\delta\psi}{\psi} \sim \frac{m}{M_p} \ll 1 , \tag{13.22}$$

since $mc^2 \sim 10^{15}$ GeV if m is the X-boson, while $M_p c^2 \sim 10^{19}$ GeV. We conclude that the assumption that all energy resides in the scalar field implies its near-homogeneity. It is therefore reasonable to put $\nabla\psi = 0$ in the derivation of (13.19) and (13.20), and we take (13.22) as the initial condition of ψ at $t = t_p$. According to (13.19), ψ is a harmonic oscillator with frequency m and damping $3H/2$. For weak damping ($H \ll m$) ψ will oscillate. But the damping turns out to be strong ($H \gg m$), and ψ approaches zero only very slowly. In that case the inertia term $\ddot{\psi}$ can be neglected. We assume that:

$$H \gg m \quad \text{and} \quad \dot{\psi} \ll m\psi . \tag{13.23}$$

Furthermore we omit the curvature k/S^2 term. These approximations will be justified later. We are then left with:

$$3H\dot{\psi} = -m^2\psi ; \tag{13.24}$$

$$H^2 = \frac{4\pi m^2}{3M_p^2}\psi^2 . \tag{13.25}$$

The nature of the solution is rather obvious. Relation (13.25) says that $H (:) \psi$, and then (13.24) says that $\dot{\psi}$ is a negative constant which turns out to be small, so that ψ is also approximately constant. Hence, according to (13.25), $H = \dot{S}/S \sim$ constant, i.e. exponential expansion. As ψ slowly decreases, so does H until the weak damping limit is attained. The explicit solution for the strong damping case is obtained by solving (13.24) and (13.25) for $\dot{\psi}$ and $H = \dot{S}/S$:

$$\dot{\psi} = -\frac{mM_p}{\sqrt{12\pi}} ; \tag{13.26}$$

$$\frac{\dot{S}}{S} = -\frac{4\pi}{M_p^2}\psi\dot{\psi} . \tag{13.27}$$

These equations may be integrated:

13.3 Chaotic inflation

$$\psi(t) = \psi_p - \frac{mM_p t}{\sqrt{12\pi}} ; \tag{13.28}$$

$$S(t) = S_p \exp\left[\frac{2\pi}{M_p^2}\{\psi_p^2 - \psi(t)^2\}\right], \tag{13.29}$$

where $\psi_p = \psi(t_p) \simeq M_p^2/m$ and $S_p = S(t_p) \simeq L_p$.

An exercise invites the reader to show that the approximations (13.23) are valid as long as $\psi \gg M_p/\sqrt{3\pi}$. The range of validity of the solution is therefore

$$M_p/\sqrt{3\pi} < \psi \lesssim \frac{M_p^2}{m} . \tag{13.30}$$

With the help of (13.28) the exponent in (13.29) may be expanded as $(2\pi/M_p^2) \cdot [\psi_p^2 - (\psi_p + \dot\psi t)^2] \simeq (2\pi/M_p^2)(-2\dot\psi\psi_p t) = \sqrt{4\pi/3}\, M_p t$, as long as $t \ll -\psi_p/\dot\psi$. The expansion is therefore exponential,

$$S(t) = S_p \exp(H_p t) ; \quad H_p = \sqrt{\frac{4\pi}{3}} M_p \simeq 2M_p , \tag{13.31}$$

as long as $t \ll t_e$ where

$$t_e = -\frac{\psi_p}{\dot\psi} = \sqrt{12\pi}\, \frac{M_p}{m^2} = \sqrt{12\pi}\left(\frac{M_p}{m}\right)^2 t_p . \tag{13.32}$$

For large t, $S(t)$ reaches the final value

$$\frac{S_e}{S_p} \sim \exp\left\{\frac{2\pi}{M_p^2}\psi_p^2\right\} \sim \exp\left\{2\pi\left(\frac{M_p}{m}\right)^2\right\} . \tag{13.33}$$

The numerical value of (13.33) is very uncertain because the boson mass m is unknown. But since $m \ll M_p$ it is clear that the scale factor S is blown up by a huge amount, possibly as large as $\sim 10^{(10^8)}$ if $mc^2 \sim 10^{15}$ GeV. This number is so large that even astronomers, not known to be easily impressed by large numbers, are baffled.

And it all happens in a very brief time span. The time when the inflation terminates can be estimated by requiring $\psi(t_e) = 0$ in (13.28), and this leads again to (13.32). For $m/M_p = 10^{-4}$ we have $t_e \sim 6 \times 10^8 t_p \sim 3 \times 10^{-35}$ s. When ψ becomes of the order of $M_p/\sqrt{3\pi}$ it is no longer possible to ignore $\ddot\psi$, and $\psi(t)$ becomes oscillatory. This is the reheating phase, during which the energy in the scalar field is converted into matter, which is not included in the equations. The subsequent expansion of the universe to its present-day size, by a factor of $\sim 10^{27}$, is relatively modest with respect to (13.33).

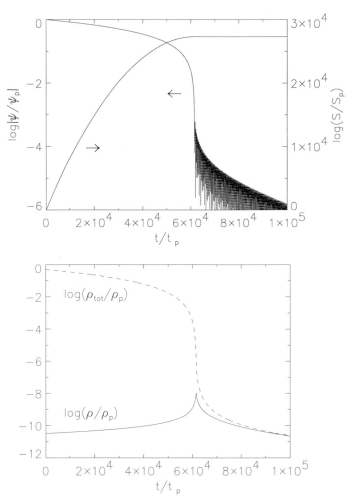

Fig. 13.6. Numerical solution of eqs. (13.34) - (13.36). Top: scale factor and field amplitude. Bottom: the energy density of the radiation and the total energy density in units of the Planck energy density $\rho_p c^2$. Note that $H \propto (\rho_{\rm tot})^{1/2}$. All logarithms are base 10. Parameters: $m/M_{\rm p} = 0.01$, $\gamma = 0.01\, m$, $\psi_{\rm p} = M_{\rm p}^2/m$, timestep $\Delta t = 0.25\, t_{\rm p}$. The equations require a small timestep because they are stiff, and we did not bother to use a special integration routine. This renders integration for more realistic parameters such as $m/M_{\rm p} = 10^{-4}$ difficult.

Toy model

To demonstrate the inflation and reheating in some detail we add an equation for relativistic matter (that is, radiation) to eqs. (13.19) and (13.20):

$$\ddot{\psi} + (3H + \gamma)\dot{\psi} + m^2\psi = 0 , \tag{13.34}$$

$$H^2 = \frac{8\pi}{3M_p^2} \left(\tfrac{1}{2}\dot{\psi}^2 + \tfrac{1}{2}m^2\psi^2 + \rho\right) , \tag{13.35}$$

$$\dot{\rho} + 4H\rho = \gamma\dot{\psi}^2 . \tag{13.36}$$

These equations are obtained as follows. The interaction of the scalar field with other fields is modelled by a damping term $\gamma\dot{\psi}$ in (13.34). This interaction is initially not important since H is large, but as ψ and H decrease, the damping of ψ by coupling with matter fields becomes more important than expansion. In physical terms, the energy in the scalar field is converted into particles, for example X-bosons, that subsequently decay into quarks and leptons. The matter has an energy density ρ, which has been added to the total energy density in (13.35). The curvature term k/S^2 has been dropped as it is soon unimportant. For $\gamma = 0$, (13.36) says that ρS^4 is constant (relativistic matter). The choice $\gamma\dot{\psi}^2$ for the matter source term is motivated by the fact that it makes the interaction between matter and scalar field energy-conserving. We verify that by computing the time derivative of the total energy density $\rho_{\text{tot}} = \tfrac{1}{2}\dot{\psi}^2 + \tfrac{1}{2}m^2\psi^2 + \rho$:

$$\frac{d}{dt}\left(\tfrac{1}{2}\dot{\psi}^2 + \tfrac{1}{2}m^2\psi^2 + \rho\right) = -H(3\dot{\psi}^2 + 4\rho) , \tag{13.37}$$

which is independent of the rate γ at which energy is exchanged between ψ and ρ. The conversion of the scalar field into matter proceeds without loss of energy, since ρ_{tot} decreases only insofar expansion dilutes the energy density of the scalar field and the matter.

These phenomenological equations nicely illustrate the key features of inflation and reheating, see Fig. 13.6. There is a huge expansion as long as the scalar field dominates. In this phase the matter evolves quasistationary, $\rho \simeq \gamma\dot{\psi}^2/4H$ ($\dot{\rho} \simeq 0$), and is energetically unimportant. The end of the inflation phase is correctly predicted by (13.32), and the simulations confirm that the scalar field has by then collapsed to a value $\psi \sim M_p/\sqrt{3\pi}$, see (13.30). The field becomes oscillatory thereafter, with a period $\sim m^{-1} = (M_p/m)t_p$, which is hardly resolved in Fig. 13.6. The expansion continues but slows down to $S \propto t^{1/2}$, which is no longer visible on the logarithmic scale of the figure. The total expansion factor is well reproduced by relation (13.33). The matter energy density surges to an estimated peak value of $\rho/\rho_p \sim \gamma m/12\pi M_p^2$

(without proof), and scales as $\rho \propto S^{-4} \propto t^{-2}$ soon thereafter: the beginning of a hot big bang.

Exercise 13.3: Verify that the assumptions in (13.23) are correct as long as $\psi \gg M_p/\sqrt{3\pi}$.

Hint: ψ in (13.19) is a damped harmonic oscillator. As long as the damping $3H/2$ is supercritical (i.e. $3H/2 > m$) one may neglect $\ddot{\psi}$. With (13.25) $\rightarrow$ $\psi > M_p/\sqrt{3\pi}$; (13.26) $\rightarrow |\dot{\psi}| < m\psi$.

Exercise 13.4: Prove that during the inflation phase the Hubble constant is $\sim 10^{61}$ times larger than it is today.

Hint: $H_p \sim M_p = M_p t_p / t_p = 1/t_p$ and $H_0 \sim 1/t_0 \rightarrow H_p/H_0 \sim t_0/t_p$.

Exercise 13.5: Show that the scalar field ψ is equivalent to a density ρ and a negative pressure p:

$$\left.\begin{array}{l}\rho = \tfrac{1}{2}\dot{\psi}^2 + \tfrac{1}{2}m^2\psi^2\ ;\\[4pt] p = \tfrac{1}{2}\dot{\psi}^2 - \tfrac{1}{2}m^2\psi^2\ .\end{array}\right\} \qquad (13.38)$$

Hint: Set $H = \dot{S}/S$ in (13.20), multiply with S^2, differentiate, and eliminate $\ddot{\psi}$ with (13.19):

$$\frac{\ddot{S}}{S} = -\frac{4\pi}{3M_p^2}(2\dot{\psi}^2 - m^2\psi^2)\ . \qquad (13.39)$$

Compare with (9.40) $\rightarrow \rho + 3p = 2\dot{\psi}^2 - m^2\psi^2$ in Planck units, and ρ from (13.20) and (9.36); Λ is completely negligible. The pressure is negative on account of (13.23).

13.4 Discussion

The horizon problem has disappeared because if we compute once more d_{ho} using (13.31) and (13.33) we find, for $t \gg t_e$:

$$d_{\mathrm{ho}} = cS\int_0^t \frac{dt}{S} > cS_e\int_0^{t_e} \frac{dt}{S} \simeq e^{H_p t_e}\int_0^{t_e} e^{-H_p t}dt$$

$$\sim \frac{c}{H_{\rm p}} \, {\rm e}^{H_{\rm p} t_{\rm e}} \sim L_{\rm p} \exp\left\{ 2\pi \left(\frac{M_{\rm p}}{m}\right)^2 \right\}, \qquad (13.40)$$

because $c/H_{\rm p} \sim L_{\rm p}$. After the inflation phase the value of $d_{\rm ho}$ is by any measure enormous. On the other hand, as long as $t > t_{\rm e}$, we know[4] that $d_{\rm lk}$ is at most of the order of ct_0. In other words, the inequality $d_{\rm ho}/d_{\rm lk} \ll 1$ that caused all the problems has now been reversed to $d_{\rm ho}/d_{\rm lk} \gg 1$.

Since H is approximately constant while S grows rapidly to huge proportions, the term k/S^2 in eq. (13.20) is soon negligible. Inflation thus implies that the universe is flat. Likewise it is correct to ignore the cosmological constant term $\Lambda/3$ in (13.20) as $\Lambda \sim H_0^2 \ll H^2$ (in Planck units). Our visible universe is a tiny fraction of the original quantum fluctuation and is therefore homogeneous, regardless how inhomogeneous the initial fluctuation was.

Current status

The inflation concept was originally introduced by Guth[5] in 1981, and presently there are a number of different models on the market.[6] Because of its many achievements inflation has become a paradigm in cosmology that is likely to stay – even though the nature of the scalar field that does the magical trick is unknown. To this comes that all models have loose ends, and none is wholly accepted. For example, there is no explanation for the fine-tuning problem of the cosmological constant Λ. The model expounded here has one advantage over others: it seems not to depend strongly on the (unknown) details of the particle physics, although self-interaction of the field is ignored. It produces an inflation factor much larger than other models do, and also much larger than is needed (see exercise). On the other hand, the fact that under certain conditions a quantum fluctuation would inflate may be regarded as an instability of the vacuum, and it remains to be seen if a complete quantum theory of gravity permits such a phenomenon. Another serious objection is the fact that the model uses a semi-classical formulation right after the Planck time.

What drives it?

The clue is that the energy density of the scalar field is *not* diluted by expansion like ordinary radiation ($\propto S^{-4}$). This very counterintuitive property is confirmed by eq. (13.37): for $\rho = 0$ the energy density of the scalar field decreases as $-3H\dot{\psi}^2$ which is of the order of $(m/M_{\rm p})^2$ Planck energy density $\rho_{\rm p} c^2$ per Planck time. In the present context this is very small, so H

---

[4] See Table 11.1; $d_{\rm lk} \equiv d$.
[5] Guth, A.H., *Phys. Rev. D 23* (1981) 347.
[6] For more information on inflation theory see Börner (1988) Ch. 9; Kolb and Turner (1990) Ch. 8; and Peacock (1999) Ch. 11.

remains roughly constant according to (13.35). And that means that exponential expansion continues unabated. Another way of saying this is relation (13.38): the scalar field is equivalent to a density $\rho \simeq \frac{1}{2}m^2\psi^2$ and a negative pressure $p \simeq -\frac{1}{2}m^2\psi^2$ since $\frac{1}{2}\dot\psi^2$ is small. During inflation spacetime behaves approximately as a vacuum with a large cosmological constant $\Lambda \propto \frac{1}{2}m^2\psi^2$, and we conclude that the initial quantum fluctuation is blown up by the huge anti-gravity associated with the scalar field. All inflation models have in common that the inflation takes place very early – after $\sim 10^{-30}$ s at most it is all over. Inflation may therefore be regarded as a physical mechanism that creates the homogeneous, isotropic, hot, expanding and flat FRW universe whose existence we took for granted in earlier chapters.

Seed fluctuations

The exponential expansion creates an event horizon. Events further away than $c/H \gtrsim L_\mathrm{p}$ cannot communicate with the observer, who will experience the universe during the inflation phase as a kind of black hole turned inside out. Although this is merely an analogy, there is one consequence that carries over: the creation of quantum fluctuations in the scalar field, which turn out to have an r.m.s. amplitude $\delta\psi/\psi \sim m/(M_\mathrm{p}\sqrt{3\pi})$ per wavelength decade. These fluctuations in ψ are eventually converted into density fluctuations $\delta\rho/\rho$ and have the right spectrum to serve as the seeds for structure formation if $m/M_\mathrm{p} \sim 10^{-4}$. This is an important reason for believing that the scalar field of mass m may correspond to the supermassive X-bosons of grand unified theories.

Energetics

The energetics of inflation is an elusive problem. We start with a total energy $\sim M_\mathrm{p}c^2$ at $t = t_\mathrm{p}$ and at the end of the reheating we have $\sim S_\mathrm{e}^3 \times$ the energy density at that time, plus the gravitational energy, which is negative one would say. However, this is the reasoning of an external observer, § 11.2. An observer *in* the universe faces a different situation, as he has to perform an integration over the past lightcone. A proper calculation is called for, but then we run into the problem that an invariant definition of the gravitational energy in a volume does not exist in GR. It does only in the special cases of asymptotically flat or stationary spacetimes, neither of which applies to the FRW universe. So, when we make statements about the global energetics of the expanding universe we cannot be sure to avoid artifacts due to the choice of the co-ordinates! As a result of these problems no clear answer exists, which is very unsatisfactory.

Philosophical issues

Speculations on the origin of the universe concern issues that are often impossible to verify, which gives them a metaphysical twist where, depending

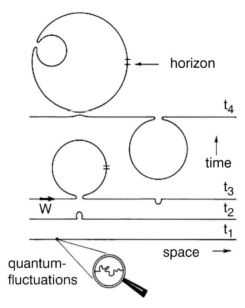

Fig. 13.7. Chaotic inflation may take the form of a hierarchical process creating many interconnected or decoupled universes. Observer W would just notice a small defect in his spacetime of size $\sim L_\mathrm{p}$ that may or may not disappear again. Our visible universe is a very small section of spacetime, for instance the space indicated by the two markers. Since observations beyond the horizon are impossible, we can only make 'reasonable' assumptions about what lies outside (such as the cosmological principle). This underlines the highly speculative character of ideas such as these.

on the temperament of the author, sometimes pretty wild extrapolations are made.[7] The model treated here is called chaotic inflation because it begins whenever a sufficiently large quantum fluctuation materializes in spacetime, and the standard lore about chaotic inflation runs something like this. Suppose that happens close to an observer W. To W such a region would appear to be a small defect in space – a kind of black hole with radius $\sim L_\mathrm{p}$, see Fig. 13.7. Inside, however, the geometry is 'redefined' in a drastic manner, as it contains an entire universe. This idea of inflation being a sudden redefinition of the geometry in a very small patch of spacetime may be helpful. W may nevertheless hold the defect, and thereby an entire universe in his hand (where perhaps other students study their cosmology books). The figure is an attempt to visualize the geometrical structure by letting space 'bulge out' into a flat embedding space, suggesting that one universe is 'next to' or 'below' another – which is of course not so because the embedding space does not exist. Our universe may likewise be enclosed inside another spacetime.

---

[7] For example Tegmark, M., *Sci. Am.* May 2003.

In this vision creation is a stochastic process that continues forever. Spacetime would have a kind of hierarchical structure without a beginning in time. Our universe may be one of many, and the constants of nature and therefore the physics would be different in each universe. The constants of nature in our universe must have the values they have because we would not exist if they were much different. This type of reasoning is called the *anthropic principle*, originally introduced by Carter.[8] The so-called weak anthropic principle maintains that what we can expect to observe is restricted by the conditions necessary for our existence as observers, see Barrow and Tipler (1986) for more details.

Perhaps the most stunning perspective offered by inflation theory is the idea that the entire universe as we know it originates from a tiny part of an already tiny quantum fluctuation. Is this a dazzling show of the power of scientific reason, or rather a figment of the mind – a modern Tower of Babel? Don't say too soon that we shall never know, and recall the example of Auguste Comte, who argued in earnest in 1835 that it would be forever impossible to determine the temperature and the internal state of stars.[9] And then of course came spectroscopy.

Exercise 13.6: The solution of the horizon problem requires a minimum inflation factor of $S_e/S_p \sim (t_0/t_p)(T_0/T_e) \sim 10^{61} 10^{-28} = 10^{33} \sim e^{76}$.

Hint: Require that the present horizon distance (about ct_0) rescaled to $t = t_p$, i.e. $ct_0(S_p/S_0)$, is equal to the horizon size L_p at $t = t_p$; then $S_e/S_p = (S_e/S_0) \cdot (S_0/S_p) \sim (S_e/S_0)(ct_0/L_p) \sim (T_0/T_e)(t_0/t_p)$ (use $T \propto S^{-1}$); estimate T_e from $\kappa T_e \sim mc^2 \to T_e \sim 10^{28}$ K. This is a very rough estimate; the energy of the scalar field at the end of the inflation may not be completely spent on reheating and then T_e is smaller than 10^{28} K.

Exercise 13.7: Verify that the expansion factor of the universe from the end of inflation to now is $\sim 10^{27}$.

Hint: The scale factor S goes as $t^{1/2}$ from $t_e \sim 6 \times 10^8 \, t_p \sim 3 \times 10^{-35}$ s to $t_{mat} \simeq 5.5 \times 10^4$ yr ($\to$ expansion factor 2.4×10^{23}). Then as $t^{2/3}$ until now, 14×10^9 yr (expansion factor 4000).

---

[8] Carter, B., in *Confrontation of Cosmological Theories with Observational Data*, ed. M.S. Longair (Reidel 1974), p. 291.
[9] Comte, A.: 1835, *Philosophie Première (cours de philosophie positive)*, 19ᵉ Leçon, ed. Hermann (Paris 1975).

A
Bibliography

Elementary texts

Adams, F. and Laughlin, G.: 1999, *The Five Ages of the Universe*, The Free Press.

Berry, M.: 1978, *Principles of Cosmology and Gravitation*, Cambridge U.P.

Geroch, R.: 1978, *General Relativity from A to B*, University of Chicago Press.

Hogan, C.J.: 1998, *The Little Book of the Big Bang: A Cosmic Primer*, Springer-Verlag.

Silk, J.: 1980, *The Big Bang*, Freeman and Co.

Weinberg, S.: 1977, *The First Three Minutes*, Basic Books, Inc.

Introductory textbooks

Adler, R., Bazin, M. and Schiffer, M.: 1965, *Introduction to General Relativity*, McGraw-Hill.

Dirac, P.A.M.: 1975, *General Theory of Relativity*, Wiley-Interscience.

Foster, J. and Nightingale, J.D.: 1989, *A short course in General Relativity*, Longman.

Frank, J., King, A.R. and Raine, D.J.: 1992, *Accretion Power in Astrophysics*, Cambridge U.P.

Kenyon, I.R.: 1990, *General Relativity*, Oxford U.P.

Landau, L.D. and Lifshitz, E.M.: 1971, *The Classical Theory of Fields*, Pergamon Press.

Linder, E.V.: 1997, *First Principles of Cosmology*, Addison-Wesley.

Price, R.H.: 1982, *General Relativity Primer*, Am. J. Phys. **50**, 300.

Rindler, W.: 2001, *Relativity, Special, General and Cosmological*, Oxford U.P.

Robertson, H.P. and Noonan, T.W.: 1969, *Relativity and Cosmology*, Saunders.

Schutz, B.F.: 1985, *A First Course in General Relativity*, Cambridge U.P.

Advanced textbooks on specific topics

Blair, D.G. (ed.): 1991, *The Detection of Gravitational Waves*, Cambridge U.P.

Börner, G.: 1988, *The Early Universe (Facts and Fiction)*, Springer-Verlag.

Chen, Y.T. and Cook, A.: 1993, *Gravitational Experiments in the Laboratory*, Cambridge U.P.

Kolb, E.W. and Turner, M.S.: 1990, *The Early Universe*, Addison-Wesley.

Misner, C.W., Thorne, K.S. and Wheeler, J.A.: 1971, *Gravitation*, Freeman and Co.

Padmanabhan, T.: 1993, *Structure Formation in the Universe*, Cambridge U.P.

Peacock, J.A.: 1999, *Cosmological Physics*, Cambridge U.P.

Peebles, P.J.E.: 1993, *Principles of Physical Cosmology*, Princeton U.P.

Saulson, P.R.: 1994, *Fundamentals of Interferometric Gravitational Wave Detectors*, World Scientific.

Schneider, P., Ehlers, J. and Falco, E.E.: 1992, *Gravitational Lenses*, Springer-Verlag.

Shapiro, S.L. and Teukolsky, S.A.: 1983, *Black Holes, White Dwarfs and Neutron Stars*, Wiley-Interscience.

Wald, R.M.: 1984, *General Relativity*, University of Chicago Press.

Weinberg, S.: 1972, *Gravitation and Cosmology*, John Wiley and Sons, Inc.

Will, C.M.: 1993, *Theory and Experiment in Gravitational Physics*, Cambridge U.P.

Foundations and history

Barrow, J.D. and Tipler, F.J.: 1986, *The Anthropic Cosmological Principle*, Clarendon Press.

Bless, R.C.: 1995, *Discovering the Cosmos*, University Science Books.

Dijksterhuis, E.J.: 1969, *The Mechanization of the World Picture*, Oxford U.P.

Evans, J.: 1998, *The History and Practice of Ancient Astronomy*, Oxford U.P.

Friedman, M.: 1983, *Foundations of Space-Time Theories*, Princeton U.P.

Koestler, A.: 1959, *The Sleepwalkers - A History of Man's changing Vision of the Universe*, Hutchinson.

Pais, A.: 1982, *'Subtle is the Lord...', the Science and Life of A. Einstein*, Oxford U.P.

Pannekoek, A.: 1989, *A History of Astronomy*, Dover.

Study reports

LISA System and Technology Study Report, Reinhard, R. and Edwards, T. (eds.), ESA-SCI(2000)11, July 2000.

Exercises

Lightman, A.P., Press, W.H., Price, R.H. and Teukolsky, S.A.: 1975, *Problem book in Relativity and Gravitation*, Princeton U.P. (general relativity and cosmology).

Taylor, E.F. and Wheeler, J.A.: 1966, *Spacetime Physics*, Freeman and Co. (special relativity).

B
Useful numbers

Table B.1. Physical and astronomical constants [a]

electron mass	9.109×10^{-28}	g	(511.0 keV)
proton mass	1.673×10^{-24}	g	(938.3 MeV)
neutron mass	1.675×10^{-24}	g	(939.6 MeV)
electron charge e	4.803×10^{-10}		c.g.s. (esu)
speed of light c	2.998×10^{10}		cm s^{-1}
Boltzmann constant κ	1.381×10^{-16}		erg K^{-1}
radiation constant σ	5.670×10^{-5}		erg cm^{-2} K^{-4}s^{-1}
Planck constant $\hbar = h/2\pi$	1.055×10^{-27}		g cm^{2}s^{-1}
gravitational constant G	6.674×10^{-8}		cm^3 g^{-1}s^{-2}
Planck mass $M_\mathrm{p} = (\hbar c/G)^{1/2}$	2.18×10^{-5}		g
Planck length $L_\mathrm{p} = (\hbar G/c^3)^{1/2}$	1.62×10^{-33}		cm
Planck time $t_\mathrm{p} = (\hbar G/c^5)^{1/2}$	5.39×10^{-44}		s
Planck density $\rho_\mathrm{p} = c^5/\hbar G^2$	5.16×10^{93}		g cm^{-3}
1 AU	1.496×10^{13}	cm	
1 light year (lyr)	9.461×10^{17}	cm	
1 parsec (pc)	3.086×10^{18}	cm	3.262 lyr
microwave background temperature	2.725 ± 0.002	K	
Hubble constant H_0	$100\,h$ km s^{-1} Mpc^{-1}		$3.24 \times 10^{-18}\,h$ s^{-1}
h	0.71 ± 0.04		
Hubble time $1/H_0$	$3.09 \times 10^{17} h^{-1}$	s	$9.79\,h^{-1}$ Gyr
Hubble radius c/H_0	$9.25 \times 10^{27}\,h^{-1}$	cm	$3.00\,h^{-1}$ Gpc

[a] http://physics.nist.gov/constants

Table B.2. Sun and Earth

	Sun	Earth
mass (g)	1.99×10^{33}	5.98×10^{27}
radius (km)	6.96×10^5	6.37×10^3
Schwarzschild radius $2GM/c^2$	2.95 km	0.887 cm
luminosity (erg s^{-1})	3.83×10^{33}	

C

Euler-Lagrange equations

In GR and other fields one often encounters the following problem. Given a function $L(y_1(p), \dot{y}_1(p), y_2(p), \dot{y}_2(p), \cdots) \equiv L(\{y_i, \dot{y}_i\})$, where $\dot{} = d/dp$. For which functions $y_i(p)$ is the value of the integral $I = \int_a^b L\,dp$ an extremum? This well known problem is handled by considering the difference between the value of I for a neighbouring function set $y_i + \delta y_i$ and the original value of I. We compute the difference to first order in δy_i:

$$\delta I = \int_a^b L(\{y_i + \delta y_i, \dot{y}_i + \delta \dot{y}_i\})\,dp - \int_a^b L(\{y_i, \dot{y}_i\})\,dp$$

$$\simeq \int_a^b \left(\frac{\partial L}{\partial y_i}\delta y_i + \frac{\partial L}{\partial \dot{y}_i}\delta \dot{y}_i\right) dp, \tag{C.1}$$

with a summation over double indices i as usual. We have

$$\delta \dot{y}_i = \delta \frac{dy_i}{dp} = \frac{d}{dp}\delta y_i. \tag{C.2}$$

δ and d/dp commute, and that enables us to partially integrate the second term:

$$\delta I \simeq \int_a^b \left[\frac{\partial L}{\partial y_i} - \frac{d}{dp}\left(\frac{\partial L}{\partial \dot{y}_i}\right)\right]\delta y_i\,dp. \tag{C.3}$$

The stock term $(\partial L/\partial \dot{y}_i)\delta y_i|_a^b$ vanishes because $\delta y_i(a) = \delta y_i(b) = 0$. The end points are held fixed. The requirement that I is an extremum implies that $\delta I = 0$ for arbitrary δy_i. It follows that

$$\frac{\partial L}{\partial y_i} = \frac{d}{dp}\left(\frac{\partial L}{\partial \dot{y}_i}\right). \tag{C.4}$$

These are the famous Euler-Lagrange differential equations from which the functions $y_i(p)$ may be solved. Note that the derivation of (C.4) clearly shows that $\partial L/\partial y_i$ and $\partial L/\partial \dot{y}_i$ should be computed as if y_i and $\dot{y}_i$ are independent variables. Note, too, that we obtain the functions for which L is an extremum, a wider class than the functions for which I is a maximum or minimum.

Example

Let $L = y^2\dot{y} + \dot{y}^2$. Then $\partial L/\partial y = 2y\dot{y}$ and $\partial L/\partial \dot{y} = y^2 + 2\dot{y}$. After insertion in (C.4) we get $2y\dot{y} = 2y\dot{y} + 2\ddot{y}$, or $\ddot{y} = 0$. Hence $\int_a^b(y^2\dot{y} + \dot{y}^2)\,dp$ has an extremal value when $y(p)$ is a linear function of p connecting the end points $y(a)$ and $y(b)$.

D

Pressure of a photon gas

The pressure P is the force per unit area due to photons bouncing, say, off a reflecting mirror, see Fig. D.1. The force on the surface element dA is $P\,dA$, and this is also equal to the rate of change of momentum of the reflected photons:

$$P\,dA = d\,\text{momentum}/dt$$

$$= \iint n(\nu)d\nu \cdot \frac{dV}{dt} \cdot \Delta p \cdot \frac{d\Omega}{4\pi} \,. \tag{D.1}$$

Here $n(\nu)d\nu$ is the number density of photons in a frequency band $d\nu$ centered on ν, and $dV/dt = c\cos\theta\,dA$ is the volume 'swept out' by the photons per unit time (Fig. D.1); Δp is the momentum change per photon, and $d\Omega/4\pi$ the fraction of the solid angle. The integrations are over frequency and solid angle. If p is the photon's momentum, its energy is $E = pc$. It follows that $pc = h\nu$ or $p = h\nu/c$, and the momentum change equals $\Delta p = 2p_x = 2(h\nu/c)\cos\theta$. The solid angle element, finally, is $d\Omega = 2\pi\sin\theta\,d\theta$. Inserting everything yields

$$P = \int n(\nu)h\nu\,d\nu \int_0^{\pi/2} \cos^2\theta \sin\theta\,d\theta = \tfrac{1}{3}\epsilon \,. \tag{D.2}$$

The first integral is equal to the photon energy density ϵ and the second integral equals $\tfrac{1}{3}$. Note that the energy distribution of the photons is immaterial, but isotropy is essential.

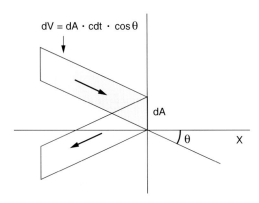

Fig. D.1. The volume dV swept out per unit time by photons impinging on a mirror under an angle θ.

E

The angular power spectrum of the CMB

We derive here three relations from § 11.4. First we consider (11.35), then (11.36), after which (11.34) is trivial. Write down (11.32) twice and take the ensemble average:

$$\langle a_{\ell m} a^*_{\ell' m'} \rangle = \iint d\Omega_1 d\Omega_2 \, Y^*_{\ell m}(\boldsymbol{n}_1) Y_{\ell' m'}(\boldsymbol{n}_2) \, \langle \Delta T(\boldsymbol{n}_1) \Delta T(\boldsymbol{n}_2) \rangle \, . \quad \text{(E.1)}$$

Assuming spherical symmetry, the autocorrelation function $\langle \Delta T(\boldsymbol{n}_1) \Delta T(\boldsymbol{n}_2) \rangle$ can only depend on θ_{12}, where $\cos\theta_{12} = \boldsymbol{n}_1 \cdot \boldsymbol{n}_2$. Accordingly, it should be possible to expand the autocorrelation function in Legendre polynomials P_n:

$$\langle \Delta T(\boldsymbol{n}_1) \Delta T(\boldsymbol{n}_2) \rangle = \sum_n \text{const}_n P_n(\cos\theta_{12}) \, . \quad \text{(E.2)}$$

The addition theorem of the spherical harmonics,

$$\sum_j Y_{nj}(\boldsymbol{n}_1) Y^*_{nj}(\boldsymbol{n}_2) = \frac{2n+1}{4\pi} P_n(\cos\theta_{12}) \, , \quad \text{(E.3)}$$

allows one to express $P_n(\cos\theta_{12})$ in terms of $\boldsymbol{n}_1$ and $\boldsymbol{n}_2$. Insert (E.2) in (E.1) then make use of (E.3) and rename $4\pi \, \text{const}_n/(2n+1) \equiv C_n$. These constants are the same as those in (11.35). As a result of these operations we find

$$\langle a_{\ell m} a^*_{\ell' m'} \rangle = \sum_{nj} C_n \int d\Omega_1 \, Y^*_{\ell m}(\boldsymbol{n}_1) \, Y_{nj}(\boldsymbol{n}_1) \int d\Omega_2 \, Y_{\ell' m'}(\boldsymbol{n}_2) Y^*_{nj}(\boldsymbol{n}_2)$$

$$= \sum_{nj} C_n \, \delta_{\ell n} \delta_{mj} \delta_{\ell' n} \delta_{m'j} = C_\ell \, \delta_{\ell \ell'} \delta_{mm'} \, . \quad \text{(E.4)}$$

In the second line we have twice made use of the orthogonality of the spherical harmonics:

$$\int d\Omega \, Y_{\ell m} Y^*_{\ell' m'} = \delta_{\ell \ell'} \delta_{mm'} \, . \quad \text{(E.5)}$$

It follows that

$$\langle |a_{\ell m}|^2 \rangle = C_\ell \, , \quad \text{(E.6)}$$

which is relation (11.35). It shows that $\langle |a_{\ell m}|^2 \rangle$ is indeed independent of m. The next step is that we may now write (E.1) as

$$C_\ell = \iint d\Omega_1 d\Omega_2 \, Y^*_{\ell m}(\boldsymbol{n}_1) Y_{\ell m}(\boldsymbol{n}_2) \, \langle \Delta T(\boldsymbol{n}_1) \Delta T(\boldsymbol{n}_2) \rangle \, . \quad \text{(E.7)}$$

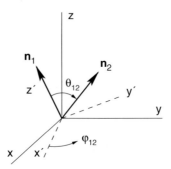

Fig. E.1. The integration over Ω_2 in (E.8) is performed first, using spherical co-ordinates in the $x'y'z'$-frame at fixed $\boldsymbol{n}_1$. In this frame the co-ordinates of $\boldsymbol{n}_2$ are θ_{12}, φ_{12}. Because the integrand is axially symmetric around the z'-axis we have $\int d\Omega_2 = 2\pi \sin\theta_{12} d\theta_{12}$, and the result is (E.9). Spherical symmetry renders the remaining integrand independent of $\boldsymbol{n}_1$, so that $\int d\Omega_1$ produces just a factor 4π.

Summing this relation over m produces a factor $2\ell + 1$ on the left, while on the right we invoke the addition theorem (E.3):

$$(2\ell+1)C_\ell = \frac{2\ell+1}{4\pi} \iint d\Omega_1 d\Omega_2 \, P_\ell(\cos\theta_{12}) \, \langle \Delta T(\boldsymbol{n}_1)\Delta T(\boldsymbol{n}_2)\rangle \; . \quad \text{(E.8)}$$

We now exploit the fact that the integrand depends only on θ_{12}. As explained in Fig. E.1, the result is

$$C_\ell = \tfrac{1}{2} \int d\Omega_1 \int_0^\pi \sin\theta_{12} \, d\theta_{12} \, P_\ell(\cos\theta_{12}) \cdot$$

$$\langle \Delta T(\boldsymbol{n}_1)\Delta T(\boldsymbol{n}_2)\rangle \,|\, \boldsymbol{n}_1 \cdot \boldsymbol{n}_2 = \cos\theta_{12} \quad \text{(E.9)}$$

$$\equiv 2\pi \int_0^\pi C(\theta) P_\ell(\cos\theta) \sin\theta \, d\theta \; , \quad \text{(E.10)}$$

where we have dropped the index 12. This proves relation (11.36).

Finally, expand $C(\theta)$ in Legendre polynomials, $C(\theta) = \sum_n A_n P_n(\cos\theta)$, insert that in (E.10) and use the orthogonality of the Legendre polynomials,

$$\int_{-1}^1 P_n(x) P_m(x) \, dx = \frac{2}{2n+1} \delta_{nm} \; , \quad \text{(E.11)}$$

to find that $A_\ell = (2\ell+1)C_\ell/4\pi$, or $C(\theta) = (4\pi)^{-1}\sum_n (2n+1)C_n P_n(\cos\theta)$, which is (11.34).

E The angular power spectrum of the CMB

The position of the maxima

In real life the peak positions are found by processing temperature maps through (11.32) and (11.35). Here we are forced to follow a simpler approach, and we employ the angles (10.47) between directions of maximal temperature difference:

$$\theta_n \simeq \frac{\Delta}{(2n+1)d} \left(\Omega_m + \Omega_\Lambda\right)^{1/2} . \tag{E.12}$$

We now focus on the position of the first peak:

$$\theta_0 = \frac{\Delta}{d} \left(\Omega_m + \Omega_\Lambda\right)^{1/2} = 1.64 \times 10^3 \, H_0 t_m \left(\Omega_m + \Omega_\Lambda\right)^{1/2}$$

$$\simeq 1.12 \times 10^{-2} \cong 0.65° \tag{E.13}$$

The distance Δ travelled by the $be\gamma$ mode at recombination is the horizon distance $9ct_m$ at recombination, see below (11.20), divided by $\sqrt{3}$, to allow for a signal speed of $c/\sqrt{3}$. And according to Table 11.1 the distance d to the last scattering surface at recombination is $\simeq 3.3 \times 10^{-3} c t_0 = 3.3 \times 10^{-3} \times 0.96 \, c/H_0$. The numerical value of θ_0 follows by inserting $t_m = 9.4 \times 10^4$ yr, see below (10.24), and $\Omega_m + \Omega_\Lambda = 1$.

To estimate the maximum in the angular power spectrum we argue that according to (11.32) C_ℓ will be maximal if the grid of $+$ and $-$ signs laid out on the sphere by $Y_{\ell m}$ is commensurate with that of ΔT. There are maximally 2ℓ zeros on the equator, hence $2\ell \, \theta_n = 2\pi$. The position of the first peak is therefore expected at

$$\ell_0 \simeq \pi/\theta_0 \simeq 277 , \tag{E.14}$$

while the observations give $\ell_0 = 220 \pm 1$. The origin of the discrepancy is that we have ignored two important effects that alter the value of Δ and therefore of θ_0. In the first place we have tacitly assumed that the $be\gamma$ modes are *free*, but in fact they are driven by the gravity perturbation $\delta\phi$ generated by the dark matter modes. This turns out to enhance their effective speed of propagation and hence also the value of Δ, by an amount that depends on the wavelength λ, i.e. on n. In the second place we have assumed that the signal speed is $c/\sqrt{3}$, but in reality baryon loading reduces the speed, in particular at late times, and that in turn diminishes Δ. Although the two effects partly cancel, we cannot hope our result (E.14) to be very accurate.

Index

accretion, 14, 16, 110
  onto black hole, 110
  onto neutron star, 95
  onto white dwarf, 93
active galactic nuclei (AGNs), 16, 110
anthropic principle, 270

bar detectors, 143
  MiniGRAIL, 145
  ongoing projects, 145
baryogenesis, 241
baryon to photon ratio, 203
baryon-electron-photon fluid, 206
baryons, 238
  dark, 170
  luminous, 170
Big Bang, 173, 192
  misconception of, 222
Big Crunch, 193, 258
Big Emptiness, 195
binary pulsars, 16, 81, 93
  and gravitational radiation, 141
binding energy
  of neutron star, 100
  of nucleons, 93, 100
Binet's method, 75
Birkhoff's theorem, 70
black hole, 16, 89, 109
  and future light-cone, 115
  elementary properties, 113
  entropy, 129
  evaporation, 129
  event horizon, 114

  growth, 116
  hole in spacetime, 124
  in galaxy, 112
  microscopic, 130
  no hair, 110
  observations, 110
  of intermediate mass, 112
  primordial, 110, 130
  rotating, 125
  specified by M, L and Q, 109
  supermassive, 110
  temperature, 129
brown dwarfs, 89

calendar, 1
cataclysmic variables, 93
Chandrasekhar limit, 96
chaotic inflation, 261
  and beginning of time, 270
  duration, 263
  expansion factor, 263
  toy model, 265
  transition to hot big bang, 266
charge
  colour, 238
  electric, 238
Christoffel symbol, 28
  computation of, 32, 39
  in freely falling frame, 47
  nontensor, 29, 36
  properties, 29
  Robertson-Walker metric, 181
  Schwarzschild metric, 67

Index

classical tests of GR, 77
clock paradox, 8
closed universe, 179, 192
CMB
  angular power spectrum, 209, 229, 230
  dipole anisotropy, 172, 229
  discovery, 17
  energy density, 172
  evolution, 197
  temperature, 172
  temperature fluctuations, 205, 228
CMB (Cosmic Microwave Background), 172
co-ordinate distance, 214, 215
co-ordinate picture, 20, 49, 175, 214, 219
co-ordinates
  Gaussian, 176
  harmonic, 57
  in Riemann space, 19
  Kruskal-Szekeres, 121
  meaning, 43
  Schwarzschild, 70
  spatial, 45
  time, 44
cold dark matter (CDM) models, 207
conservation of mass, 50
conserved quantities
  baryon number, 240
  electric charge, 240
constant of the motion, 32
continuity equation, 51
Cosmic Background Explorer (COBE), 172, 228
Cosmic Microwave Background (CMB), 172
cosmic variance, 174, 230
cosmological constant, 60, 183, 227
  and dark energy, 170, 183
cosmological parameters, 171
  determination, 230
cosmological principle, 2, 174
covariant derivative
  of higher rank tensor, 34
  of vector, 33
critical density, 169, 170, 191
curvature, 11
  and Riemann tensor, 36
  Gaussian, 37
  total, 38
cyclist analogy
  and photon propagation, 150, 219

d'Alembert operator, 56
dark energy, 170, 185
dark matter, 83, 170
  baryons, 170
  cold, 205
  nonbaryonic, 170
decoupling, 240
density fluctuations
  and inflation, 207
  at recombination, 199, 205
distance
  co-ordinate, 215
  geometrical, 213, 214
  luminosity, 224
distribution of matter
  evolution effects, 171
  filaments and voids, 169
  isotropy, 171

Eddington limit, 111
Einstein clock, 9
Einstein tensor, 38
  Robertson-Walker metric, 182
  Schwarzschild metric, 69
Einsteinturm, 78
electron-positron annihilation, 245
elementary particles
  overview, 237
energetics of inflation, 268
energy
  of test particle, 58
entropy
  black hole, 129
ephemeris, 1
equation of state (EOS)
  cold matter, 104, 105, 250
  hot matter, 189, 250
ergosphere, 127
Euler-Lagrange equations, 32, 67
event, 3, 43
event horizon
  and Hawking radiation, 128
  and Unruh effect, 128
evolution of universe, 171

overview, 242
expansion of universe, 171
   adiabatic, 183
   swelling of space, 222

Fermi energy, 95
Fermi-Walker transport, 155, 156
field equations
   basic idea, 52
   classical limit, 57
   general form, 54, 60
   structure, 61
   vacuum, 53
   weak field, 57
flat universe, 178, 192
flatness problem, 253
four-momentum, 47, 50
four-velocity, 32, 47, 50
frame-dragging effect, 13, 125
   and Mach's principle, 126
   LAGEOS satellites, 126, 163
freeze-out, 240
Friedmann-Robertson-Walker (FRW) models, 189
FRW models
   failures, 253
   matter-dominated, 192
   radiation-dominated, 197
   successes, 253
FRW reference model, 194

Galilean transformation, 4
Gamma-ray bursts, 16
gauge bosons, 238
Gauss's theorem, 51
Gaussian co-ordinates, 176
general covariance, 13
   how to use it, 60
geodesic
   extremal property, 32
   null, 31
   timelike, 31
geodesic deviation, 40, 52
geodesic equation, 30
geodesic motion, 46
geodesic precession, 29, 161
   and binary pulsar, 163
geometrical distance, 213, 214, 255
geometrical picture, 20, 175, 219

Grand Unified Theories, 243
gravitational deflection of light, 11, 14, 78
gravitational lensing, 17, 82
   arcs, 83
   Einstein ring, 83
   macrolensing, 85
   microlensing, 85
   of neutron star image, 83, 86
gravitational mass, 10
gravitational redshift, 11, 14, 49, 79
   in solar spectrum, 14, 79
   of neutron star surface, 106
gravitational time delay, 14, 79
gravitational waves, 16, 133
   detectors, 143, 145
   dispersion relation, 134
   effect on test masses, 136
   energy flux density, 140
   generation, 138
   metric tensor, 135
   polarization, 137
   quadrupole radiation, 139
   TT-gauge, 135
gravity
   Newtonian, 10
   SR theories, 10
   weak, 47, 56
Gravity Probe A, 79
Gravity Probe B, 6, 16, 162, 164
   and geodesic precession, 164
   and Lense-Thirring effect, 164

hadrons, 238
Hawking radiation, 128
   and event horizon, 128
helium synthesis, 247
High mass X-ray binaries, 95, 113
Hipparcos satellite, 81
horizon
   event, 114, 223
   in cosmology, 220
   particle, 115, 220, 223
horizon problem, 221, 254
   and causality, 256
   in closed universe, 257
   origin, 256
   remedy, 257
hot dark matter (HDM) models, 207

Hubble constant, 171
Hubble flow, 171
    cold, 171, 185
Hubble radius, 175, 220
Hubble relation, 171, 214
    general form, 225
Hubble time, 173, 195
hyperbolic universe, 179

index
    contraction, 24
    dummy, 24
    lowering, 22
    raising, 22
inertial frame
    global, 4, 11
    local, 11, 12
inertial mass, 10
inflation, 17, 60, 195, 221
    and cosmological constant, 268
    and creation of FRW universe, 268
    and density fluctuations, 268
    and flatness problem, 267
    and horizon problem, 267
    and scalar field, 258
    basic idea, 257
    current status, 267
    energetics, 268
    equations, 260
    first and second phase, 195
    loose ends, 267
    philosophical issues, 268
    what drives it, 267
interferometer detectors, 145
    LIGO, 146, 147, 152
    LISA, 147, 151, 152
    ongoing projects, 152
    operational principle, 149
    signal on detector, 150
    VIRGO, 152
interval, 6
isotropy of universe, 171

Jeans instability, 203

Kamiokande facility, 243
Kepler's second law, 74
Kerr metric, 125
Klein-Gordon equation, 259

Kruskal-Szekeres co-ordinates, 121

Legendre polynomials, 230
Lense-Thirring effect, 163
lepton era, 243
leptons, 238
leptoquarks
    asymmetric decay, 243
light-cone, 7
    past, 215
light-element synthesis, 17, 246
    and observations, 249
Lorentz gauge, 57
Lorentz transformation, 8
    of rest mass density, 54
Low mass X-ray binaries, 95, 113
luminosity distance, 216, 224

Mach's principle, 13
MACHOs, 86
mass limit
    neutron star, 96, 105
    white dwarf, 96
mass transfer, 93
matter
    definition in cosmology, 189
matter era, 191
    models, 217
matter-antimatter asymmetry, 242
maximum mass
    neutron stars, 95, 99, 105
    white dwarfs, 95
mesons, 238
metric tensor
    covariant derivative, 35
    experimental determination, 45
    for weak field, 58
    gravitational waves, 135
    in general relativity, 12
    in special relativity, 6
    Kerr metric, 125
    of space, 45
    Riemann space, 20
    Robertson-Walker metric, 181
    Schwarzschild metric, 66, 70
Minkowski spacetime, 3

neutrino
    decoupling, 244

oscillations, 238
neutrino background, 173, 245
neutron, 238
  beta-decay, 240, 243, 247
neutron drip, 104
neutron stars, 89
  bare mass, 99
  binding energy, 100
  constant density model, 101
  discovery, 15
  equation of state, 104
  gravitational acceleration, 120
  maximum mass, 95, 96, 99, 105
  measured mass and radius, 106
  minimum radius, 101
  physical mass, 99
  realistic models, 103
neutron-to-proton ratio, 245
nova, 93
nuclear fusion
  in early universe, 237, 246
  in stars, 90, 248
nuclear reactions
  in early universe, 237
  inverse beta-decay, 91
number density of compact objects, 93, 112

Olbers's paradox, 232
  remedy, 233
open universe, 179, 192
orbit classification
  in Schwarzschild metric, 74

parallel transport
  formal definition, 29
  intuitive definition, 27
  on a sphere, 39
parameters
  of characteristic objects, 49
past light-cone
  cyclist analogy, 219
  integrations over, 231
  shape, 218
  volume, 232
perihelium precession, 14, 77, 82
Planck density, 260
Planck length, 113, 260
Planck mass, 260

PLANCK mission, 231
Planck time, 260
planetary nebula, 90
Pound-Rebka-Snider experiment, 12, 49, 79
pressure
  degeneracy, 90
  dual role, 99
  source of gravity, 60, 99, 183
principle
  anthropic, 270
  cosmological, 2
  general covariance, 13, 55, 60, 184, 259
  Mach, 13
  relativity, 3
  strong equivalence, 13, 46
  weak equivalence, 10
proper time, 7, 12, 44
  minor role in cosmology, 176
proper volume, 51
proton, 238
  lifetime, 243
pulsars, 15, 81, 94
  X-ray, 95

QPOs, 106
quadrupole moment of the Sun, 78
quality factor, 143
quark-gluon plasma, 240
  in laboratory, 241
quarks, 238
quasars, 16, 78, 110, 166, 227
quasi-periodic oscillations (QPOs), 106

radiation
  definition in cosmology, 189
radiation era, 197
  time evolution, 199
radio astronomy, 14
re-ionization, 199
recombination, 199
redshift, 214
  and astronomical jargon, 215
  and scale factor, 215
  and tired light, 216
  not additive, 72, 120
  of De Broglie wavelength, 215
reference frame

freely falling, 11, 12, 46
global, 3, 10
quasar, 166
reheating
    after inflation, 266
    in matter era, 199
rest, 44, 174
rest-frame
    global, 5, 7
    local, 12
Ricci tensor, 38
    Robertson-Walker-metric, 182
    Schwarzschild metric, 68
Riemann space, 19
    definition, 19
    embedding, 19, 20
Riemann tensor, 34
    and curvature, 36, 37
    and tidal forces, 53
    in freely falling frame, 47
Robertson-Walker metric, 178
    co-ordinates, 178
    Einstein tensor, 182
    geodesics, 185
    Ricci tensor, 182
    scale factor, 178
rotating black hole, 125
    ergosphere, 127
    static limit, 127
rotation
    galactic, 112

Sachs-Wolfe effect, 210
scalar (tensor of rank 0), 24, 25
scalar field
    and inflation, 258
scale factor, 178
    evolution equation, 182
Schwarzschild metric, 70
    Einstein tensor, 69
    geodesics, 72
    orbit classification, 74
    orbit equation, 75
    Ricci tensor, 68
    singularity, 113
Schwarzschild radius, 70
Shapiro effect, 79
sign convention, 6, 38
signature, 6

simultaneity, 4, 5
singularity
    in cosmology, 192
    of Schwarzschild metric, 113
spacetime
    curvature, 11
    Minkowski, 3
spherical harmonics, 228
spherical universe, 179
standard candle, 224
    type Ia supernovae, 227
stellar evolution, 89
    binary systems, 93
    main sequence, 89
    mass loss, 90, 93
    neutrino losses, 92
    nova, 93
    nuclear fusion, 90
    red giant, 90
    supernova, 92
stress-energy tensor, 54, 55, 60, 97, 182
    of cold dust, 54
    of matter, 60
    of scalar field, 259
    of vacuum, 184
strong equivalence principle, 13, 46
structure formation, 203
    and dark matter, 205
    imprints on CMB, 209
summation convention, 6, 23
Sunyaev-Zeldovich effect, 227
supernova, 15, 92
    type Ia, 93, 227
Supernova Cosmology Project, 227
supernova remnant, 94
supersoft X-ray sources, 93

tangent space
    and embedding, 21
    base vectors, 21
    preferred metric, 21
temperature of universe, 245
tensor
    contravariant representation, 24
    covariant representation, 24
    Einstein, 38
    of higher rank, 24
    quotient theorem, 25
    Ricci, 38

Riemann, 34
  stress-energy, 54, 55
  unit, 25
thermal equilibrium
  in early universe, 239
Thomas precession, 5, 159
Thomson scattering, 198, 206
tidal forces, 11
  and curvature, 53
time dilation, 9
Tolman-Oppenheimer-Volkoff (TOV) equation, 97
transport of accelerated vector, 155
transverse traceless gauge, 135

ultra-luminous X-ray sources, 112
universe
  age, 195, 201
  age indicators, 195
  closed, 179
  evolution, 171
  expansion, 171
  flat, 178
  future development, 196
  homogeneity, 174
  hyperbolic, 179
  isotropy, 171, 174
  open, 179
  scale model, 218
  spherical, 179
  temperature, 245
  thermal history, 198
  visible, 221
Unruh effect, 128

UrQMD, 241

variational calculus, 32, 67
vector, 4
  contravariant, 24
  covariant, 24
  null, 7
  spacelike, 7
  tensor of rank 1, 24
  timelike, 7
virial theorem, 89, 96
visible universe, 221
VLBI, 78, 166
Vulcan, 78

weak equivalence principle, 10
white dwarfs, 89, 90
  maximum mass, 95, 96
white hole, 123
Wilkinson Microwave Anisotropy Probe (WMAP), 17, 170, 228
WIMPs (weakly interacting massive particles), 170
world model
  geocentric, 2
  Greek, 1
  heliocentric, 2
  Hindu, 1
  Ptolemy's, 3
worldline, 3
wormhole, 66

X-boson, 243, 258
X-ray astronomy, 14
X-ray binaries, 14, 95, 105, 106, 112

ASTRONOMY AND ASTROPHYSICS LIBRARY

Series Editors: G. Börner · A. Burkert · W. B. Burton · M. A. Dopita
A. Eckart · T. Encrenaz · B. Leibundgut · J. Lequeux
A. Maeder · V. Trimble

The Stars By E. L. Schatzman and F. Praderie

Modern Astrometry 2nd Edition
By J. Kovalevsky

The Physics and Dynamics of Planetary Nebulae By G. A. Gurzadyan

Galaxies and Cosmology By F. Combes, P. Boissé, A. Mazure and A. Blanchard

Observational Astrophysics 2nd Edition
By P. Léna, F. Lebrun and F. Mignard

Physics of Planetary Rings Celestial Mechanics of Continuous Media
By A. M. Fridman and N. N. Gorkavyi

Tools of Radio Astronomy 4th Edition
By K. Rohlfs and T. L. Wilson

Tools of Radio Astronomy Problems and Solutions 1st Edition, Corr. 2nd printing By T. L. Wilson and S. Hüttemeister

Astrophysical Formulae 3rd Edition (2 volumes)
Volume I: Radiation, Gas Processes and High Energy Astrophysics
Volume II: Space, Time, Matter and Cosmology
By K. R. Lang

Galaxy Formation By M. S. Longair

Astrophysical Concepts 2nd Edition
By M. Harwit

Astrometry of Fundamental Catalogues
The Evolution from Optical to Radio Reference Frames
By H. G. Walter and O. J. Sovers

Compact Stars. Nuclear Physics, Particle Physics and General Relativity 2nd Edition
By N. K. Glendenning

The Sun from Space By K. R. Lang

Stellar Physics (2 volumes)
Volume 1: Fundamental Concepts and Stellar Equilibrium
By G. S. Bisnovatyi-Kogan

Stellar Physics (2 volumes)
Volume 2: Stellar Evolution and Stability
By G. S. Bisnovatyi-Kogan

Theory of Orbits (2 volumes)
Volume 1: Integrable Systems and Non-perturbative Methods
Volume 2: Perturbative and Geometrical Methods
By D. Boccaletti and G. Pucacco

Black Hole Gravitohydromagnetics
By B. Punsly

Stellar Structure and Evolution
By R. Kippenhahn and A. Weigert

Gravitational Lenses By P. Schneider, J. Ehlers and E. E. Falco

Reflecting Telescope Optics (2 volumes)
Volume I: Basic Design Theory and its Historical Development. 2nd Edition
Volume II: Manufacture, Testing, Alignment, Modern Techniques
By R. N. Wilson

Interplanetary Dust
By E. Grün, B. Å. S. Gustafson, S. Dermott and H. Fechtig (Eds.)

The Universe in Gamma Rays
By V. Schönfelder

Astrophysics. A New Approach 2nd Edition
By W. Kundt

Cosmic Ray Astrophysics
By R. Schlickeiser

Astrophysics of the Diffuse Universe
By M. A. Dopita and R. S. Sutherland

The Sun An Introduction. 2nd Edition
By M. Stix

Order and Chaos in Dynamical Astronomy
By G. J. Contopoulos

Astronomical Image and Data Analysis
By J.-L. Starck and F. Murtagh

ASTRONOMY AND ASTROPHYSICS LIBRARY

Series Editors: G. Börner · A. Burkert · W. B. Burton · M. A. Dopita
A. Eckart · T. Encrenaz · B. Leibundgut · J. Lequeux
A. Maeder · V. Trimble

The Early Universe Facts and Fiction
4th Edition By G. Börner

The Design and Construction of Large Optical Telescopes By P. Y. Bely

The Solar System 4th Edition
By T. Encrenaz, J.-P. Bibring, M. Blanc,
M. A. Barucci, F. Roques, Ph. Zarka

General Relativity, Astrophysics, and Cosmology By A. K. Raychaudhuri,
S. Banerji, and A. Banerjee

Stellar Interiors Physical Principles, Structure, and Evolution 2nd Edition
By C. J. Hansen, S. D. Kawaler, and V. Trimble

Asymptotic Giant Branch Stars
By H. J. Habing and H. Olofsson

The Interstellar Medium
By J. Lequeux

Methods of Celestial Mechanics (2 volumes)
Volume I: Physical, Mathematical, and Numerical Principles
Volume II: Application to Planetary System, Geodynamics and Satellite Geodesy
By G. Beutler

Solar-Type Activity in Main-Sequence Stars
By R. E. Gershberg

Relativistic Astrophysics and Cosmology
A Primer By P. Hoyng

DATE DUE

SCI QB 462.65 .H69 2006

Høyng, Peter.

Relativistic astrophysics
and cosmology